METHODS IN MOLECULAR BIOLOGY

Series Editor
John M. Walker
School of Life and Medical Sciences
University of Hertfordshire
Hatfield, Hertfordshire, UK

For further volumes:
http://www.springer.com/series/7651

For over 35 years, biological scientists have come to rely on the research protocols and methodologies in the critically acclaimed *Methods in Molecular Biology* series. The series was the first to introduce the step-by-step protocols approach that has become the standard in all biomedical protocol publishing. Each protocol is provided in readily-reproducible step-by-step fashion, opening with an introductory overview, a list of the materials and reagents needed to complete the experiment, and followed by a detailed procedure that is supported with a helpful notes section offering tips and tricks of the trade as well as troubleshooting advice. These hallmark features were introduced by series editor Dr. John Walker and constitute the key ingredient in each and every volume of the *Methods in Molecular Biology* series. Tested and trusted, comprehensive and reliable, all protocols from the series are indexed in PubMed.

Ethylene Sensing and Signaling

Edited by

Devesh Shukla

Tissue Culture and Transformation Facility, Plant Biotechnology, CSIR Central Institute of Medicinal and Aromatic Plants, Lucknow, Uttar Pradesh, India; Academy of Scientific and Innovative Research (AcSIR), CSIR-HRDC Campus, Ghaziabad, Uttar Pradesh, India

Humana Press

Editor
Devesh Shukla
Tissue Culture and Transformation Facility
Plant Biotechnology
CSIR Central Institute of Medicinal and Aromatic Plants
Lucknow, Uttar Pradesh, India

Academy of Scientific and Innovative Research (AcSIR)
CSIR-HRDC Campus
Ghaziabad, Uttar Pradesh, India

ISSN 1064-3745 ISSN 1940-6029 (electronic)
Methods in Molecular Biology
ISBN 978-1-0716-4649-6 ISBN 978-1-0716-4650-2 (eBook)
https://doi.org/10.1007/978-1-0716-4650-2

This Humana imprint is published by the registered company Springer Science+Business Media, LLC, part of Springer Nature.
The registered company address is: 1 New York Plaza, New York, NY 10004, U.S.A.

Preface

My fascination with ethylene began during my early research years at CSIR-NBRI, Lucknow, where I was involved in isolating ripening-related fruit-specific promoters. A key role of ethylene in fruit ripening rapidly became clear. Later, during my postdoctoral research at Western Kentucky University Bowling Green, KY (USA), I encountered ethylene again. This time, I found it as a regulatory factor in the excess phosphate response (EPR), which affects root system architecture. Historically, ethylene was first noted when plants near leaking gas lines displayed accelerated senescence and petal abscission. Richard Gane later recognize ethylene as an endogenous plant growth regulator. Because it induces tissue degradation, ethylene was once labeled as the "death hormone." Yet it also shapes fruit ripening, stress adaptation, apical hook formation in dark grown seedlings, and auxin dependent cell expansion. Ethylene further affects root hairs that anchor plants and several developmental processes. This wide role makes ethylene an ideal topic for the protocols collected in this volume.

This volume, part of the Methods in Molecular Biology series, aims to present a set of protocols focused on ethylene sensing and signaling. Covering methods from detection and quantification of ethylene to postharvest applications and molecular assays, each chapter offers readers a hands-on approach to investigate ethylene in plant growth, development and ripening.

To facilitate navigation, the book has been organized into four distinct sections:

1. **Part I: Ethylene Detection and Quantification Techniques**
   Protocols for measuring ethylene production, gas chromatography methods, and quantification of ethylene-mediated changes in plant tissues (Chaps. 1, 2, 3, 4, and 5).
2. **Part II: Role of Ethylene in Postharvest Biology and Fruit Ripening**
   Chapters examine ethylene function in fruit and floral senescence, ripening-specific promoters, and postharvest management strategies (Chaps. 6, 7, 8, and 9).
3. **Part III: Tools and Methods for Investigating Ethylene Action**
   Novel assays and molecular biology protocols ranging from gene expression analyses to bioinformatics workflows at cellular and whole-plant levels (Chaps. 10, 11, 12, 13, and 14).
4. **Part IV: Ethylene Signaling in Plant Development and Stress**
   In-depth studies highlight regulatory capacity of ethylene in plant development, root system modifications, and adaptation to diverse biotic and abiotic stressors (Chaps. 15, 16, 17, and 18).

Taken together, these chapters provide reliable methodologies for investigating the central function of ethylene in plant biology. I hope that the protocols and perspectives provided here will not only accelerate progress in ethylene research but also illuminate applications in crop improvement, postharvest management, and beyond.

I am grateful to all chapter authors for their expert contributions. My special thanks go to Prof. John E. Walker, whose invitation to edit the book and steady guidance made this volume possible. I remain grateful to my early mentors, Dr. P. K. Trivedi and Dr. P. Nath, for sparking my interest in ethylene biology. Finally, I thank the editorial team at Springer Nature for their professionalism and support.

Lucknow, Uttar Pradesh, India ***Devesh Shukla***

Contents

Part III Tools and Methods for Investigating Ethylene Action

Part IV Ethylene Signaling in Plant Development and Stress

Contributors

YUKI ACHIHA • *Graduate School of Bioresource Sciences, Nihon University, Fujisawa, Kanagawa, Japan*
YOSEF AL SHOFFE • *Horticulture Section, School of Integrative Plant Science College of Agriculture and Life Sciences, Cornell University, Ithaca, NY, USA*
SHAHRE ALAM • *CSIR-National Botanical Research Institute (CSIR-NBRI), Lucknow, Uttar Pradesh, India*
ULBIO EDUARDO ALCÍVAR-CEDEÑO • *Departamento de Procesos Químicos, Alimentos y Biotenología, Facultad de Ingeniería y Ciencias Aplicadas, Universidad Técnica de Manabí, Portoviejo, Manabí, Ecuador*
MEHAR HASAN ASIF • *CSIR-National Botanical Research Institute (CSIR-NBRI), Lucknow, Uttar Pradesh, India; Academy of Scientific and Innovative Research (AcSIR), CSIR- Human Resource Development Centre, (CSIR-HRDC) Campus, Ghaziabad, Uttar Pradesh, India*
ARKADIPTA BAKSHI • *Department of Botany, University of Wisconsin-Madison, Madison, WI, USA*
GABRIEL ALFONSO BURGOS-BRIONES • *Departamento de Procesos Químicos, Alimentos y Biotenología, Facultad de Ingeniería y Ciencias Aplicadas, Universidad Técnica de Manabí, Portoviejo, Manabí, Ecuador*
CARLOS ALFREDO CEDEÑO-PALACIOS • *Departamento de Procesos Químicos, Alimentos y Biotenología, Facultad de Ingeniería y Ciencias Aplicadas, Universidad Técnica de Manabí, Portoviejo, Manabí, Ecuador*
HSING-YI CHO • *Department of Molecular Biology and Center for Computational and Integrative Biology, Massachusetts General Hospital, Boston, MA, USA; Department of Genetics, Harvard Medical School, Boston, MA, USA*
MEI-YI CHOU • *Agricultural Biotechnology Research Center, Academia Sinica, Taipei, Taiwan*
ASIS DATTA • *National Institute of Plant Genome Research, New Delhi, India*
MARÍA HIPATIA DELGADO-DEMERA • *Departamento de Veterinaria, Facultad de Ciencias Veterinarias, Universidad Técnica de Manabí, Portoviejo, Manabí, Ecuador*
YOGESHWAR VIKRAM DHAR • *CSIR-National Botanical Research Institute (CSIR-NBRI), Lucknow, Uttar Pradesh, India; Academy of Scientific and Innovative Research (AcSIR), CSIR- Human Resource Development Centre, (CSIR-HRDC) Campus, Ghaziabad, Uttar Pradesh, India*
LUIS DANIEL GARCÍA-PARRALES • *Carrera de Ingeniería Química, Facultad de Ingeniería y Ciencias Aplicadas, Universidad Técnica de Manabí, Portoviejo, Manabí, Ecuador*
KAPUGANTI JAGADIS GUPTA • *National Institute of Plant Genome Research, New Delhi, India*
RAHUL KUMAR GUPTA • *Division of Crop Production and Protection, Central Institute of Medicinal and Aromatic Plants, Lucknow, India; Academy of Scientific and Innovative Research (AcSIR), Ghaziabad, India*
SHWETA GUPTA • *Tissue culture and Transformation Facility, Plant Biotechnology Division, CSIR-Central Institute of Medicinal and Aromatic Plants, Lucknow, Uttar Pradesh, India*

DONALD J. HUBER • *Horticultural Sciences, University of Florida, Gainesville, FL, USA*
MOHAMMAD IRFAN • *Plant Biology Section, School of Integrative Plant Science, Cornell University, Ithaca, NY, USA; National Institute of Plant Genome Research, New Delhi, India*
SHIN ISHIKAWA • *College of Bioresource Sciences, Nihon University, Fujisawa, Kanagawa, Japan*
REKHA JAISWAL • *National Institute of Plant Genome Research, New Delhi, India*
YUSUKE KAMIYOSHIHARA • *College of Bioresource Sciences, Nihon University, Fujisawa, Kanagawa, Japan; Graduate School of Bioresource Sciences, Nihon University, Fujisawa, Kanagawa, Japan*
RAVI KESARI • *Department of Molecular Biology and Genetic Engineering, Bihar Agricultural College, Bihar Agricultural University, Sabour, Bihar, India*
HARRY J. KLEE • *Horticultural Sciences, University of Florida, Gainesville, FL, USA*
PANKAJ KUMAR • *National Institute of Plant Genome Research, New Delhi, India; Department of Biotechnology, Dr. Y.S. Parmar University of Horticulture and Forestry, Solan, Himachal Pradesh, India*
POONAM KUMARI • *Division of Crop Production and Protection, Central Institute of Medicinal and Aromatic Plants, Lucknow, India*
LEI LI • *Peking University Institute of Advanced Agricultural Sciences, Shandong Laboratory of Advanced Agricultural Sciences in Weifang, Shandong, China*
PRIYAMVADA MISHRA • *CSIR-National Botanical Research Institute (CSIR-NBRI), Lucknow, Uttar Pradesh, India; Academy of Scientific and Innovative Research (AcSIR), CSIR- Human Resource Development Centre, (CSIR-HRDC) Campus, Ghaziabad, Uttar Pradesh, India*
SHINJI MIZUNO • *College of Bioresource Sciences, Nihon University, Fujisawa, Kanagawa, Japan; Graduate School of Bioresource Sciences, Nihon University, Fujisawa, Kanagawa, Japan*
HITOSHI MORI • *Institute for Glyco-core Research (iGCORE), Nagoya University, Nagoya, Aichi, Japan*
PRAVENDRA NATH • *CSIR-National Botanical Research Institute, Lucknow, Uttar Pradesh, India*
SUMYA PATHAK • *PME, CSIR Central Institute of Medicinal & Aromatic Plants, Lucknow, India*
AFSANA PRAVEEN • *National Institute of Plant Genome Research, New Delhi, India*
POONAM RAKWAL • *Council of Scientific and Industrial Research-Central Institute of Medicinal and Aromatic Plants (CSIR-CIMAP), Lucknow, India; Academy of Council of Scientific and Industrial Research (ACSIR), Ghaziabad, India*
CH. RATNASEKHAR • *Council of Scientific and Industrial Research-Central Institute of Medicinal and Aromatic Plants (CSIR-CIMAP), Lucknow, India; Academy of Council of Scientific and Industrial Research (ACSIR), Ghaziabad, India*
LEONARDO ANDRES SALTOS-MEZA • *Carrera de Ingeniería Química, Facultad de Ingeniería y Ciencias Aplicadas, Universidad Técnica de Manabí, Portoviejo, Manabí, Ecuador*
GARIMA SAXENA • *CSIR-National Botanical Research Institute (CSIR-NBRI), Lucknow, Uttar Pradesh, India; Academy of Scientific and Innovative Research (AcSIR), CSIR-Human Resource Development Centre, (CSIR-HRDC) Campus, Ghaziabad, Uttar Pradesh, India*
MING-CHE SHIH • *Agricultural Biotechnology Research Center, Academia Sinica, Taipei, Taiwan*

DEVESH SHUKLA • *Tissue culture and Transformation Facility, Plant Biotechnology Division, CSIR-Central Institute of Medicinal and Aromatic Plants, Lucknow 226015, Uttar Pradesh, India; Academy of Scientific and Innovative Research (AcSIR), CSIR-HRDC campus, Ghaziabad, Uttar Pradesh, India*

AKANKSHA SINGH • *Division of Crop Production and Protection, Central Institute of Medicinal and Aromatic Plants, Lucknow, India; Academy of Scientific and Innovative Research (AcSIR), Ghaziabad, India*

AKIRA TATEISHI • *College of Bioresource Sciences, Nihon University, Fujisawa, Kanagawa, Japan; Graduate School of Bioresource Sciences, Nihon University, Fujisawa, Kanagawa, Japan*

PRABODH KUMAR TRIVEDI • *Tissue Culture and Transformation Facility, Plant Biotechnology Division, CSIR-Central Institute of Medicinal and Aromatic Plants, Lucknow, Uttar Pradesh, India*

ANOOP KUMAR VERMA • *Council of Scientific and Industrial Research-Central Institute of Medicinal and Aromatic Plants (CSIR-CIMAP), Lucknow, India; Jawaharlal Nehru University, New Delhi, India*

YANZHI YANG • *Peking University Institute of Advanced Agricultural Sciences, Shandong Laboratory of Advanced Agricultural Sciences in Weifang, Shandong, China*

Part I

Ethylene Detection and Quantification Techniques

Chapter 1

Estimation of 1-Aminocyclopropane-1-Carboxylic Acid (ACC) Deaminase Enzyme in Bacteria

Poonam Kumari, Rahul Kumar Gupta, and Akanksha Singh

Abstract

Ethylene at a low concentration is essential for the germination of seeds and the development of seedlings in the plant species. However, the same ethylene at a higher concentration inhibits the elongation of roots, seed germination, and seedling development. In plants, 1-aminocyclopropane-1-carboxylic acid (ACC) serves as the direct precursor for the production of ethylene. The enzyme ACC deaminase produced by certain bacteria plays a pivotal role in reducing ethylene levels in plants by breaking down ACC into α-ketobutyrate and ammonia, thus positively influencing various physiological processes. Bacteria can utilize these compounds for their own growth and development. This chapter provides a comprehensive overview of quantitative as well as qualitative methods including colorimetric assay, solid plate, and PCR-based assays for readily screening unknown ACC deaminase–producing bacterial strains.

Key words ACC deaminase, Bacteria, Ethylene, α-Ketobutyrate, Plant growth

1 Introduction

Ethylene is a gaseous hormone that is naturally produced by all higher plants, and it plays a crucial role in regulating various beneficial plant responses even at low concentrations (as low as 10 ppm). However, the high concentrations of ethylene up to 25 ppm can have adverse effects on the growth and development. Under stress conditions, the higher production of ethylene has various adverse effects like elongation of roots and shoots, leaf expansion suppression, and induction of epinasty [1]. The production of the enzyme 1-aminocyclopropane-1-carboxylic acid (ACC) synthase is responsible for converting S-adenosyl-L-methionine (SAM) into ACC, which serves as the direct precursor for the synthesis of ethylene. Fortunately, certain bacteria play a pivotal role in mitigating plant stressors like drought, salt, heavy metal toxicity, and extreme temperatures by producing the ACC deaminase enzyme. This enzyme breaks down ACC into ammonia and α-ketobutyrate, thereby

Devesh Shukla (ed.), *Ethylene Sensing and Signaling*, Methods in Molecular Biology, vol. 2945,
https://doi.org/10.1007/978-1-0716-4650-2_1,

reducing ethylene-mediated responses triggered by the stress [2, 3]. Furthermore, beyond reducing the ethylene levels, ACC deaminase–producing bacteria offer additional benefits to the plants, such as enhancing nutrient uptake, increasing root growth, and suppressing the plant pathogens. For instance, in an investigation, the ACC deaminase–producing rhizospheric bacterium *Pseudomonas putida* UW4 was found to effectively reduce the submergence-triggered ethylene-mediated responses *in Rumex palustris* [4]. Moreover, studies have highlighted the beneficial effects of ACC deaminase–producing bacteria in the alleviation of various other stresses. For example, four strains of rhizobacteria *P. fluorescens* have been shown to ameliorate the detrimental impacts of drought stress and enhanced productivity in sweet corn (*Zea mays* L. var. saccharata) [5]. Additionally, inoculation with bacteria such as *Aneurinibacillus aneurnilyticus* and *Paenibacillus* sp. exhibited ACC deaminase activity by degrading ACC into α-ketobutyrate. Treatment with consortia containing these bacterial strains was found to effectively alleviate the negative effects of salinity stress and promote the growth of *Phaseolus vulgaris* seedlings. The treatment also led to increment in root length, shoot length, root biomass, shoot biomass, and total chlorophyll content [6]. Therefore, using ACC deaminase–producing bacteria in plants is looked upon as a sustainable and eco-friendly approach for improving plant performance under stress conditions (Fig. 1).

2 Materials

2.1 Media and Reagents

1. Luria-Bertani Broth (LB) medium: 10 g tryptone, 5 g yeast extract, 10 g NaCl, and 1 L of distilled water; adjust the pH to 7.0 with 1 N NaOH. Autoclave the mixture for 25 min at 120 °C.
2. Luria Agar (LA) medium: 10 g tryptone, 5 g yeast extract, 10 g NaCl, 15 g Agar, and 1 L of distilled water; pH 7.0. Autoclave the mixture for 25 min at 120 °C.
3. DF (Dworkin & Foster) salt minimal media: 0.2 g $MgSO_4{\cdot}7H_2O$, 6.0 g Na_2HPO_4, 4.0 g KH_2PO_4, 2.0 g glucose, 2.0 g citric acid, and 2.0 g gluconic acid. Additionally, trace elements 10 μg H_3BO_3, 11.2 μg $MnSO_4{\cdot}H_2O$, 1 mg $FeSO_4{\cdot}7H_2O$, 78.22 μg $CuSO_4{\cdot}5H_2O$, 124.6 μg $ZnSO_4{\cdot}7H_2O$, 10 μg MoO_3 (pH 7.2), and 2.0 g $(NH_4)_2SO_4$ act as a N_2 source [7] per liter water. Autoclave the mixture for 25 min at 120 °C.
4. 3 mM and 50 mM 1-aminocyclopropane-1-carboxylic acid (ACC) were sterilized using a nylon filter with a pore size of 0.22 μm.

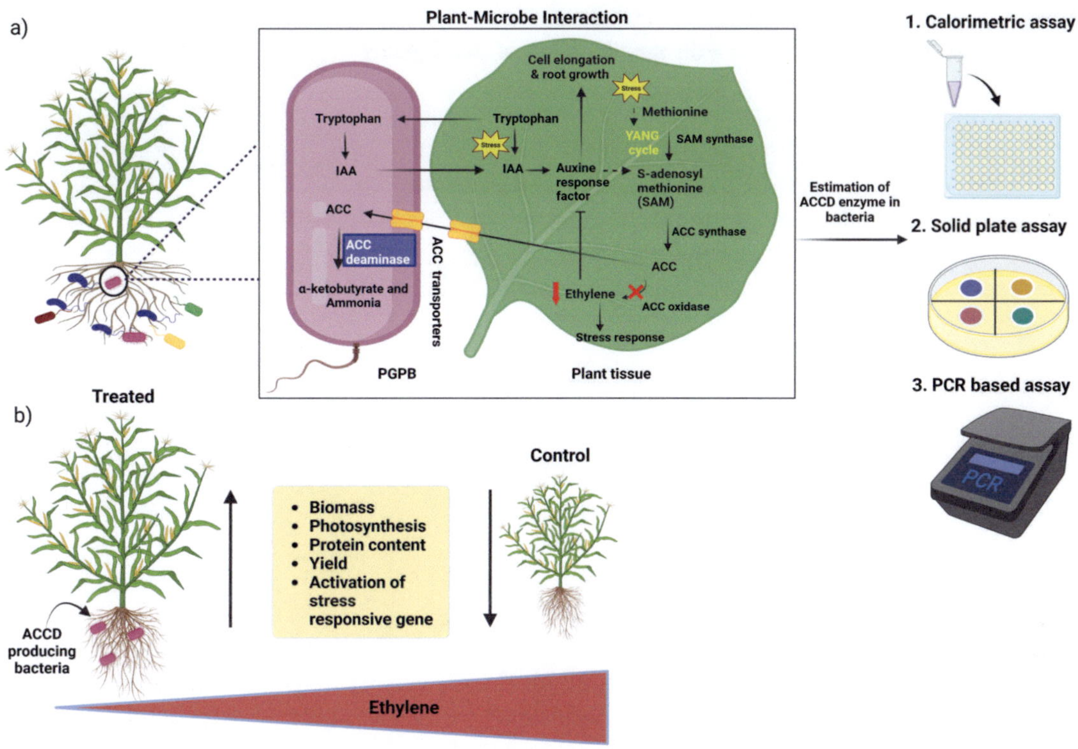

Fig. 1 (**a**) A model illustrating how plant growth–promoting bacteria (PGPB) can promote plant growth through the production of 1-aminocyclopropane-1-carboxylate (ACC) deaminase and indole-3-acetic acid (IAA). The enzyme ACC synthase converts S-adenosyl methionine (SAM) to ACC, which is further converted to ethylene by ACC oxidase. PGPB containing ACC deaminase activity can break down ACC, leading to the production of α-ketobutyrate and ammonia, thereby reducing the synthesis of plant growth–inhibitory ethylene. (**b**) A model depicting the role of ACC deaminase–producing bacteria in promoting plant growth through various mechanisms

5. DF-ACC medium: 0.2 g $MgSO_4 \cdot 7H_2O$, 6.0 g Na_2HPO_4, 4.0 g KH_2PO_4, 2.0 g glucose, 2.0 g citric acid, and 2.0 g gluconic acid. Additionally, trace elements 10 μg H_3BO_3, 11.2 μg $MnSO_4 \cdot H_2O$, 1 mg $FeSO_4 \cdot 7H_2O$, 78.22 μg $CuSO_4 \cdot 5H_2O$, 124.6 μg $ZnSO_4 \cdot 7H_2O$, 10 μg MoO_3 (pH 7.2), and 3.0 mM ACC act as a N_2 source per liter water.
6. 1 N and 2 N sodium hydroxide (NaOH) solution: 40 g and 80 g NaOH per 1 L H_2O, respectively.
7. 100 mM α-ketobutyrate.
8. 0.1 M Tris–HCl buffer (pH 7.6 and 8.5); 1X TAE buffer (pH 8.0).
9. 0.56 N HCl.
10. 2 M HCl.
11. 70% ethanol.
12. 5% toluene.

Table 1
Primers that have been used for PCR amplification of *acdS* genes

Primer set	Forward primer	Reverse primer	References
Set 1	5′-ATGAATCTGAATCGTTTT GAAC-3′	5′-TCAGCCGTTGCGGAACAG-3′	[11]
Set 2	5′-CAGCAGGAAAAGGATTT GGG-3′	5′-ACTCCACTGAATTGAAC CCG-3′	[12]

13. 0.2% ammonium sulfate.
14. 0.2% 2,4-dinitrophenylhydrazine (DNF).
15. 10 pM PCR primers specific to the ACC deaminase structural gene (*acdS*) (*see* Table 1).
16. PCR-grade water.
17. 1× PCR buffer.
18. 2.5 mM Nucleotides (dNTPs).
19. 1 U Taq polymerase.
20. 25 ng DNA template.
21. 1% Agarose gel: 1 g agarose was added to 100 mL of 1× Tris–acetate (TAE) buffer (*see* **Note 5**).

All media and reagents used were of analytical grade and obtained from Sigma-Aldrich Co., St. Louis, MO, USA, and Thermo Fisher Scientific, USA, respectively. All the media and solutions were freshly prepared as required before performing the assay, using double-distilled deionized water from Millipore, Bedford, MA.

2.2 Equipment

1. Sterile culture tubes (Borosil, India Ltd.).
2. Nylon filter, 0.20 μm (Sigma-Aldrich Co., St. Louis, MO, USA).
3. Autoclave (LabGard, India Pvt., Ltd.).
4. Rotary shaker (Thermo Fisher Scientific, USA).
5. Pipettes (Eppendorf, India).
6. Sterile tips (Tarson, India).
7. Centrifuge (Sigma, Germany).
8. Microcentrifuge tubes (Tarson, India).
9. Spectrophotometer (Thermo Fisher Scientific, USA).
10. 96-well microplate (Thermo Fisher Scientific, USA).
11. Sterile petri dish (Tarson, India).
12. BOD incubator (28 ± 1 °C) (LabGard, India Pvt., Ltd.).

13. Fume hood (LabGard, India Pvt., Ltd.).
14. Thermocycler (Applied Biosystems, USA).
15. Electrophoresis equipment and reagents (Genei, India Pvt., Ltd.).
16. UV transilluminator (Genei, India Pvt., Ltd.).
17. Inoculation loop.
18. Hand gloves.

3 Methods

3.1 Determination of ACC Deaminase Activity Through Colorimetric Assay

1. Inoculate a sterile culture tube comprising 15 mL of LB medium by a single colony of the bacterial culture and incubate it overnight at 28 °C temperature with shaking at 200 rpm (*see* **Notes 6** and **11**).
2. After overnight incubation, harvest 1 mL of the bacterial culture through centrifugation at 8000 g for 10 min at 4 °C, discarding the supernatant.
3. Rinse the cell pellet using 5 mL of DF salt minimal media and resuspend the same in 7.5 mL of DF-ACC medium.
4. Place the culture on a shaker and incubate it at 28 °C for an additional 36–72 h.
5. Harvest the bacterial cells through centrifugation at 8000 g for 10 min at 4 °C. After rinsing the cells twice with 0.1 M of Tris–HCl (pH 7.6), resuspend them in 600 μL of 0.1 M Tris–HCl (pH 8.5).
6. To toluenize the cells, add 5% toluene (volume by volume) and vortex for 30 s (*see* **Notes 9** and **10**).
7. Add 200 μL of the toluenized cell suspension along with 20 μL of 0.5 M ACC in a microcentrifuge tube and incubate it at 30 °C for 15 min.
8. Thoroughly mix the samples by vortexing with 1 mL of 0.56 N HCl and remove the debris of the cell by centrifugation at 16,000 g for 5 min.
9. Pour 1 mL of the supernatant to a glass test tube and then mix it with 800 μL of 0.56 M HCl and 300 μL of a solution containing 0.2% 2,4-dinitrophenylhydrazine (DNF) in 2 M HCl, and incubate the resulting mixture at 30 °C for 30 min.
10. Before measuring the absorbance at 540 nm, add 2 mL of 2 N NaOH to the sample.
11. The α-ketobutyrate concentration in each sample can be determined through comparing it to a standard curve that is prepared using α-ketobutyrate as a reference (*see* **Notes 3** and **4**) [8].

3.2 Determination of ACC Deaminase Activity Through Solid Plate Assay

1. Prepare LA plates and allow them to cool and solidify.
2. Inoculate the LA plates with the bacterial cultures using a sterile inoculation loop (*see* **Notes 6** and **11**).
3. Incubate the LA plates at 28 °C for 48 h.
4. Once the bacterial cultures have grown on the LA plates, select a single colony and inoculate it into the LB medium.
5. Keep the LB tubes at 28 °C in an incubator and shake them at 200 rpm for 24 h.
6. Prepare minimal medium plates and add either ammonium sulfate (2 g per liter) or 3 mM ACC as a nitrogen source.
7. Using a pipette, spot a small amount of the 48-h-old bacterial culture onto the prepared minimal media plates supplemented with 3 mM ACC (*see* **Note 8**).
8. Place the plates in an incubator at a temperature of 28 °C and allow them to grow for a period of 2–5 days.
9. After the incubation period, observe the plates on a daily basis for the bacterial growth.
10. Measure the diameter of bacterial growth on ACC-supplemented plates for different bacterial isolates.
11. The bacterial cultures growing well on ACC-supplemented media plates reveal that the bacterium can utilize ACC as a nitrogen source [9].

3.3 Determination of ACC Deaminase Activity in Bacteria Through PCR-Based Assay

1. Obtain genomic DNA from ACC deaminase–producing bacteria by utilizing established protocols such as the phenol/ chloroform extraction method [10] or other standard protocols (*see* **Note 11**).
2. Utilize the extracted genomic DNA as a template for PCR amplification of the *acdS* gene with primers specific to the gene.
3. Perform PCR reactions by mixing 25 μL of the reaction mixture containing 1× PCR reaction buffer, a mixture of 2.5 mM dNTPs, 10 pM of each specific primer, 1 unit of Taq DNA polymerase, and 25 ng of template DNA (*see* **Note 7**).
4. Conduct the PCR reaction in a thermal cycler using the following conditions: start with an initial denaturation step at 95 °C for 5 min, followed by 35 cycles of denaturation at 95 °C for 15 s, annealing at 55 °C for 15 s, and elongation at 72 °C for 30 s. Conclude the reaction with a final extension step at 72 °C for 5 min.
5. After amplification, load the PCR products onto a 1% agarose gel and visualize the bands. Once the bands are visualized, purify the desired band from the gel and then proceed with DNA sequencing.

6. Compare the obtained sequence with the existing *acdS* gene sequences available in the database for final confirmation of the result [11].

4 Notes

1. It is essential to use appropriate controls, including a negative control and a positive control, to validate the assay and ensure accurate measurement of ACC deaminase activity in bacterial cells.
2. Prepare a negative control by making a 200 μL toluenized cell suspension without ACC, and prepare a blank by mixing 0.1 M Tris–HCl (pH 8.5) along with 20 μL of 0.5 M ACC.
3. A 100 mM stock solution of α-ketobutyrate is prepared in 0.1 M Tris–HCl buffer (pH 8.5) and kept at 4 °C for storage. Make a graph that shows the concentration of α-ketobutyrate ranging from 0.1 to 1.0 μmol to use it as a reference during the assay.
4. α-Ketoglutarate is a stable chemical, and it is safe to store it at 2–8 °C temperature for long-term use. Fresh alpha-ketobutyrate should be clear and colorless. If the solution appears cloudy, contains particles, or has changed color (e.g., yellow or brown), it may indicate degradation.
5. TAE buffer is commonly prepared as a 50× stock solution for laboratory use. To prepare this solution, dissolve 242 g of Tris base in water, add 57.1 mL of glacial acetic acid, and combine with 100 mL of a 500 mM EDTA solution adjusted to pH 8.0. Make up the final volume to 1 L. This stock solution can then be diluted 49:1 with water to create a 1× working solution. The resulting 1× solution will contain 40 mM Tris, 20 mM acetic acid, and 1 mM EDTA.
6. All the steps should be carried out aseptically using sterile equipment to avoid contamination.
7. It is recommended to utilize newly prepared solutions for conducting PCR.
8. It is recommended to take necessary precautions to reduce errors while pipetting.
9. Make sure to wear hand gloves and use sterile techniques when handling laboratory materials. Toluene is a flammable chemical, so keep it away from open flames and sparks. Additionally, remember to wear lab coats and use fume hoods when working with volatile substances such as toluene.

10. Proper disposal methods for hazardous materials should also be followed according to the Resource Conservation and Recovery Act (RCRA). These precautions can help to avoid any disruptions in the protocol and ensure your safety in the laboratory.
11. The methods outlined in the manuscript are applicable to a range of bacterial strains that have ACC deaminase enzyme gene in their genome.

Acknowledgments

The team is thankful to the Director, CSIR CIMAP, Lucknow, India for facilitating with the laboratory aids. PK immensely acknowledges the CSIR-UGC SRF funding body for financial assistance through Senior Research Fellowship. The publication number provided by CSIR-CIMAP is CIMAP/PUB/2024/41.

References

1. Glick BR (2005) Modulation of plant ethylene levels by the bacterial enzyme ACC deaminase. FEMS Microbiol Lett 251(1):1–7. https://doi.org/10.1016/j.femsle.2005.07.030
2. Honma M, Shimomura T (1978) Metabolism of 1-aminocyclopropane-1-carboxylic acid. Appl Biol Chem 42(10):1825–1831. https://doi.org/10.1080/00021369.1978.10863261
3. Glick BR, Penrose DM, Li J (1998) A model for the lowering of plant ethylene concentrations by plant growth-promoting bacteria. J Theor Biol 190(1):63–68. https://doi.org/10.1006/jtbi.1997.0532
4. Ravanbakhsh M, Sasidharan R, Voesenek LA, Kowalchuk GA, Jousset A (2017) ACC deaminase-producing rhizosphere bacteria modulate plant responses to flooding. J Ecol 105(4):979–986. https://doi.org/10.1111/1365-2745.12721
5. Zarei T, Moradi A, Kazemeini SA, Akhgar A, Rahi AA (2020) The role of ACC deaminase producing bacteria in improving sweet corn (*Zea mays* L. var saccharata) productivity under limited availability of irrigation water. Sci Rep 10(1):20361. https://doi.org/10.1038/s41598-020-77305-6
6. Gupta S, Pandey S (2019) ACC deaminase producing bacteria with multifarious plant growth promoting traits alleviates salinity stress in French bean (*Phaseolus vulgaris*) plants. Front Microbiol 10:1506. https://doi.org/10.3389/fmicb.2019.01506
7. Dworkin M, Foster J (1958) Experiments with some microorganisms which utilize ethane and hydrogen. J Bacteriol 75(5):592–603
8. Penrose DM, Glick BR (2003) Methods for isolating and characterizing ACC deaminase-containing plant growth-promoting rhizobacteria. Physiol Plant 118(1):10–15. https://doi.org/10.1034/j.1399-3054.2003.00086.x
9. Khandelwal A, Sindhu SS (2013) ACC deaminase containing rhizobacteria enhance nodulation and plant growth in clusterbean (*Cyamopsis tetragonoloba* L.). J Microbiol 3(3):117–123
10. Sobieralski CA, Stanley DM (1994) DNA extraction strategies for amplified fragment length polymorphism analysis. J Forensic Sci 39(5):1254–1269
11. Ali SZ, Sandhya V, Venkateswar Rao L (2014) Isolation and characterization of drought-tolerant ACC deaminase and exopolysaccharide-producing fluorescent *Pseudomonas* sp. Ann Microbiol 64(2):493–502. https://doi.org/10.1007/s13213-013-0680-3
12. Wang C, Knill E, Glick BR, De'fago G (2000) Effect of transferring 1-aminocyclopropane-1-carboxylic acid (ACC) deaminase genes into *Pseudomonas fluorescens* strain CHA0 and its gac A derivative CHA96 on their growth-promoting and disease-suppressive capacities. Can J Microbiol 46(10):898–907. https://doi.org/10.1139/w00-071

Chapter 2

Method for the Measurement of Ethylene During Hypoxia in Rice Plants

Rekha Jaiswal, Kapuganti Jagadis Gupta, and Afsana Praveen

Abstract

Ethylene is a versatile phytohormone that is involved in the regulation of both growth and development such as senescence, and also it can act as a signaling hormone during hypoxia. Ethylene acts alone or in interaction with different phytohormones and proteins to regulate numerous cellular processes. Accumulating evidence suggest that endogenous ethylene production and emission into atmosphere are modulated by various biotic and abiotic stresses. Since it is a gaseous hormone, a precise detection, particularly under low-oxygen (hypoxic) conditions, is important for understanding its role in regulatory processes and stress signaling pathways. Currently, measurement practices such as gas chromatography, electrochemical sensing, and optical sensing are widely employed to detect ethylene. These methods are distinct from each other in terms of sensitivity, time response, selectivity, and cost. However, each method has its own advantages and limitations. Gas chromatography (GC) is one of the best techniques that is applied for the separation and measurement of ethylene due to its volatile and supersensitive nature. In this chapter, we describe a detailed GC-based procedure specifically optimized for measuring ethylene levels during hypoxic stress application in (*Oryza sativa*) rice plants.

Key words Ethylene, Gas chromatography, Hypoxia, Rice

1 Introduction

For the maintenance of growth, development, and to counter numerous biotic and abiotic stresses, plants modulate the biosynthesis of various hormones, among which ethylene is gaining importance. Measurement of ethylene is important for understanding the molecular and signaling mechanisms and its downstream processes. Gas chromatography (GC) is one of the widely utilized methods for the precise quantification of ethylene. In order to segregate and investigate volatile compounds of extreme sensitivity, GC is an often-used method; therefore, it emerged as an important tool for ethylene research as well. A complex mixture of compounds present in volatile form or that can be vaporized at a particular temperature can be examined through GC [1, 2]. Basically, the

Devesh Shukla (ed.), *Ethylene Sensing and Signaling*, Methods in Molecular Biology, vol. 2945,
https://doi.org/10.1007/978-1-0716-4650-2_2,

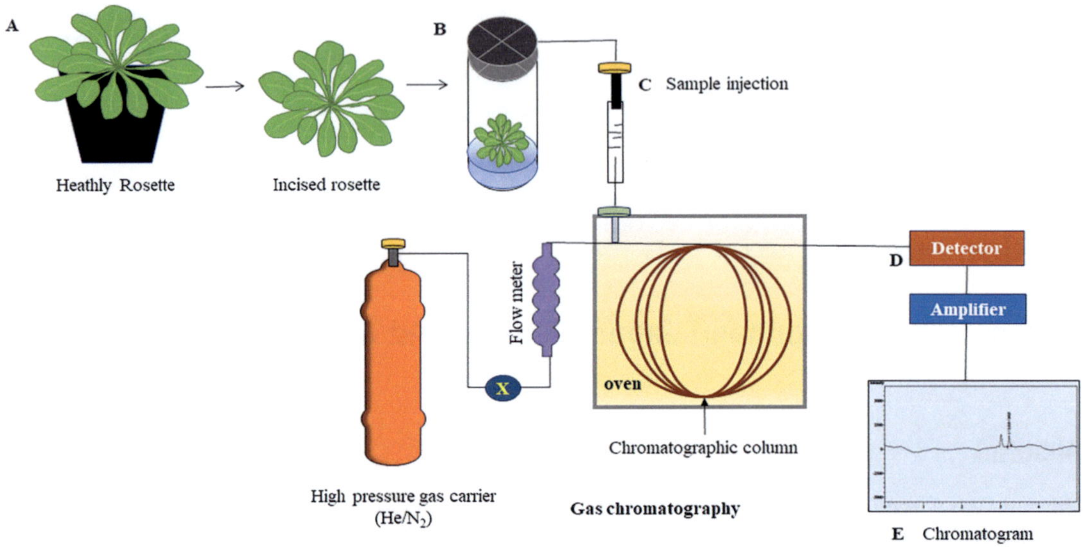

Fig. 1 A schematic diagram depicting the workflow of gas chromatography. A sealed incubation system used to trap ethylene released by intact plants. The example plant shown is *Arabidopsis thaliana*; the same arrangement can be applied to *Oryza sativa* seedlings for hypoxia assays. (**a**) Arabidopsis rosette incised at the root base. (**b**) Incised rosette in a glass vial with 2 mL of water. (**c**) Gas from the vial injected into the gas chromatography inlet. (**d**) Gas passed through the column is detected. (**e**) Chromatogram develops after 5 min run and peak develops after 3.2 min

function of GC is to distinguish constituents from a composite mixture through a specified column, analyze, and measure them separately (Fig. 1). The detailed procedure regarding the functioning of GC is already described by Cristescu et al. [3].

Frequent changes in environmental conditions occur due to climate change, negatively impacting crop yield and productivity, thus creating a threat to food security worldwide. Flooding/waterlogging, submergence, unseasonal rainfalls, and flash floods are among other abiotic stresses causing huge damage to crops in various climatic regions. Excess of water stress frequently creates hypoxia or anoxia, leading to energy crisis and causing concomitant effects such as poor germination, loss of seed vigor, and seedling establishment [4]. Rice is comparatively more tolerant to flooding in comparison to other cereal crops [5]. In order to cope with hypoxia/anoxia, plants have developed a variety of adaptive responses among ethylene-induced aerenchyma formation, and hyponastic responses are the most prevalent adaptive responses [6]. During the flooding stress, the growth of rice seedlings is promoted by ethylene, and this flooding-mediated elongation facilitates the plant's survival via hypoxia escape mechanisms and enables them to access air from atmosphere [7]. In rice plants, the overall elongation and growth induced by ethylene are restricted to not only internodes but also other developmental structures such as roots. In addition to proper gaseous exchange in shoots with their surroundings, ethylene also arbitrates adventitious root development and accelerates the energy metabolism required for better survival under hypoxia [8]. Accumulating evidence suggest that the production of ethylene during hypoxia and reoxygenation exposure is considered as one of the indicators of oxygen sensing and survival of plants. Ethylene response factor group VII (ERF VII) transcription factor family primarily regulates low oxygen sensing and survival through the N-end rule pathway of targeted proteolysis. In this pathway, cysteine residues present at the N-terminal of ERF VII are degraded in normoxia by plant cysteine oxidase (PCO) via the N-end rule pathway during the normal growth and development of plants. However, under hypoxia, the activity of PCOs is highly inhibited, resulting in the stabilization of ERF VII and leading to the enhancement of hypoxia-responsive gene expression [9]. Past studies reported that there is a wider cross talk between ethylene signaling and oxygen sensing and signaling mechanisms. For instance, in low oxygen escape syndrome and low oxygen quiescent syndrome, which occur in flooding-tolerant plants, ethylene along with other hormones such as nitric oxide plays a major role [1, 10]. In the submergence-resilient rice genotypes such as Swarna SUB1, China IRRI, Sonachur, Pravat, Pankaj, and Ranjit SUB1, a considerable enhancement in ethylene production was observed during the submergence exposure. During the submergence exposure in rice seedlings, genes involved in ethylene synthesis such as 1-aminocyclopropane-1-carboxylic acid (*ACC*), ACC synthase (*ACS*), and ACC oxidases (*ACOs*) with limited isoform transcripts become abundant in the culm and roots [11].

1.1 Measurement of Ethylene from Rice Plants during Hypoxia

In this chapter, we presented a detailed description of the ethylene measurement technique using GC. In this method, rice seeds (MTU-1010) are kept in dark for 2 days for inducing germination. After germination, seedlings are allowed to grow in petri dishes for a week. Thereafter, the rice seedlings are transferred to a green house in plastic pots kept inside customized polyacrylic box and divided them into two groups. The control group is kept in the box with the lid open and supplied with sufficient water for growth, while the other box is tightly closed with a lid to maintain hypoxia conditions and flooding stress (Fig. 2). After 1 week of growth in these conditions, the plants are used for the measurement of ethylene using GC (Fig. 3). In our experiments, during the flooding stress, the level of ethylene increased significantly in rice plants in comparison to control (normoxia) plants (Fig. 4). Ethylene facilitates plant escape mechanisms by recovering the air contact. It allows proper gaseous exchange via the development of aerenchyma or hyponasty for better plant survival during hypoxia.

Fig. 2 Complete procedures for plant growth under control and hypoxia conditions: (**a**) MTU-1010 seeds kept for germination. (**b**) Seedlings grown for 1 week. (**c**) Seedlings transferred to the green house in plastic pots filled with soil. Seedlings kept in customized polyacrylic box under control (normoxia) conditions, and the lid is open. (**d**) For maintaining hypoxia, the box is closed with a lid, and water is supplied in a larger amount for flooding. (**e**) Seedlings grown under these conditions for 10 more days. Plants grown under control conditions. (**f**) Flooding or hypoxia-treated plants

2 Materials

1. Rice plants (MTU-1010).
2. Nitrile gloves.
3. 0.1% Mercuric chloride.
4. Autoclave distilled water.
5. Beakers (100 mL).
6. Petri dishes.
7. Blotting papers.
8. Customized polyacrylic airtight box (20 cm × 20 cm × 20 cm).
9. Plastic pots filled with soil.
10. Plant growth chamber with optimum conditions for light, temperature, and humidity.

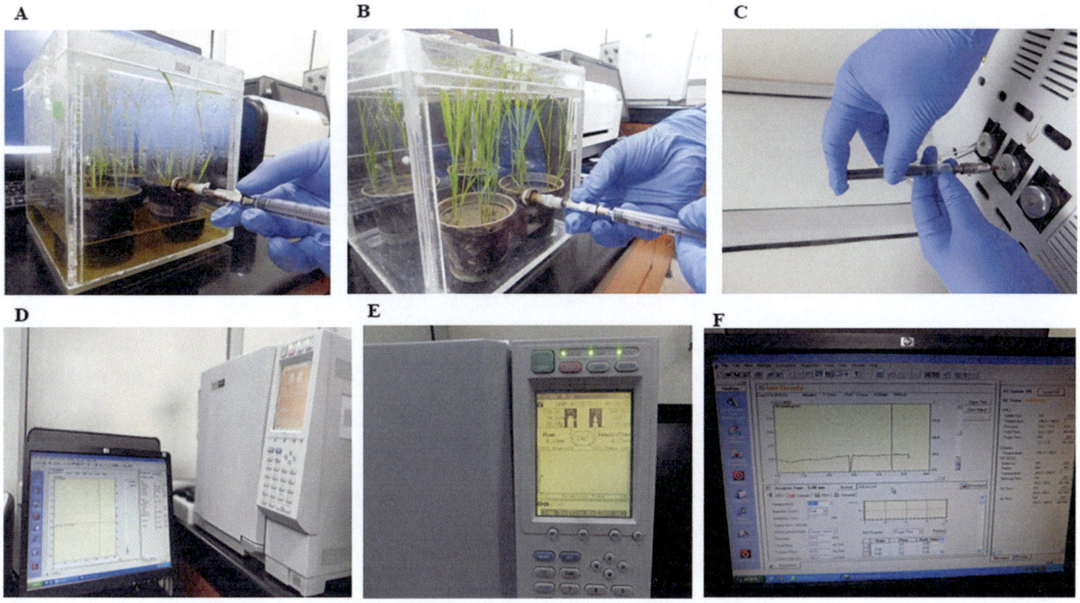

Fig. 3 Detailed procedures for ethylene measurement using gas chromatography (GC) in rice plants. (**a**) Withdrawal of 1 mL of headspace gas from hypoxia-treated plants. (**b**) Withdrawal of 1 mL of headspace gas from control plants. (**c**) Injection of samples into the GC column. (**d**) GC equipped with HP-AL/S 50 m, 0.53 mm column, and an attached monitor for analysis. (**e**) GC display screen showing detector, column, and injector temperatures. Analysis time is also displayed. (**f**) Monitor showing the chromatogram

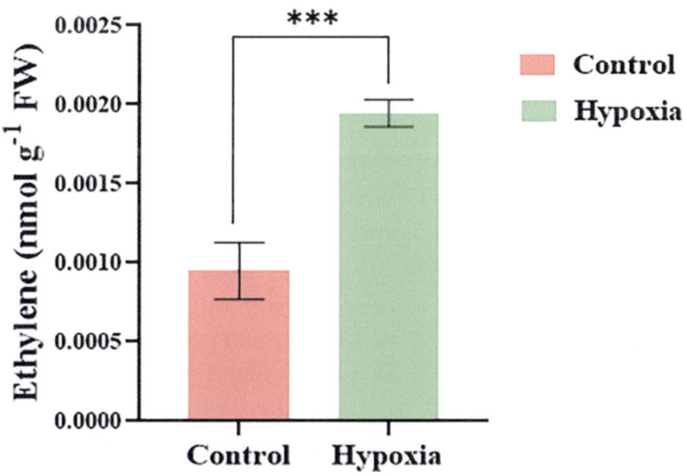

Fig. 4 Ethylene emission from rice plants under control (normoxia) and hypoxia conditions. The graph represents ethylene emission in nmoL g^{-1} FW. All the experiments were carried out in triplicate. A bar depicts the mean value with a standard error. Asterisks indicate values that are significant at $P \leq 0.001$ (***)

11. Weighing balance (Citizen, Cx 120).
12. Fine needle (2 mL syringe).
13. Gas chromatography (GC-2010,) equipped with HP-AL/S (HP-PLOT Al_2O_3 Standard column) 50 m, 0.53 mm column.

3 Methods

1. Hand-pick healthy rice seeds (50 g).
2. Surface-sterilize seeds with 0.1% mercuric chloride ($HgCl_2$) solution for 5 min and then discard the solution (*see* **Note 1**).
3. Use autoclaved distilled water for washing the seeds up to four to five times.
4. Cut the blotting papers according to the size of the petri dishes and place them inside. Dampen the blotting papers by spraying autoclaved distilled water.
5. For seed germination, keep sterilized seeds on the blotting papers with forceps.
6. Cover the petri dishes with lids (initially) to maintain the moisture of seeds.
7. Transfer the petri dishes to the dark phase and place them for 2 days at 23 °C.
8. After germination, transfer the petri dishes in the growth chamber under optimal conditions of temperature and light.
9. The optimum conditions for rice growth are 12–14 h light, and temperature should be between 25 °C and 30 °C.
10. Prepare the plastic pots for plant growth.
11. Grow the seedlings for 1 week, and divide them into control and treated groups.
12. Keep the 1-week-old rice seedlings in pots filled with 1:1 ratio of sterile soil and commercial potting mix (such as peat, coconut coir, wood products, vermiculite, sand, etc.). Place the pots in a customized polyacrylic box, one with the lid and another without the lid in a green house.
13. Maintain the optimum conditions in the greenhouse for rice growth (12–14 h of light and additional lightening during winters. The temperature of the day should be 25–30 °C, and night temperature should be 20–25 °C.
14. Add water to the pots, and keep the box open for control.
15. Supply water in excess amount for flooding (Fig. 2) and keep the box closed for maintaining the hypoxia conditions.
16. Grow the seedlings under the above conditions for 10 days.

17. Use triplicates for both control and hypoxia seedlings.
18. Withdraw 1 mL of headspace gas for measuring the ethylene production.
19. Weigh the seedlings to be used for ethylene measurement (*see* **Note 2**).
20. Calibrate the GC instruments (*see* **Note 3**).
21. Before injecting the sample into the GC chamber, set the parameters in the system attached to it.
22. Set the column temperature at 120 °C, detector temperature at 200 °C, injector temperature at 180 °C, and analysis time for 5 min (*see* **Note 4**).
23. Carrier gases present in the column are nitrogen and air gases, and for flame ionization detectors, hydrogen and air gases are used.
24. Carefully inject samples into a GC system equipped with an HP-AL/S 50 m, 0.53 mm column (*see* **Notes 5** and **6**).
25. Select the peak with RT ~3.2 on the X-axis which is displayed on screen.
26. Calculate the mean and standard error of the mean for three biological replicates.
27. Normalize the ethylene release with the seedling weight and ethylene standard gas used as controls.
28. Measure ethylene with independent biological replicates.
29. Express ethylene in terms of nmoL g^{-1} FW.

4 Notes

1. After the sterilization of seeds, proper discard of the used $HgCl_2$ is required as per safety practices.
2. Always use fresh plant samples for ethylene measurement.
3. It is advisable that for proper functioning of instruments, GC should be calibrated.
4. It is often recommended to set the temperature of injector, column, and detector for the separation of ethylene through GC.
5. Two facets that need caution and attention in GC are the sampling process and the injection of samples into columns.
6. A small volume of gas sample is required for inserting into the column (generally having an inner diameter of 5 mm for packed columns and 250 mm for capillary columns).

7. Always inject accurate samples, and any overflow results in defective disassociation.
8. To prevent the leakage of ethylene or to get the actual amount of its release, samples should be kept in an airtight container.

Acknowledgments

This work was financially supported by the MK Bhan Young Researcher Fellowship Programme HRD-12/4/2020-AFS-DBT-Part (1) (13815) from the Department of Biotechnology, Ministry of Science and Technology, Government of India. RJ acknowledges the Council of Scientific & Industrial Research (CSIR), India, for providing senior research fellowships (grant no. 09/803(0162)/2019-EMR-I).

References

1. Hartman S, Sasidharan R, Voesenek LACJ (2021) The role of ethylene in metabolic acclimations to low oxygen. New Phytol 229:64–70. https://doi.org/10.1111/NPH.16378
2. Hu B, Sun DW, Pu H, Wei Q (2019) Recent advances in detecting and regulating ethylene concentrations for shelf-life extension and maturity control of fruit: a review. Trends Food Sci Technol 91:66–82. https://doi.org/10.1016/J.TIFS.2019.06.010
3. Cristescu SM, Mandon J, Arslanov D, De Pessemier J, Hermans C, Harren FJM (2013) Current methods for detecting ethylene in plants. Ann Bot. https://doi.org/10.1093/aob/mcs259
4. Usman B, Derakhshani B, Jung KH (2023) Recent molecular aspects and integrated omics strategies for understanding the abiotic stress tolerance of rice. Plants (Basel) 12:2019. https://doi.org/10.3390/PLANTS12102019
5. Behera PK, Kumar V, Sharma SS, Lenka SK, Panda D (2023) Genotypic diversity and abiotic stress response profiling of short-grain aromatic landraces of rice (Oryza sativa L. Indica). Curr Plant Biol 33:100269. https://doi.org/10.1016/J.CPB.2022.100269
6. Li M, Xu G, Li H (2022) Simultaneous improvement of lodging resistance and grain yield by stacking dep1 with elite genes in rice. https://doi.org/10.21203/RS.3.RS-1726367/V1
7. Panda D, Barik J, Sarkar RK (2021) Recent advances of genetic resources, genes and genetic approaches for flooding tolerance in rice. Curr Genomics 22:41–58. https://doi.org/10.2174/1389202922666210114104140
8. Ma M, Cen W, Li R, Wang S, Luo J (2020) The molecular regulatory pathways and metabolic adaptation in the seed germination and early seedling growth of rice in response to low O2 stress. Plan Theory 9:1363. https://doi.org/10.3390/PLANTS9101363
9. Haque MA, Rafii MY, Yusoff MM, Ali NS, Yusuff O, Arolu F, Anisuzzaman M (2023) Flooding tolerance in rice: adaptive mechanism and marker-assisted selection breeding approaches. Mol Biol Rep 50(3):2795–2812. https://doi.org/10.1007/S11033-022-07853-9
10. Jagadis Gupta K, George Ratcliffe R, Mur LAJ, Wany A, Kumari A, Fernie AR (2020) The role of nitrite and nitric oxide under low oxygen conditions in plants. New Phytol 225(3):1143–1151. https://doi.org/10.1111/nph.15969
11. Neves M, Correia S, Cavaleiro C, Canhoto J (2021) Modulation of organogenesis and somatic embryogenesis by ethylene: an overview. Plan Theory 10:1208. https://doi.org/10.3390/PLANTS10061208

Chapter 3

Method for the Measurement of Ethylene During Pathogen Infection in Arabidopsis

Rekha Jaiswal, Kapuganti Jagadis Gupta, and Afsana Praveen

Abstract

Ethylene is a gaseous phytohormone that plays an important role as a signaling molecule during pathogen attack, influencing disease resistance and defense responses in plants. A precise measurement of ethylene production upon pathogen challenge is essential to elucidate its role in plant–pathogen interactions. Gas chromatography (GC) is among the most accurate and sensitive techniques for detecting and quantifying ethylene emissions due to its selectivity and effectiveness with gaseous molecules. In this chapter, we provide a detailed procedure employing GC specifically adapted for measuring ethylene levels during pathogen infection (*Botrytis cinerea*) in *Arabidopsis* leaflets. Arabidopsis leaflets infected by the necrotrophic pathogen *Botrytis cinerea* exhibit increased ethylene emission, facilitating the activation of defense pathways and secondary metabolites such as camalexin. The present GC method captures ethylene dynamics at early infection stages, ensuring precise quantification critical for dissecting the molecular mechanisms of plant immunity.

Key words Ethylene, Gas chromatography, Hypoxia, *Botrytis cinerea*, Arabidopsis

1 Introduction

Biotic stress triggers a plant immune system where several hormones, including ethylene, play a major role in certain interactions [1]. *Botrytis cinerea* is a necrotrophic fungus that belongs to the Sclerotiniaceae (ascomycetes) family and causes gray mold disease in several fruit crops such as tomatoes, grapes, strawberries, and flower crops such as rose. *Botrytis cinerea* uses various infection strategies for disease development to successfully invade the host, primarily through the secretion of cell wall–degrading enzymes, phytotoxins (polyketide botcinin, sesquiterpene botrydial), cell death-inducing proteins, carbohydrate-active enzymes (CAZyme), modulation of cellular pH, and altering the redox metabolism of the host [2]. Plants deploy ethylene and jasmonic acid as the second layer of defense to survive under stress conditions by reprogramming their metabolic activity [3]. In tomato, *Botrytis* accelerates the

Devesh Shukla (ed.), *Ethylene Sensing and Signaling*, Methods in Molecular Biology, vol. 2945,
https://doi.org/10.1007/978-1-0716-4650-2_3,

fruit-ripening process via inducing high ethylene production and activation of cell wall–degrading enzymes such as pectin-degrading enzymes which facilitate the fungus to colonize the fruit surface via necrotrophy, leading to postharvest losses [4]. The necrotrophic nature of *Botrytis* instigates the ethylene or jasmonic acid and Mitogen-activated protein kinase (MAPK) signaling pathways, which are synergistically responsible for resistance in plants by activating camalexin biosynthesis [5]. Low oxygen also aids *Botrytis* colonization over the host surface, so it induces local hypoxia around the inoculated region dissected by the expression of the ERF-VII protein RAP2.12 [6]. In rose and carnation flowers, it was found that ethylene induces susceptibility against *Botrytis* [7]. A study reveals that iron deficiency in *Arabidopsis* can lead to resistance against *Botrytis* and activates ethylene emission [8].

1.1 Measurement of Ethylene Levels in Arabidopsis Against Botrytis cinerea

In this protocol, we describe the method to measure the ethylene levels in *Arabidopsis* rosette in response to *Botrytis cinerea* infection using the GC method (Fig. 1). Ethylene is a gaseous signaling molecule which can be detected by using the sensor in a GC system and represented as a peak in the chromatogram (Fig. 1m). The area under the chromatogram is considered equivalent to the ethylene emission under stress conditions. In our experiment, we observed the elevation in ethylene emission in response to *Botrytis* after 1 and 2 dpi (Fig. 2). For ethylene measurement, a healthy rosette of Col 0 and fresh *Botrytis* culture are required. After infection with an

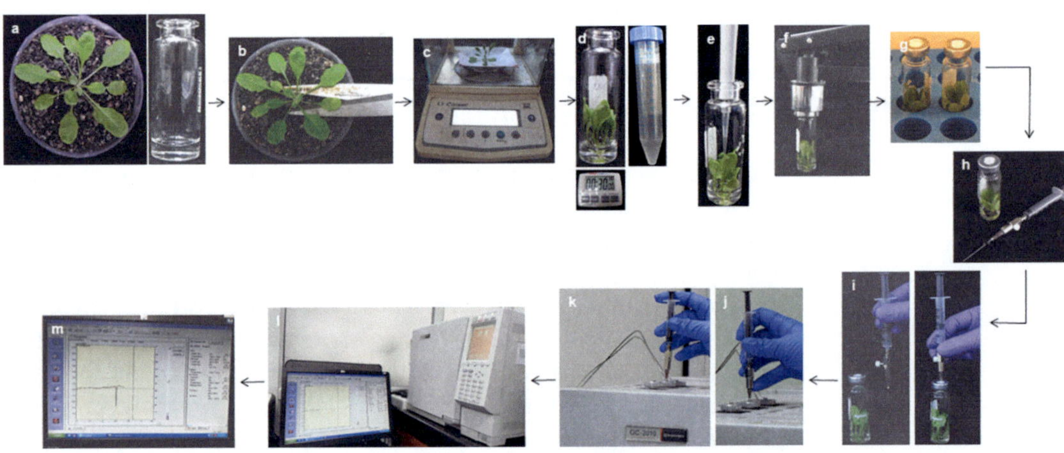

Fig. 1 Steps in ethylene estimation using GC. (**a**) *Arabidopsis* plant and glass vials with 2 mL of autoclaved water in them. (**b**) Rosette incised at the base. (**c**) Incised rosette weighed, and weight was noted. (**d**) Rosette kept in a glass vial, and the vial opened for 30 min, and *Botrytis cinerea* suspension of 1×10^6 concentration was kept in a falcon. (**e**) Three to four leaves of a rosette inoculated with 10 μL of spore suspension. (**f**) Immediately after infection, the glass vial is sealed with an aluminum cap. (**g**) Incubation. (**h**) Samples and 2 mL syringe. (**i**) A syringe which is placed inside the vial and 1 mL of air is taken. (**j**) Syringe taken into the column. (**k**) Air injected inside the column. (**l**) Run the start through the column. (**m**) Chromatograms develop after 5 min run

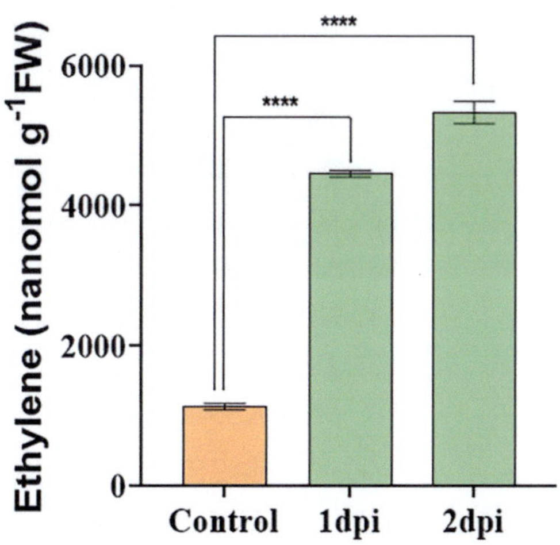

Fig. 2 Graph showing ethylene emission (nanomole g^{-1} FW) in response to *Botrytis cinerea* in Col-0 plants after 1 and 2 days post infection (dpi). Each bar represents the mean value of three different replicates with a standard error. Asterisks indicate $P \leq 0.0001$ (****), $P \leq 0.001$ (***), $P \leq 0.01$ (**), and $P \leq 0.05$ (*)

optimized concentration of *Botrytis*, over (1×10^6) leaflets were enclosed in 20 mL glass vials and incubated for the required 1–2 days. Upon infection, the necrotrophic nature of *Botrytis* elevates ethylene emission, which was taken out from the glass vial with the help of a syringe and injected into the chromatography column to get the peak for ethylene. The area under the peak at the retention time of 3.2 min was taken, and ethylene levels were calculated in nano mol g^{-1} FW. Although demonstrated here at an early-stage growth, the protocol can be adapted to many crops and tissue types. We have demonstrated protocols in early-stage growth, but this protocol can be well suitable for any crop and tissue type.

2 Materials

1. Four weeks old *Arabidopsis* rosette (Col-0) (grown under 8/16 h (day/night) photoperiod, 22/18 °C temperature, and 100 μmoL m^{-2} s^{-1} light intensity).
2. *Botrytis cinerea* culture (grown under 22 °C temperature, 21 days old).
3. Autoclaved water.
4. 20 mL glass vial.
5. Hemocytometer.
6. Coverslip.
7. Weighing balance.

8. Scissors.
9. Forceps.
10. Timer.
11. 25 mL falcon.
12. 10 μL pipette and tips.
13. Aluminum cap.
14. Glass vial sealer and opener.
15. 2 mL syringe.
16. Microscope.
17. GC instrument (Shimadzu, GC-2010, equipped with a HP-AL/S 50 m, 0.53 mm column, Agilent Technologies).

3 Methods

3.1 Plant Growth

1. Take Col-0 seeds in a microcentrifuge tube (1.5 mL) and sterilize using 70% ethanol and 0.1% TritonX-100.
2. Wash the seeds three to four times with 100% ethanol. Air-dry the seeds by placing them on a sterile filter paper.
3. Sow the seeds in an Agropeat:Soilrite (1:1) mixture and keep in a cold room at 4 °C for 2 days for stratification in the dark and then transfer to the growth chamber.
4. After 10 days, transfer the plants into individual pots and keep them into 8/16 h (day/night) photoperiod, 22/18 °C temperature, and 100 μmoL m^{-2} s^{-1} light intensity in the growth chamber.
5. After 4 weeks, use rosette-stage plants to perform the experiment.

3.2 Botrytis cinerea (B05.10) Growth

1. Prepare potato dextrose agar (PDA) media.
2. Pour them on petri dishes.
3. Inoculate the media with a mycelium plug of *Botrytis* inside a laminar air flow.
4. Keep plates in the growth chamber at 22 °C temperature for 21 days.
5. After spore development, take the plates out of the chamber.
6. Harvest spores from the *Botrytis* plates using sterile water and collect them in a 50 mL falcon tube.
7. Prepare a dilution of spores by taking 100 μL of spores in 5 mL of sterile water.
8. Take 10 μL of spores and load them into both grids of the hemocytometer.

9. Count spores in bright field at 40× magnification of the microscope.
10. Calculate spore concentration using the formula:

$$\text{Total number of spores}\,\text{mL}^{-1} = \frac{\text{Total number of spores counted}}{\text{Number of squares in hemocytometer}} \times 10^4 \times V_i \times D$$

Vi = Initial Volume, d = dilution factor.

11. Prepare the optimized concentration of 1 × 10^6 by diluting the stock of *Botrytis* spore suspension using the formula: Concentration (Stock spore count) × volume taken from the stock = Concentration required (1 × 10^6 spores) × volume required (C1V1 = C2V2).

3.3 Protocol Steps

1. Take a healthy rosette (Col-0).
2. Take 2 mL of sterile water in a 20 mL glass vial.
3. Cut the rosette at the base.
4. Weigh the rosette and note it.
5. Transfer the rosette into a glass vial with water using forceps (Fig. 1).
6. Keep the vial open for 30 min to remove wound-induced ethylene (*see* **Note 1**).
7. After that, inoculate three to four leaflets with 10 μL of *Botrytis* spores (1 × 10^6 spore concentration) (*see* **Note 2**).
8. Immediately, seal the vial with an aluminum cap using a clip sealer (*see* **Note 3**).
9. Keep the vial for incubation.
10. After a specific time point, take the samples.
11. Take out 1 mL of gas from the vial using a 2 mL syringe.
12. Inject it directly into the column (*see* **Note 4**).
13. Run the protocol for an analysis time of 5 min, with splitless inlets (SPL) at 180 °C, column at 120 °C, and flame ionizer detector (FID) at 200 °C.
14. After a 5 min run, the screen will display a chromatogram.
15. The retention time for ethylene is at 3.2 min.
16. Calculate the area under the peak.
17. Calculate the ethylene level in nanomol g^{-1}FW after comparing with the standard (0, 0.001, 0.01, 0.1, 10, and 100% of ethylene volume and 1 mL of 100% of ethylene calculated to be 41,160 nanomoL).
18. Calculate the ethylene level in nanomoL g^{-1} after comparing with the standard.
19. Plot the graph for ethylene levels for control at 1 and 2 days postinfection.

4 Notes

1. During ethylene measurement in *Arabidopsis* rosette, avoid opening the control samples for more than 30 min; this will cause an exorbitant loss of ethylene.
2. Samples inoculated with spores should be sealed immediately after the infection to prevent cross-contamination.
3. A glass vial should be sealed properly and incubated for appropriate time points to cause pathogenicity.
4. 1 mL of gas from the vial should be taken and carefully injected into the column to avoid extra quantity of gas.

Acknowledgments

This work was financially supported by the MK Bhan Young Researcher Fellowship Programme HRD-12/4/2020-AFS-DBT-Part (1) (13815) from the Department of Biotechnology, Ministry of Science and Technology, Government of India. RJ acknowledges the Council of Scientific & Industrial Research (CSIR), India, for providing senior research fellowships (grant no. 09/803(0162)/2019-EMR-I).

References

1. Bhar A, Chakraborty A, Roy A (2021) Plant responses to biotic stress: old memories matter. Plants (Basel) 11:84. https://doi.org/10.3390/PLANTS11010084
2. Bi K, Liang Y, Mengiste T, Sharon A (2023) Killing softly: a roadmap of Botrytis cinerea pathogenicity. Trends Plant Sci 28:211–222. https://doi.org/10.1016/J.TPLANTS.2022.08.024
3. Pérez-Llorca M, Pollmann S, Müller M (2023) Ethylene and jasmonates signaling network mediating secondary metabolites under abiotic stress. Int J Mol Sci 24:5990. https://doi.org/10.3390/IJMS24065990
4. Silva CJ, Adaskaveg JA, Mesquida-Pesci SD, Ortega-Salazar IB, Pattathil S, Zhang L, Hahn MG, van Kan JAL, Cantu D, Powell ALT, Blanco-Ulate B (2023) Botrytis cinerea infection accelerates ripening and cell wall disassembly to promote disease in tomato fruit. Plant Physiol 191:575–590. https://doi.org/10.1093/PLPHYS/KIAC408
5. Cox KL (2022) Stronger together: ethylene, jasmonic acid, and MAPK signaling pathways synergistically induce camalexin synthesis for plant disease resistance. Plant Cell 34:2829–2830. https://doi.org/10.1093/PLCELL/KOAC155
6. Valeri MC, Novi G, Weits DA, Mensuali A, Perata P, Loreti E (2021) Botrytis cinerea induces local hypoxia in Arabidopsis leaves. New Phytol 229:173–185. https://doi.org/10.1111/NPH.16513
7. Bika R, Baysal-Gurel F, Jennings C (2020) Botrytis cinerea management in ornamental production: a continuous battle. 43:345–365. https://doi.org/10.1080/07060661.2020.1807409
8. Lu C, Liang G (2023) Fe deficiency-induced ethylene synthesis confers resistance to Botrytis cinerea. New Phytol 237(2023):1843–1855. https://doi.org/10.1111/nph.18638

Chapter 4

Determination of the Effect of Ethylene Treatment on the Ripening Process, Weight Loss, Free Humidity, Colorimetry, and Drying Kinetics of *Solanum lycopersicum L* (Tomato)

Gabriel Alfonso Burgos-Briones, María Hipatia Delgado-Demera, Leonardo Andres Saltos-Meza, Luis Daniel García-Parrales, Ulbio Eduardo Alcívar-Cedeño, and Carlos Alfredo Cedeño-Palacios

Abstract

This chapter examines the effects of ethylene application on the maturation, weight reduction, moisture content, color characteristics, and drying behavior of *Solanum lycopersicum* L. (tomato). Employing a 100 ppm ethephon (an ethylene source) concentration, a notable acceleration of the ripening process occurs, shortening the usual 20-day period to only 4 days. This method adhered to stringent criteria for selecting, preparing, and analyzing samples, emphasizing reproducibility and compliance with international quality guidelines like the Codex Stan and national regulations including NMX-FF-031 and NTE INEN 2832. Color changes in tomatoes are monitored using the following colorimetric parameters, L^*, a^*, and b^*, and weight loss is shown in a controlled environment of 23 °C and 80% humidity. Moisture content is monitored using a thermobalance device, and acidity levels are evaluated using titratable acidity techniques. The values of the drying process at different temperatures (105 °C, 115 °C, and 130 °C) show better results, especially at 130 °C. This thorough chapter highlights the role of ethylene in improving tomato ripening and sheds light on the resultant alterations in physical and chemical attributes, conforming to international quality standards and suggesting applications for agricultural and food processing sectors.

Key words Ethylene treatment, Tomato ripening, Drying kinetics, Colorimetry, *Solanum lycopersicum*

1 Introduction

Domesticated tomato (*Solanum lycopersicum L.*), a prominent member of the Solanaceae family and genus Solanum L., has

Electronic Supplementary Material: The online version of this chapter (https://doi.org/10.1007/978-1-0716-4650-2_4) contains supplementary material, which is available to authorized users.

Devesh Shukla (ed.), *Ethylene Sensing and Signaling*, Methods in Molecular Biology, vol. 2945,
https://doi.org/10.1007/978-1-0716-4650-2_4,

significant biological and economic value within its family. Globally recognized for its nutritional and culinary attributes, tomato, a fruiting vegetable, is widely consumed in its red-ripe and fresh state and is possibly the largest documented economic plant species in the family [1, 2]. Its widespread use is especially notable in areas such as Ecuador, underscoring its global importance as a culinary staple [3, 4]. In 2020, Ecuador planted approximately 2579 ha of tomatoes, achieving an average yield of around 14.9 t/ha and a total production of around 38,438 t ha^{-1} [5]. The Sierra Region emerges as the epicenter of this activity, mainly covering some 1976 ha. In terms of provinces, Imbabura, Manabí, and Pichincha stood out as the main producers, with areas of 971, 579, and 263 ha, respectively [6].

Tomatoes are relatively simple to grow but face production problems, especially under varying environmental conditions [7]. The predominant challenges in tomato cultivation in Ecuador involve inadequate water management and fertigation [8], as well as the presence of viral diseases [9, 10] and pests such as the tomato moth (Tuta absoluta) [11]. In addition, it faces challenges such as low soil fertility or contamination, water quality problems, and the constant threat of drought, all of which impact not only tomato production but agriculture also in general [12]. For example, increased ambient temperature can reduce fruit yields, generating a marked decrease in production, which underscores the need for efficient cultivation and ripening methods [13, 14].

Ethylene, a plant hormone, plays a key role in the ripening of agricultural products, including tomatoes. It regulates essential developmental stages leading to aging. This natural process can lead to a decrease in nutritional and commercial value [15–17]. The complex interplay between ethylene synthesis and response, changes in cell wall metabolism, and the impact of environmental elements such as light on ripening has attracted a great deal of scientific interest [18]. Understanding these processes is vital, considering the role of fruits in human nutrition and their unique biological transformations [19, 20].

After harvest, the transit time from farm to consumer can last for weeks, during which phytochemical reactions to environmental conditions can significantly modify the biological and medicinal properties of the fruits; this period is crucial to preserve product quality and safety [21]. Factors such as weight loss and free moisture are key quality indicators at this stage. Free moisture, which reflects water evaporation, is affected by the intrinsic properties of the food and storage or processing environment [22, 23].

In the postharvest life of tomatoes, the stability of lycopene, a pigment subject to oxidative degradation and isomerization during drying and heating, is also critical [24]. The main objective in drying produce such as tomatoes is to reduce the moisture content for safe and prolonged storage. Reducing water activity not only prolongs shelf life but also prevents pathogen growth, slows enzymatic activities, and reduces moisture-related adverse reactions [25–27].

Tomato skin color is an excellent indicator of maturity. To accurately assess these changes, methods such as acoustic impact and colorimetry techniques are suggested [28–30]. These nondestructive tests offer a quantitative and effective way to determine the ripening stages of tomatoes, improving on traditional sensory methods [31, 32].

In this chapter, we present an exhaustive procedure for ripening *Solanum lycopersicum L.* (tomato) using ethephon as a source of ethylene treatment. We aim to clarify the impact of this treatment on aspects like weight loss, free humidity, colorimetry, and drying kinetics, offering a methodology that can be replicated and utilized in various agricultural and research contexts.

2 Materials

2.1 Equipment

1. Colorimeter.
2. Digital electric scale.
3. Thermobalance.
4. Digital pH meter.
5. Blender.
6. Magnetic stirrer.
7. Thermo-hygrometer.
8. Tunnel dryer.
9. Mortar and pestle.
10. Storage containers.
11. Beaker (150 mL).
12. Metal tweezers.
13. Sample holder.
14. Newspaper.
15. Cutter (knife).
16. Pipette (10 mL) and burette.
17. Erlenmeyer flask (250 mL).
18. Magnetic tablet.
19. Flask (50 mL).
20. Storage container (1 L).
21. Syringe (3 mL/cc).
22. Sponge (100 × 75 × 30 mM).

2.2 Solutions

1. Calibration buffers.
   (a) Buffer 4.
   (b) Buffer 7.
   (c) Buffer 10.
2. Distilled water.
3. Phenolphthalein.
4. Ethanolic KOH solution.
5. Ethylene releaser.

2.3 Fresh Tomatoes

Fresh tomatoes are the primary subject of this study and will undergo various analyses to assess the impact of ethylene on their ripening process.

3 Methods

3.1 Ripening with Ethylene

3.1.1 Selection of Tomatoes

Start by choosing healthy tomatoes, making sure they are devoid of holes, decay, or other imperfections. Record the weight of each tomato to track any alterations throughout the ripening period.

3.1.2 Preparation of Ethephon Solution

1. Place a sample of approximately 840 g of the chosen tomatoes in a container.
2. Create a diluted ethephon solution (a source of ethylene) as follows: Obtain a commercial ethephon from a standard bottle designated for experimental use. Utilizing a precise syringe, extract 0.14 mL of ethephon. Introduce this liquid into a sealed container holding 1 L of distilled water. This container should have a special inlet valve to facilitate the safe and precise addition of ethephon into water, resulting in a diluted ethephon solution with a concentration of 100 ppm ethephon.

   Concentration of 100 ppm of (ethephon)

   $$\text{ppm} = \frac{\text{Miligrams solute (mg)}}{\text{Liter solution (L)}}$$

   $$720\frac{\text{g}}{\text{L}} * \frac{1000\text{mg}}{1\text{g}} = 720,000\frac{\text{mg}}{\text{L}}$$

   For the volume of the solution, the known data are used

   $$C1 * V1 = C2 * V2$$

   $$720{,}000 * X = 100 \text{ ppm} * 1 \text{ L}$$

   $$X = \frac{100\text{ppm} * 1\text{L}}{720,000\text{ppm}}$$

$$X = 1.38888 \times 10^{-4} L$$

$$1.3888 \times 10^{-4}\ \mathrm{L} * \frac{1000\ \mathrm{mL}}{1\ \mathrm{L}} = 0.138\ \mathrm{mL}$$

0.138 mL ethephon diluted to 1 liter of water. (*see* **Note 1**)

3. Gently mix this solution to achieve uniformity.
4. Immerse a clean sponge in the ethephon solution and softly apply it over the surface of each tomato to ensure uniform coverage.

3.1.3 Ripening Procedure

1. Place the ethylene-treated tomatoes in a fresh container lined with newspaper. This lining aids in absorbing surplus moisture and keeps the environment clean.
2. Hermetically seal the container, creating an airtight space to prevent ethylene from escaping and to maintain a stable internal environment.
3. Keep the sealed container in a controlled setting where the temperature is held between 22 °C and 24 °C, an ideal range for ethylene-induced ripening.
4. Set the ripening period to last for 4 days (*see* **Note 3**).

3.1.4 Monitoring and Inspection

1. Regularly inspect the tomatoes during the 4-day period, observing changes in color, texture, and other visible signs of ripening.
2. Consistently use a thermo-hygrometer to monitor and record the environmental conditions of temperature and humidity within the storage area, ensuring stable conditions for accurate and consistent ripening.
3. Ethephon acquired had a concentration of 100 ppm; then, applying calculations, it was considered to place 0.14 mL of ethephon in a volume of 1 L of distilled water.

3.2 Colorimetry

3.2.1 Introduction to Colorimetry

1. In this context, colorimetry involves using L^*, a^*, and b^* parameters to evaluate the color changes in tomatoes during ripening.
2. L^* (Lightness) ranges from 0 (black) to 100 (white), indicating the sample's lightness or darkness.
3. a^* and b^* represent color dimensions, with a^* indicating red/-magenta to green (positive values for red/magenta, negative for green) and b^* indicating yellow to blue (positive values for yellow, negative for blue).
4. Saturation and Hue: a^* and b^* together assess color saturation (color purity) and hue (perceived color).

Table 1
Tomato color parameters

Tomato type	L^*	a^*	b^*	C^*	h^*
A0	27.10	41.80	64.10	76.52	56.89
A1	25.20	38.30	58.00	69.50	56.56
A2	30.20	45.90	69.80	83.54	56.67
A3	23.20	38.80	59.60	71.11	56.93
A4	28.60	41.40	65.70	77.65	57.78
A5	28.10	39.30	64.90	75.87	58.80
A6	27.30	41.60	63.00	75.49	56.56

Source and preparation: Authors

3.2.2 Calculating Chroma and Hue Angle (see Table 1)

Chroma (C^*): Represents color intensity, calculated using C^*

$$\text{chromatide}\,(C^*) = \sqrt{a^{*2}} + b^{*2.}$$

Hue Angle (h^*): Represents the color tone, calculated as: $h^*=$

$$(h^*) = \text{arct}\,g\ b^* / a^*$$

3.2.3 Measurement Procedure (Adapted from Talens Oliag Ref. [33])

1. Calibrate the colorimeter, placing the measuring head on the calibration plate and activating the "Calibrate" function.
2. Switch the instrument to "Measure" mode.
3. Position the colorimeter on the sample surface, ensuring consistent conditions for accuracy.
4. Document L^*, a^*, and b^* values (*see* Table 1).

3.2.4 Factorial Design and Color Difference Calculation

1. Use factorial design ($y = A * b$) to interpret results.
2. Calculate color difference (ΔE) between the initial and final ripening stages using $\Delta E = [(L_1 - L_2)^2 + (a_1 - a_2)^2 + (b_1 - b_2)^2]^{1/2}$ (*see* Table 2) quantifying color change due to ripening and ethylene treatment (*see* **Note 2**).

3.3 Weight Loss Measurement

Quantify the tomato weight loss over 4 days at 23 °C and 80% humidity, simulating typical storage environments and assessing ethylene's impact on weight loss during ripening. For weighing:

1. Record each tomato's initial fresh weight (PIF) using a calibrated digital electric scale (Adapted from Fonseca Ref. [34]).
2. Store tomatoes in controlled conditions for 4 days.
3. Weigh each tomato to determine the final weight (PFF).

Table 2
Variations of ΔE in tomatoes

	Temperature 22 °C		Temperature 26 °C	
Reply	Time	ΔE	Time	ΔE
1	1	70.8078	1	172.6809
	2	528.9278	2	178.2784
	3	603.4003	3	114.3084
	4	1050.0478	4	56.2134
	5	1117.2553	5	105.3409
	6	1210.4753	6	84.8184
2	1	217.6944	1	208.5184
	2	376.9694	2	176.9359
	3	549.1869	3	197.7409
	4	981.5769	4	194.5759
	5	1048.7144	5	173.8159
	6	1158.2044	6	159.6934
3	1	244.3928	1	86.7853
	2	910.8478	2	298.7128
	3	932.2178	3	166.8403
	4	1477.5403	4	321.0628
	5	1544.3753	5	253.1303
	6	1684.7028	6	279.7103

Source and preparation: Authors

4. Calculate the weight loss percentage (PPF) for each tomato using the formula (*see* Fig. 1):

$$\text{PPF }(\%) = \frac{\text{PIF} - \text{PFF}}{\text{PIF}} \times 100$$

3.4 Free Humidity Measurement

Measure free humidity to determine the moisture content in tomatoes, essential for understanding ethylene's impact on water content during ripening.

3.4.1 Sample Preparation

Begin by setting the ambient temperature to either 26 °C or 27 °C, as these conditions are optimal for preparing tomato samples (Adapted from Nollet and Nollet Ref. [35]).

3.4.2 Thermobalance Setup

1. Balance and level the thermobalance, placing the tomato sample in the holder.
2. Set the thermobalance to 105, 115, and 130 °C and dry for 20 min.
3. Record the final value indicating free humidity or moisture content (*see* Table 3).

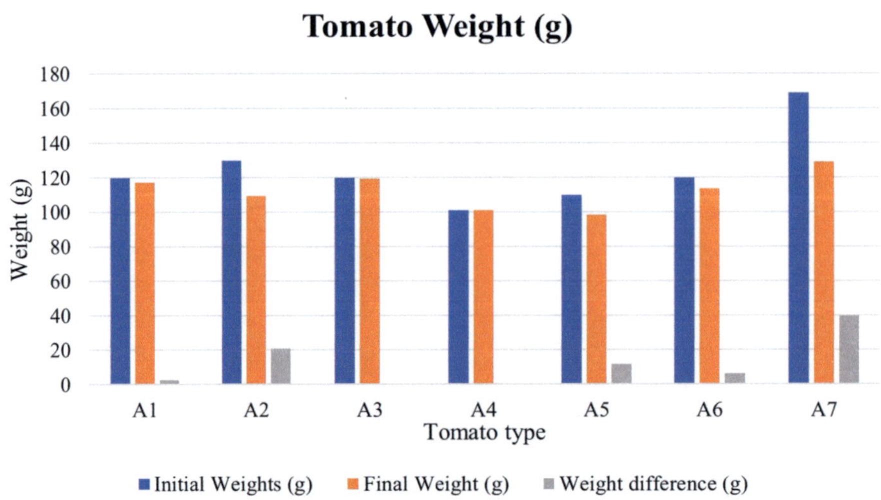

Fig. 1 Relationship between initial weight and percent weight loss (%PPF) in different types of tomatoes. And if there is a significant correlation between initial weight and %PPF, the initial and final weights of seven different types of tomatoes were recorded, and the %PPF was calculated for each type. Data were analyzed using descriptive statistics and Pearson's correlation analysis to evaluate the relationship between initial weight and %PPF. Mean initial weight: 124.29 g; mean %PPF: 8.29%; standard deviation of initial weight: 21.73 g; standard deviation of %PPF: 9.14%; Pearson's correlation between initial weight and %PPF: 0.820; *P*-value: 0.024. There is a strong and significant positive correlation between initial weight and %PPF. This suggests that tomatoes with a higher initial weight tend to have a greater percent weight loss during drying

Table 3
Free moisture present in the tomato samples

Tomato type	Humidity %
A0	94.53
A1	94.22
A2	94.50
A3	94.31
A4	94.10
A5	93.89
A6	94.60

Source and preparation: Authors

3.5 Acid Number Measurement

The aim of measuring the acid number is to determine the titratable acidity of tomatoes, to assess the ripening process and fruit quality (Adapted from Terevinto Ref. [36]).

3.5.1 Sample Preparation

1. Cut and blend tomatoes, measuring out 10 mL of juice.
2. Calibrate the pH meter; gradually add NaOH to the tomato juice until pH reaches 8.1.

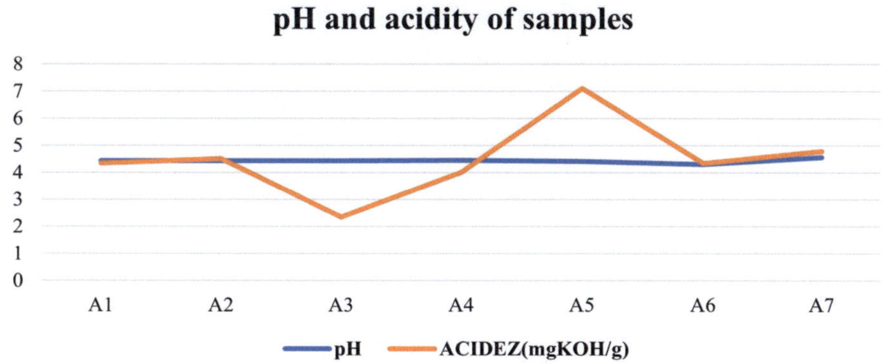

Fig. 2 Relationship between pH and acidity of different types of tomatoes and the variability in these parameters. The pH and acidity levels (expressed in mg KOH/g) of seven different types of tomatoes labeled A0 to A6 were measured. Measurements were performed using standard laboratory methods to ensure accuracy and consistency. Tomatoes have an average pH of 4.437 and an average acidity of 4.487 mg KOH/g. The variability in pH values is low (SD = 0.074), indicating the consistency in pH levels among the different tomato types. On the other hand, acidity shows greater variability (SD = 1.403), suggesting more notable differences in acidity levels between tomato types. No significant correlation was found between pH and acidity (Pearson's correlation = −0.053, *P*-value = 0.911). This implies that pH values are not linearly related to acidity values in the tomatoes analyzed. Mean pH: 4.437; mean acidity: 4.487 mg KOH/g; standard deviation of pH: 0.074; standard deviation of acidity: 1.403 mg KOH/g; Pearson's correlation: −0.053; correlation *P*-value: 0.911. The absence of a significant correlation ($P > 0.05$) between pH and acidity indicates that there is no linear relationship between these two parameters in the tomato types studied. These results suggest that both pH and acidity should be evaluated separately when considering the tomato quality and characteristics

3. Record the KOH volume required for neutralization.
4. Use the following formula to calculate TA: TA = (mL KOH × N(KOH) × Acid Meq Factor × 100)/(mL juice) (*see* Fig. 2).

3.6 Drying Process

The drying process aims to evaluate the moisture removal kinetics from tomato samples.

3.6.1 Preparation of Tomato Samples

Start by slicing tomatoes into equal-sized pieces, each weighing around 3.5 g. It is important to maintain this consistency in weight for a reliable comparison under various temperature settings.

3.6.2 Setting Drying Conditions

1. Get the thermobalance ready for the drying tests.
2. Plan to carry out the drying at three separate temperatures: 105 °C, 115 °C, and 130 °C, to explore a spectrum of drying intensities, from moderate to very high (*see* Figs. 3, 4, and 5).
3. Put the cut tomato pieces in the thermobalance.
4. Begin the drying at each of the three different temperatures, observing changes in the weight of the samples periodically.

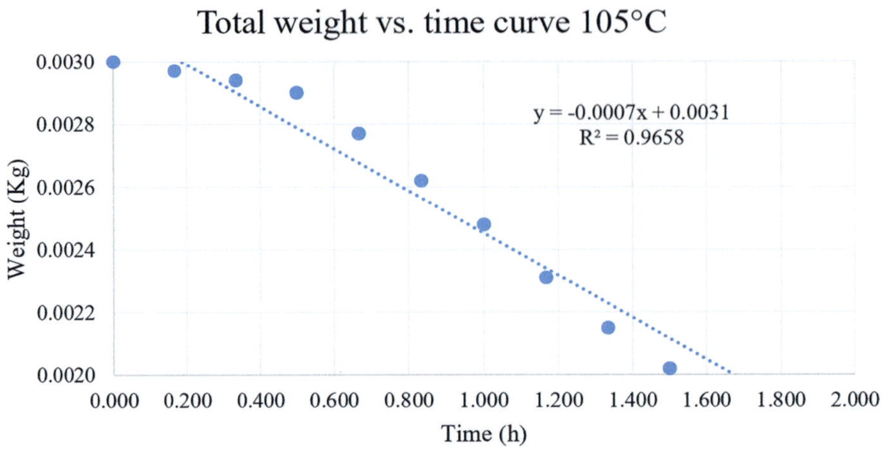

Fig. 3 Drying time curve at 105 °C showing the tomato sample weight (kg) versus time (h). Data points represent weight measurements taken every 10 min (0.167 h) over 170 min (2.833 h). A significant negative correlation (Pearson = −0.9795; $P = 2.76 \times 10^{-8}$) was observed, indicating rapid weight loss at this temperature

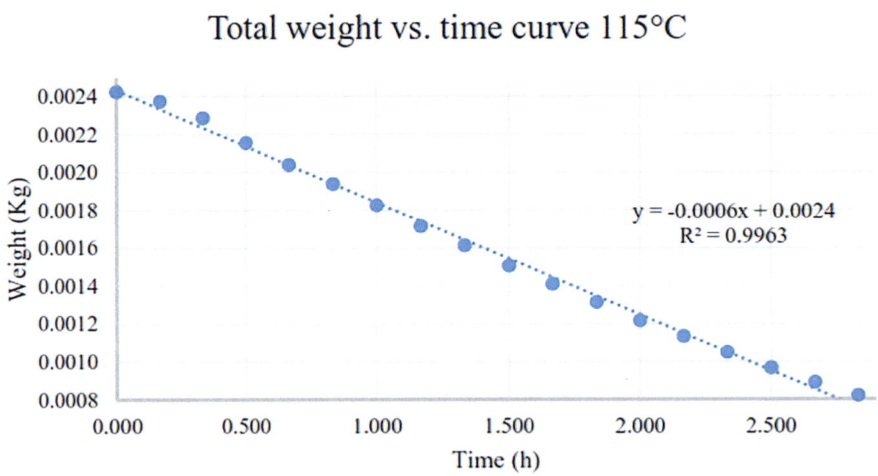

Fig. 4 Drying time curve at 115 °C showing the tomato sample weight (kg) versus time (h). Weight measurements were recorded every 10 min (0.167 h) over 170 min (2.833 h). Pearson's correlation (−0.996; $P = 4.24 \times 10^{-18}$) demonstrates a significant decrease in weight with drying time

5. Keep track of the time taken for drying until the weight stabilizes, indicating that the tomatoes have lost all their moisture (*see* **Note 4**).

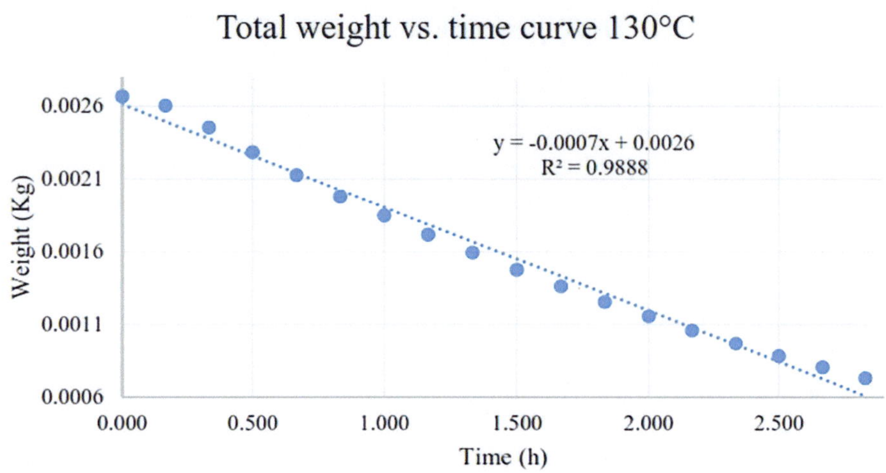

Fig. 5 Drying time curve at 130 °C showing the tomato sample weight (kg) versus time (h). Data collection was carried out at 10 min (0.167 h) intervals for 170 min (2.833 h). The strong negative correlation (Pearson = − 0.994; $P = 8.40 \times 10^{-17}$) indicates a marked reduction in weight at the highest temperature tested

3.6.3 Calculations of Kinetics

1. After completing the drying at each set temperature, utilize the Ocon and Tojo [37] equations for kinetic analysis.
2. This step is essential to gauge the speed of moisture loss and how the tomatoes dry under varying heat conditions.
3. Examine the collected data to understand the kinetics of drying in the tomato samples.
4. Evaluate and contrast the rates of drying and the total time taken for drying at different temperatures, to discern how heat affects the drying efficiency.
5. The insights from this analysis are vital to comprehend the potential effects of ethylene treatment on the drying traits of tomatoes (*see* Figs. 6, 7, and 8).

4 Notes

1. The use of ethephon as a pure source of ethylene, specifically at a working concentration of 100 ppm (diluted from stock concentration, 720,000 mg/L), has been noted to significantly hasten the ripening of tomatoes. This is a critical observation, highlighting ethylene's effectiveness in speeding up the ripening process. Such a discovery has meaningful implications in the realms of agriculture and food processing.

   Figure 1, showing the comparison of the initial and final weights of tomatoes measured in g, shows that there is a strong

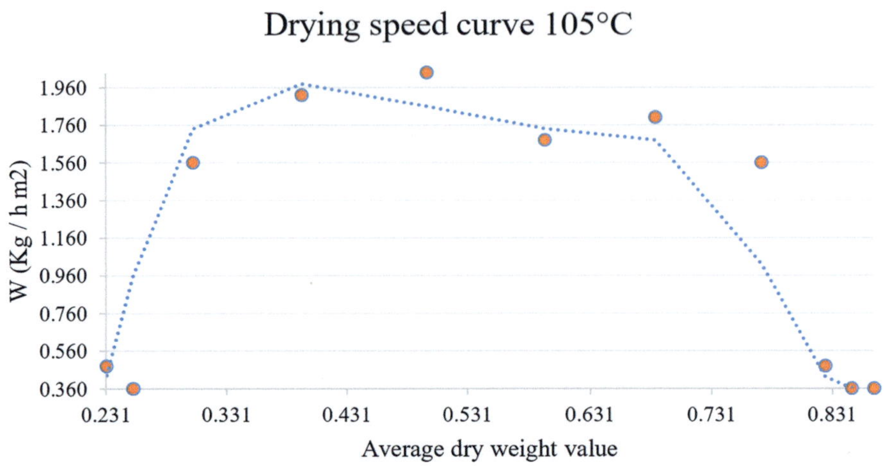

Fig. 6 Moisture content curve at 105 °C showing *w* (kg/hm^2) versus time (h). Data points were collected at defined intervals throughout the drying process. A weak negative positive correlation (r = -0.1613). However, one-way ANOVA (F = 0.0845; P = 0.9198) showed no statistically significant changes in *w* over time

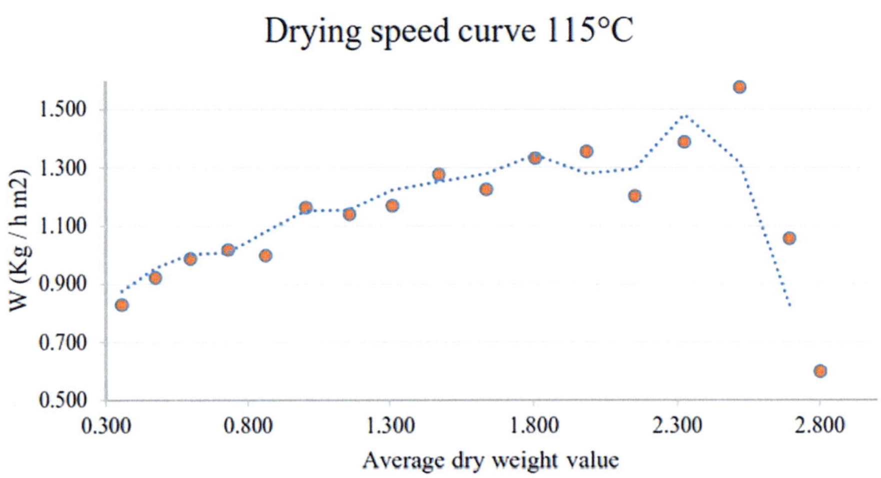

Fig. 7 Moisture content curve at 115 °C showing *w* (kg/hm^2) versus time (h). Measurements were taken periodically during the drying process. A moderate negative correlation (r = −0.388) suggests decreasing moisture content with time, consistent with the typical drying process. Yet, ANOVA (F = 0.145; P = 0.705) indicates no significant differences in *w* at this temperature

positive and significant correlation between the initial weight of tomatoes and both the weight difference and the percentage weight loss (%PPF). This indicates that initially heavier tomatoes tend to lose more weight, both in absolute and percentage terms, during the drying process. The low variability in final weights suggests consistency in the drying process, while the

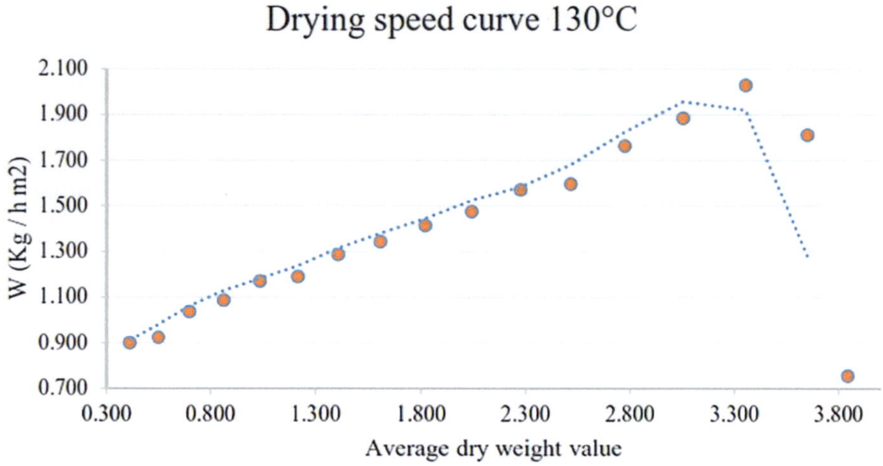

Fig. 8 Moisture content curve at 130 °C showing w (kg/hm^2) versus time (h). Data points were recorded at set intervals throughout the drying process. A moderate negative correlation ($r = -0.324$) again indicates a decrease in w over time. ANOVA ($F = 0.088$; $P = 0.768$) reveals no statistically significant variation in moisture content over the measured time points

higher variability in weight difference and %PPF indicates differences in the response of different tomato types to drying.

2. Ethylene application has been found to dramatically shorten the ripening time of tomatoes, reducing it from 20 days to merely 4 days. This significant decrease underscores the efficiency of ethylene in expediting the natural maturation process of tomatoes.
3. When assessing the quality of tomatoes ripened with ethylene, it is imperative to align with established quality standards. International guidelines like the Codex Stan [38] set forth criteria for food quality, applicable to fresh vegetables including tomatoes. Additionally, national standards such as the NMX-FF-031 and the Ecuadorian NTE INEN 2832 are referenced for ensuring the quality of non-industrialized food products [39, 40]. These standards are key in determining if ethylene-treated tomatoes meet the necessary quality levels for consumption.
4. Figure 2 evaluates the relationship between pH and acidity in different types of tomatoes, as well as analyzes the variability in these parameters. The pH and acidity values of seven types of tomatoes were measured using standard laboratory methods. The average pH of tomatoes is 4.437 with a standard deviation of 0.074, while the average acidity is 4.487 mg KOH/g with a standard deviation of 1.403. No significant correlation was found between pH and acidity ($r = -0.053$, $P = 0.911$). The

low variability in pH values suggests consistency among tomato types, while the higher variability in acidity indicates significant differences. The lack of a significant correlation between pH and acidity suggests that these parameters should be evaluated independently.

5. It has been observed that the drying rate of tomatoes treated with ethylene is notably higher at elevated temperatures, particularly at 130 °C. This indicates that ethylene's impact extends beyond ripening, potentially altering the drying behavior of tomatoes, an aspect crucial in postharvest processing and storage.
6. Figures 3, 4, and 5 show significant correlations, and the ANOVA results confirm that both time and temperature have a considerable impact on the mass of the material during drying.

Acknowledgments

Our thanks, dedication, and support are to all those who contributed to the completion of this book chapter as they have been instrumental to the success of this research.

References

1. Pérez-Díaz F, Arévalo-Galarza ML, Pérez-Flores LJ, Lobato-Ortiz R, Ramírez-Guzmán ME (2020) Crecimiento y características postcosecha de frutos de genotipos nativos de tomate (*Solanum lycopersicum L.*). Rev Fitotec Mex 43(1):89. https://doi.org/10.1186/s40529-023-00370-3
2. Kaur S, Khanal N, Dearth R, Kariyat R (2023) Morphological characterization of intraspecific variation for trichome traits in tomato (Solanum lycopersicum). Bot Stud 64(1):7. https://doi.org/10.1186/s40529-023-00370-3
3. Dascón Hurtado A, Morales Astudillo R, Pérez-Almeida I (2022) Evaluation of the agronomical performance of five new tomato varieties (*Solanum lycopersicum L.*) against two pest and disease control treatments in the province of Loja. Bioagro 34(3):311–318. https://doi.org/10.51372/bioagro343.10
4. Gabriel-Ortega J, Cevallos Gutiérrez K, Vera Velázquez R, Castro Piguave C, Narváez Campana W, Burgos López G (2022) Evaluación y selección de híbridos de tomate *Solanum lycopersicum L.* (Mill.) en Puerto la Boca, Ecuador. J Selva Andina Biosph 10(1):21–31. https://doi.org/10.36610/j.jsab.2022.100100021
5. FAOSTAT (2023) Datos sobre alimentación y agricultura. https://www.fao.org/faostat/es/#home. Accessed 26 Feb 2024
6. INEC (2023) Estadísticas agropecuarias. Instituto Nacional de Estadística y Censos. https://www.ecuadorencifras.gob.ec/estadisticas-agropecuarias-2/ Accessed 26 Feb 2024
7. Delgado-Vargas VA, Magdaleno-Villar JJ, Silva-Rojas HV, García-De los Santos G, Ayala-Garay ÓJ (2022) Calidad de semillas de tomate producidas en temperatura alta durante diferentes etapas de desarrollo. Rev Fitotec Mex 45(4):445. https://doi.org/10.35196/rfm.2022.4.445
8. Hernández Victoria MA, Ponce Hernández MS, Álvarez Montalvo S, Armas Recio M (2018) Depth, weight, and volume of tomato roots (Solanum lycopersicon) cv. Piedro in greenhouse conditions. Agrisost 24(2):110–114. https://doi.org/10.5281/zenodo.1086978

9. Melgarejo TA, Kon T, Rojas MR, Paz-Carrasco L, Zerbini FM, Gilbertson RL (2013) Characterization of a new world monopartite Begomovirus causing leaf curl disease of tomato in Ecuador and Peru reveals a new direction in Geminivirus evolution. J Virol 87(10):5397–5413. https://doi.org/10.1128/jvi.00234-13
10. Paz-Carrasco LC, Castillo-Urquiza GP, Lima ATM, Xavier CAD, Vivas-Vivas LM, Mizubuti ESG et al (2014) Begomovirus diversity in tomato crops and weeds in Ecuador and the detection of a recombinant isolate of rhynchosia golden mosaic Yucatan virus infecting tomato. Arch Virol 159(8):2127–2132. https://doi.org/10.1007/s00705-014-2046-y
11. Kutinkova H, Caicedo F, Lingren B (2016) The main pests on solanacea crops in zona 1 of Ecuador. New Knowl J Sci 5(1):72–78
12. Satama Bermeo M (2021) Assessment of farmers' perceptions of the sustainable agricultural practices in the "biocorridors for living well" program in Ecuador. https://dialnet.unirioja.es/servlet/tesis?codigo=304615. Accessed 26 Feb 2024
13. Hatfield JL, Prueger JH (2015) Temperature extremes: effect on plant growth and development. Weather Clim Extremes 10:4–10. https://doi.org/10.1016/j.wace.2015.08.001
14. Lubis WMY, Adrian M, Jadid N, Widiastuti A, Ezura H, Mubarok S et al (2023) Transcriptome dataset from *Solanum lycopersicum L.* cv. Micro-Tom; wild type and two mutants of INDOLE-ACETIC-ACID (SlIAA9) using long-reads sequencing oxford nanopore technologies. BMC Res Notes 16(1):40. https://doi.org/10.1186/s13104-023-06306-1
15. Tucker G, Yin X, Zhang A, Wang M, Zhu Q, Liu X et al (2017) Ethylene† and fruit softening. Food Qual Saf 1(4):253–267. https://doi.org/10.1093/fqsafe/fyx024
16. García Parrales LD, Saltos Meza LA, Burgos Briones GA, Delgado Demera MH, Cedeño Palacios CA (2023) Ripening process of *Solanum lycopersicum L.* (tomato) with ethylene. Afinidad 80(598):72–79. https://doi.org/10.55815/413443
17. Balaguera-López HE, Salamanca-Gutiérrez FA, García JC, Herrera-Arévalo A (2014) Etileno y retardantes de la maduración en la poscosecha de productos agrícolas. Una revisión Revista Colombiana de Ciencias Hortícolas 8(2):302–313
18. Fenn MA, Giovannoni JJ (2021) Phytohormones in fruit development and maturation. Plant J 105(2):446–458. https://doi.org/10.1111/tpj.15112
19. Muday GK, Rahman A, Binder BM (2012) Auxin and ethylene: collaborators or competitors? Trends Plant Sci 4:181–195. https://doi.org/10.1016/j.tplants.2012.02.001
20. Kumar R, Khurana A, Sharma AK (2013) Role of plant hormones and their interplay in development and ripening of fleshy fruits. J Exp Bot 16:4561–4575. https://doi.org/10.1093/jxb/eru277
21. Javanmardi J, Kubota C (2006) Variation of lycopene, antioxidant activity, total soluble solids and weight loss of tomato during postharvest storage. Postharvest Biol Technol 2: 151–155. https://doi.org/10.1016/j.postharvbio.2006.03.008
22. Monsalve J, Machado M (2023) Evaluación de dos métodos de deshidratación del tomate (Lycopersicom esculentum mill) variedad manzano. Multiciencias 7(3):256–265
23. Tahmasebi M, Emam-Djomeh Z (2021) Lycopene degradation and color characteristics of fresh and processed tomatoes under the different drying methods: a comparative study. Chem Pap 75(7):3617–3623. https://doi.org/10.1007/s11696-021-01611-0
24. Lu Z, Wang J, Gao R, Ye F, Zhao G (2019) Sustainable valorisation of tomato pomace: a comprehensive review. Trends Food Sci Technol 86:172–187. https://doi.org/10.1016/j.tifs.2019.02.020
25. Midilli A, Kucuk H (2003) Mathematical modeling of thin layer drying of pistachio by using solar energy. Energy Convers Manag 44(7): 1111–1122. https://doi.org/10.1016/S0196-8904(02)00099-7
26. Pangavhane D (2002) Design, development and performance testing of a new natural convection solar dryer. Energy 27(6):579–590. https://doi.org/10.1016/S0360-5442(02)00005-1
27. Doymaz I (2007) Air-drying characteristics of tomatoes. J Food Eng 78(4):1291–1297. https://doi.org/10.1016/J.JFOODENG.2005.12.047
28. López Camelo AF (2004) Manual for the preparation and sale of fruits and vegetables. FAO. https://www.fao.org/4/y4893e/y4893e00.htm. Accessed 14 May 2024
29. Martí R, Leiva-Brondo M, Lahoz I, Campillo C, Cebolla-Cornejo J, Roselló S (2018) Polyphenol and l-ascorbic acid content in tomato as influenced by high lycopene genotypes and organic farming at different environments. Food Chem 239:148–156. https://doi.org/10.1016/j.foodchem.2017.06.102

30. Sipos L, Orbán C, Bálint I, Csambalik L, Divéky-Ertsey A, Gere A (2017) Colour parameters as indicators of lycopene and antioxidant activity traits of cherry tomatoes (*Solanum lycopersicum L.*). Eur Food Res Technol 243(9):1533–1543. https://doi.org/10.1007/s00217-017-2862-x
31. El Hariri E, El-Bendary N, Hassanien AE, Badr A (2014) Automated ripeness assessment system of tomatoes using PCA and SVM techniques. In: Computer vision and image processing in intelligent systems and multimedia technologies, 1st edn. Kuwait University, Kuwait
32. Baltazar A, Aranda JI, González-Aguilar G (2008) Bayesian classification of ripening stages of tomato fruit using acoustic impact and colorimeter sensor data. Comput Electron Agric 60(2):113–121. https://doi.org/10.1016/J.COMPAG.2007.07.005
33. Talens Oliag P (2017) Evaluación del color y tolerancia de color en alimentos a través del espacio CIELAB. https://riunet.upv.es/handle/10251/83392. Accessed 26 Feb 2024
34. Fonseca P (2017) Informe: Las fórmulas para calcular la cantidad de materia seca. [Internet]. [cited 2024 Feb 25]. https://www.contextoganadero.com/ganaderia-sostenible/informe-las-formulas-para-calcular-la-cantidad-de-materia-seca
35. Nollet LM (2004) Handbook of food analysis-3 Volume set. CRC Press
36. Terevinto A (2013) Técnicas de análisis de materias grasas. https://docplayer.es/31247957-Tecnicas-de-analisis-de-materias-grasas.html. Accessed 12 Sep 2023
37. Ocon García J, Tojo Barreiro G (1998) Problemas de Ingenería Química Tomo II. Madrid
38. Codex Stan (2007) NORMA DEL CODEX PARA EL TOMATE (CODEX STAN 293–2007). Codex Alimentarius. https://www.fao.org/input/download/standards/11013/CXS_293s.pdf. Accessed 14 May 2024
39. Subsecretaría de Agricultura, Ganadería Mexicana (1997) Productos alimenticios no industrializados para consumo humano. Hortalizas frescas. Tomate-(Lycopersicun esculentum Mill.). Normas mexicanas. http://www.economia-nmx.gob.mx/normas/nmx/1998/nmx-ff-031-1998.pdf Accessed 14 May 2024
40. NTE INEN 2832 (2013) Norma para el tomate. Instituto Ecuatoriano de Normalización. https://docplayer.es/232508657-Nte-inen.html. Accessed 14 May 2024

Chapter 5

Ethylene Estimation of In-Vivo Cultured Plants by Gas Chromatography

Poonam Rakwal, Anoop Kumar Verma, and Ch. Ratnasekhar

Abstract

The gaseous hormone ethylene regulates different processes in plant life. Ethylene is generally considered as a stress hormone that is stimulated by biotic and abiotic stress. To understand its role in different arrays of plant life, in-vivo quantification of ethylene is essential to understand the physiological aspects of plant metabolism. Several techniques are employed for its accurate estimation; one such popular technique is gas chromatography. Gas chromatography is commonly employed for the estimation of different types of volatile compounds. The quantification of ethylene by gas chromatography is reliable and efficient as a large number of samples can be estimated simultaneously. The measurement of ethylene is accomplished by utilizing a standard curve that is prepared from certified ethylene gas used as the standard. Here, we describe a gas chromatography–flame ionization detection (GC-FID)-based method for the quantification of ethylene from live plants.

Key words Gas chromatography, Flame ionization detection (GC-FID), Ethylene, In-vivo quantification, Column, Retention time

1 Introduction

Ethylene (C_2H_4) is a gaseous phytohormone that is involved in different physiological and developmental processes throughout the life of a plant [1, 2]. The gaseous hormone is involved in regulating several processes in plants such as leaf senescence, seed germination, shoot and root growth, flower senescence, flower development, and fruit ripening [3, 4]. It is also a significant regulator of biotic and abiotic stress response in plants [5]. Ethylene production in plants is dependent on several factors like plant species as well as on the age of the plant [6]. The endogenous biosynthesis of ethylene is dependent on three characteristic enzymes, S-adenosyl methionine synthase (SAM synthase), 1-aminocyclopropane-1-carboxylate synthase (ACS), and 1-aminocyclopropane-1-carboxylate oxidase (ACO) and their respective substrates [7, 8]. The immediate precursor of endogenous ethylene

Devesh Shukla (ed.), *Ethylene Sensing and Signaling*, Methods in Molecular Biology, vol. 2945,
https://doi.org/10.1007/978-1-0716-4650-2_5,

biosynthesis, which is 1-amino-cyclopropane-1-carboxylic acid (ACC), can be quantified using gas chromatography based on an indirect method utilizing ethylene liberation [9]. However, the easiest estimation of ethylene phytohormone must be performed in gaseous phase as it requires no sample preparation [10]. This plant hormone is synthesized within cells and then diffuses from these cells to neighboring tissues, ultimately reaching the surrounding environment, such as the atmosphere or enclosed spaces [10]. For collection of this phytohormone, the plant producing ethylene is enclosed in an airtight space or chamber and then can be subjected to gas chromatography (GC) analysis [11, 12].

For understanding the role of ethylene in various developmental and physiological pathways, its quantification is required. Gas chromatography is an efficient and effective analytical technique utilized for the quantification of volatile compounds [13]. Because of its sensitivity and versatility, GC has become an attractive and versatile tool for the quantification of ethylene. Furthermore, GC is connected to a headspace unit that simplifies the sampling process and ensures that an optimum amount of sample is introduced during the injection process [14].

During the gas chromatographic analysis, the volatile sample passes through the column with the help of a carrier gas known as mobile phase which is commonly a nonreactive inert gas like nitrogen (N_2), helium (He), etc. The stationary phase constitutes a column that is packed with a layer of polymer supported with an inert solid. Based on the adsorptive properties, different molecules are retained by the column according to their retention time (RT). The column is attached to a detector that detects the molecules and generates a series of peaks known as chromatograms. For ethylene quantification, the sample peaks are compared with that of the pure standard molecule [15]. As shown in Fig. 1, ethylene emitted by plants in-vivo can be collected and subsequently analyzed using gas chromatography.

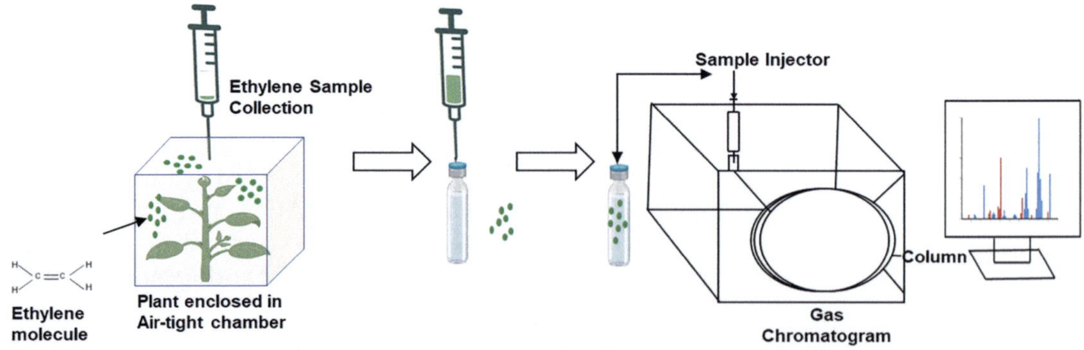

Fig. 1 Ethylene estimation in in-vivo plants by gas chromatography

2 Materials

2.1 Specific Instrumental Components and Chemicals/Reagents Required for Ethylene Measurement

1. TracePLOT TG BOND Q GC column.
2. Analytical balance.
3. Headspace vials and aluminum crimp caps with silicon/PTFE septa.
4. Headspace vial crimper.
5. Pure ethylene (standard).
6. Syringes fitted with a luer activated valve.
7. Ethanol.
8. Disposable syringes.
9. Airtight chamber or container.

3 Methods

3.1 Ethylene Collection from Plants: In-Vivo Ethylene Collection

1. Select a healthy and well-grown plant for the sampling.
2. Enclose the plant or a specific plant part (e.g., fruit, leaves) in a sealed container or airtight chamber. Once the plants are enclosed, allow the plant to naturally emit ethylene within the enclosed environment for a certain period of time (*see* **Note 1**).
3. Ensure that the gas-tight bag containing the ethylene gas is securely sealed to prevent any leakage.
4. Prepare a syringe for gas extraction. For the gas collection process, use a 2 mL syringe with a needle diameter of 0.5 mm to collect various samples of ethylene from the airtight chamber. A gas-tight syringe specifically equipped with a luer-lock is used to ensure a tight seal.
5. After gas collection, carefully remove the needle from the source and seal the syringe with a gas-tight cap or plug. Transfer the syringe to a suitable container for further analysis or storage, ensuring that it remains airtight.

3.2 Ethylene Collection from Plants: In Vitro Ethylene Collection

1. Another method to measure ethylene is the direct collection of plant tissues in airtight vials. This method is easy to follow as plant tissue is directly plucked from the plant and inserted in vials. However, this method is best suitable for the estimation of ethylene from plant parts like fruits, leaves, etc.
2. Separate the plant tissue from the main plant manually by pulling them off from the abscission layer. Immediately, insert the plant sample into the vials and close the lid properly.

3. Headspace vials along with caps are sterilized before performing the experiment. For securing gas, place a PTFE septum inside the vial caps (*see* **Note 2**).
4. Depending upon the size of the vial, add equal amount of plant sample into each vial.

3.2.1 Collection of Ethylene from the Samples

1. For in vitro ethylene collection, the collected sample generates wounds which will release ethylene into the environment. Therefore, after sample collection, vials are left open for approximately 10 min to release stress-induced ethylene.
2. After equilibration time, headspace vials are sealed with the help of suitable aluminum caps fitted with silicon or PTFE septa. To avoid ethylene leakage from vials, a hand crimper should be used to seal the vial caps (*see* **Note 3**).
3. Note down the time of vial sealing so as to avoid any false positives in the results. After sealing the vials, the samples are subjected to an incubation period. The incubation period is very crucial for measurement, as during this time, ethylene is collected into the vials.
4. Once the incubation period is over, insert the syringe in the vials for gas collection. The syringe is then injected into the GC column for further analysis.

3.3 Ethylene Estimation with a Gas Chromatography System

1. Switch on the gas chromatography and headspace sampler unit. Also, turn on the system program essential for operating the GC and headspace unit (*see* **Notes 12, 13,** and **14).**
2. The vials are properly numbered according to the sample and placed in the corresponding positions in the headspace unit of GC (*see* **Note 4**).
3. 1 mL of gas from the air chamber or vial is withdrawn and injected into the gas chromatography (GC) system equipped with a flame ionization detector (*see* **Notes 5** and **6**).
4. Injection of a large sample volume is generally avoided as it causes poor separation of peaks.
5. The sampler of GC is adjusted to 10 μL, while the given injection volume is set to 1 μL. The split ratio should be 40 mL/min, and the column is TracePLOT TG BOND Q GC (*see* **Note 15**).
6. The temperature of the column and detector is adjusted to 40 °C and 200 °C, respectively. Additionally, the initial oven temperature is set to 60 °C for the first 2 min and then heated at the rate of 30 °C/min until it reaches the set value of 240 °C that is maintained for the next approximately 1 min. The total method run time is approximately 9 min (*see* **Note 7**).

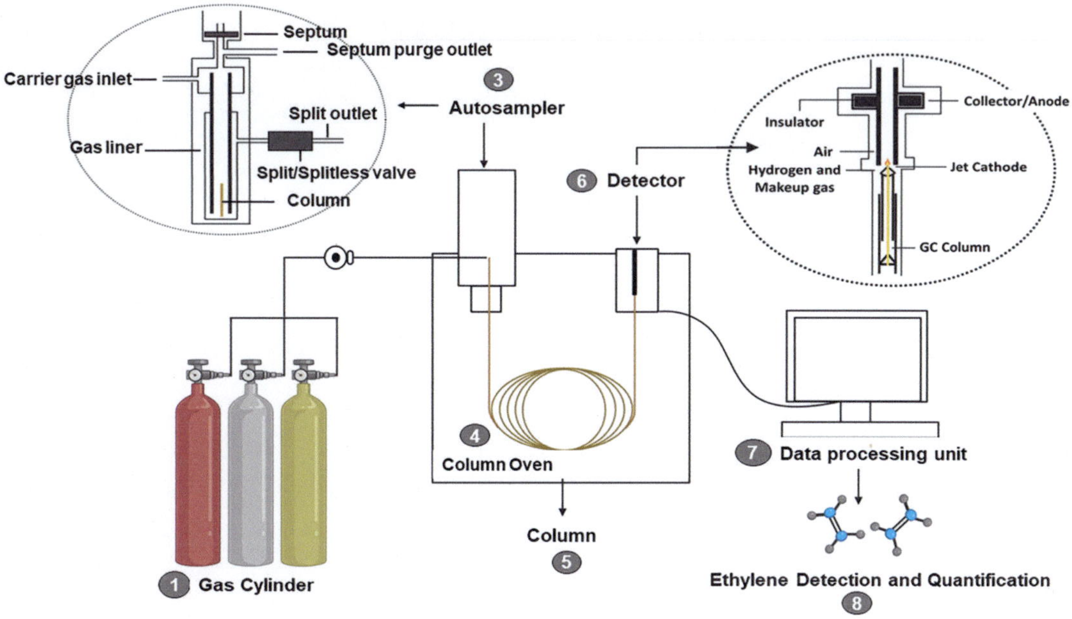

Fig. 2 The sample (in gas or liquid phase) to be analyzed is introduced into the GC system by means of an autosampler or by manually injecting with the help of a specific syringe. As the sample enters the column, the carrier gas moves the analyte through the column. As the sample exits the column, they are combusted in the presence of a hot flame (1:10, hydrogen:air). The combustion product such as hydronium ion is drawn to the collector, which generates a signal in the form of ion current that is measured by an electrometer (detector)

7. As the sample containing ethylene passes through the column and reaches the detector, a signal is detected in the form of peaks (*see* **Note 8**). A schematic overview of how the sample enters the GC column, undergoes combustion, and is detected is shown in Fig. 2.
8. For the quantification of ethylene in the plant sample and to eliminate any technical errors, a standard of ethylene having a known concentration is also parallelly run with the main sample (*see* **Note 9**).
9. The peak having the same retention time (RT) as of the ethylene standard is identified as ethylene. The RT of ethylene is around 1.64 min; deviation from RT can be observed because of GC conditions and the different setup.

3.4 Running Ethylene Standard

1. Seal around six headspace vials with airtight caps containing ambient air.
2. A cylinder with a two-stage regulator containing pure ethylene is used as the standard (*see* **Note 10**).

Table 1
Instrument parameters for running GC

Variable	Value
Carrier gas	Helium
Inlet mode	Split
Injector base temperature	100 °C
Liner	5 mM
Oven initial temperature	60 °C
Heating rate	30 °C/min
Oven final temperature	250 °C
Analog filter status	On
Hydrogen (H_2) gas	35 mL/min
Nitrogen (N_2) gas	30 mL/min
Air flow rate	350 mL/min
Syringe filling speed	50 mL/min
Needle penetration speed	25 mm/s
Injection speed	50 mL/min
Injection depth	50 mM

3. Use a gas-tight syringe to aspirate ethylene from the gas cylinder and inject it into the headspace vial. The volume of ethylene injection is dependent upon the size of the headspace vial.
4. Inject 1 mL of the ethylene standard into the GC system with the help of an autosampler.
5. Prepare a standard ethylene curve using different dilutions of the known standard. Thus, the concentration of ethylene in the plant samples can be calculated using the ethylene standard curve (*see* **Note 11**). The key instrument parameters for setting up the GC system are summarized in Table 1. The specifications of the TracePlot TG-BOND Q+ column required for ethylene analysis are presented in Table 2.

4 Notes

1. Determine the optimal duration for ethylene collection based on the specific plant species and different environmental conditions. It is important to allow sufficient ethylene accumulation in the airtight chamber. Thus, it is advisable to optimize an incubation period before conducting the main experiment.

Table 2
Column specifications required for running ethylene samples

Column specifications	Value
Column	TracePlot TG-BOND Q+
Column length (metric)	30 m
Column diameter (metric)	0.32 mM
Film thickness	10 μm
Polarity	Porous divinyl benzene

2. To prevent any contamination, it is recommended to use a sterilized syringe for gas collection. Also, before collecting the gas sample, purge the syringe by drawing and expelling the gas several times to eliminate any residual air.
3. Obtaining consistent results depends on the correct handling of the crimper. If there are leaks during the headspace vial sealing, it could lead to low levels of ethylene or inconsistent results. The cause of these leaks is probably an improperly adjusted vial crimper. When the vial seal is not crimped correctly, it can result in the leakage of ethylene from the headspace vials.
4. To minimize error during ethylene quantification, it is advisable to use at least five biological replicates of each sample. Using multiple replicates is essential to ensure the accuracy and precision in the obtained results. In the case of limited number of samples, at least three replicates must be used.
5. During the injection process, the needle may break the tissues; thus, the size of vials should be smaller.
6. The low level of ethylene may be the result of leakage of gas from the headspace vials. To ensure reproducibility of results, proper handling and storage conditions are required. Because of the large number of samples and replicates, it is not possible to analyze samples with GC after sample collection. However, the gas samples can be stored in a syringe fitted with a rubber cap for about 24 h without any change in volume.
7. During the analysis, the oven temperature can be programmed to reach a maximum temperature at a specified rate and can be held up to a specified amount of time. However, it should be noted that the maximum rate at which the oven is heated is dependent on several factors such as room temperature, detector temperature, etc. Additionally, it should be noted that each instrument has an upper and lower limit (for oven temperature, rate, etc.); thus, before running GC, follow the manufacturer's guidelines and optimize the method as required.

8. The carrier gases (nitrogen, hydrogen, and helium) used during GC analysis should be ultra-pure in nature. Using highly pure gases increases the sensitivity of results and also enhances the shelf life of the column. It is important to adjust the carrier gas flow rate for detector sensitivity. Generally, the suggested flow rate of carrier gases for packed columns is 10–60 mL/min and that for capillary column is 1–5 mL/min, but for a specific column, optimization of flow rate is required. Furthermore, the specified flow rate for detector gases like hydrogen (H2) should be set to 40 mL/min and that for air should be 450 mL/min. The fluctuation in the flow rate must be within 10% of the set value. However, if there is a significant fluctuation in the flow rate, there may be an issue with the GC flow lines or GC column which needs to be corrected prior to analysis.
9. The conditions of GC should be similar during complete analysis for all the samples belonging to each experiment/condition. The fluctuation in parameters like temperature and pressure may introduce error to the experiment. Change in parameters can shift the retention time (RT) and may produce aberrant results.
10. Ethylene gas is flammable and can be dangerous at high concentrations; thus, it is important to follow proper safety precautions while working with the standard ethylene gas. It is essential to prioritize safety by wearing appropriate protective equipment when handling an ethylene cylinder.
11. From the stock standard, standards of different concentrations are prepared and injected along with the batch of the samples. The calibration curve or standard curve can be prepared by utilizing data software or spreadsheet software like Microsoft Excel or GraphPad Prism.
12. Calibration of the GC system must be performed for ensuring accurate and reliable analysis. For this, internal and external calibrations can be performed. For the internal calibration process, an internal standard is added into the sample prior to injection. The signal/peak response corresponding to this internal standard is further utilized for correcting the variations due to sample injection or instrument response. For external calibration, a series of dilution of a known standard (certified pure standard) are prepared and injected into the GC system. The system will then generate the signals/peaks proportional to the amount of each analyte in solution. On the basis of this peak response, a standard curve will be plotted using the concentration on the X-axis and peak response on the Y-axis. Further, this standard curve will be used for defining the concentration (quantification) of the desired analyte in the sample.

13. Certain factors affect the external calibration of GC. For better accuracy, use the standard of highest purity and store the solutions and standards in an appropriate environment. Furthermore, it is advisable to check the GC components regularly or before use and follow the instructions provided by the manufacturer.
14. Before running the sample, a specific method must be established, as different analytes have different requirements such as oven temperature, low or high gas pressure, or specific run time. Thus, for each analysis, a specific method must be created and validated before running the main samples.
15. The column provides the reproducible separation of nitrogen (N_2), oxygen (O_2), and CO. The PLOT columns provide very good separation of ethylene peaks with that of nitrogen. The column is based on a nonpolar, 100% divinyl benzene stationary phase. For minimizing the particle release during the analysis, the stationary phase is very strongly bonded to the wall tubing, and it prevents damage to the detector.

Acknowledgments

The authors are thankful to Government of India for providing DST-SERB-SRG research grant (Grant number.: SRG/2021/000750-G), DBT-RRF grant (Grant No.: BT/RLF/Re-entry/21/2020) and CSIR-RDSF grant (Grant number: CSIR/RDSF/IHP24/04). They also acknowledge CSIR-CIMAP for providing necessary infrastructure, facility, and funding support from CSIR, India. Poonam Rakwal is thankful to UGC for providing Junior Research Fellowship and Anoop Kumar Verma is thankful to the DBT for providing Senior Research Fellowship (DBT-SRF).

References

1. Vaseva II, Qudeimat E, Potuschak T, Du Y, Genschik P, Vandenbussche F, Van Der Straeten D (2018) The plant hormone ethylene restricts Arabidopsis growth via the epidermis. Proc Natl Acad Sci USA 115:E4130–E4139. https://doi.org/10.1073/pnas.1717649115
2. Schaller GE (2012) Ethylene and the regulation of plant development. BMC Biol 10:9
3. Fatma M, Asgher M, Iqbal N, Rasheed F, Sehar Z, Sofo A, Khan NA (2022) Ethylene signalling under stressful environments: analyzing collaborative knowledge. Plants Theory 11:2211
4. Ceusters J, Van de Poel B (2018) Ethylene exerts species-specific and age-dependent control of photosynthesis. Plant Physiol 176: 2601–2612
5. Dubois M, Van den Broeck L, Inzé D (2018) The pivotal role of ethylene in plant growth. Trends Plant Sci 23:311–323
6. Cronje R, Jonker A (2023) Quantification of ethylene production in leaf and bud tissue of the subtropical tree crop litchi (Litchi chinensis Sonn.) using gas chromatography and flame ionization detection. Bio Protoc 13.e4636 https://doi.org/10.21769/bioprotoc.4636
7. Fišerová H., Klemš M, Mikušová Z (2008) Estimation of ethylene production and 1-aminocyclopropane-1-carboxylic acid content in plants by means of gas chromatography. Plant,

Soil and Environment,54(2),55–60. https://doi.org/10.17221/2782-PSE

8. Ecker JR (1995) The ethylene signal transduction pathway in plants. Science,268 (5211),667–675. https://doi.org/10.1126/science.7732375
9. Concepcion MA, Lizada C, Shang A, Yang FA (1979) A simple and sensitive Assay for 1-Aminocyclopropane-1-Carboxylic Acid. Analytical Biochemistry,100(1),140–145. https://doi.org/10.1016/0003-2697(79)90123-4
10. Bulens I, Van de Poel B, Hertog MLATM, De Proft MP, Geeraerd AH, Nicolaï BM (2011) Protocol: an updated integrated methodology for analysis of metabolites and enzyme activities of ethylene biosynthesis. Plant Methods 7. https://doi.org/10.1186/1746-4811-7-17
11. Makky M, Paschalidis KA, Soni P, Manganaris A, Dhima K (2014) A new rapid gas chromatographic method for ethylene, respirational, and senescent gaseous production of climacteric fruits stored in prolonged low temperature. In: International Conference on Agricultural, Environmental and Biological Sciences (AEBS-2014) April 24–25, 2014 Phuket (Thailand). International Institute of Chemical, Biological & Environmental Engineering
12. Yokotani N, Nakano R, Imanishi S, Nagata M, Inaba A, Kubo Y (2009) Ripening-associated ethylene biosynthesis in tomato fruit is autocatalytically and developmentally regulated. J Exp Bot 60:3433–3442. https://doi.org/10.1093/jxb/erp185
13. Zhou J, Lu X, Tian B, Wang C, Shi H, Luo C, Li X (2020) A gas chromatography-flame ionization detection method for direct and rapid determination of small molecule volatile organic compounds in aqueous phase. 3 Biotech 10. https://doi.org/10.1007/s13205-020-02523-8
14. Dewulf J, Van Langenhove H, Wittmann G (2002) Analysis of volatile organic compounds using gas chromatography
15. Cristescu SM, Mandon J, Arslanov D, De Pessemier J, Hermans C, Harren FJM (2013) Current methods for detecting ethylene in plants. Ann Bot 111:347–360

Part II

Role of Ethylene in Post-harvest and Fruit Ripening

Chapter 6

Measuring Internal and External Ethylene Concentrations in Fruits and Vegetables

Yosef Al Shoffe

Abstract

Ethylene, a gaseous phytohormone, plays a pivotal role in influencing plant growth, development, ripening, and senescence, as well as responses to various stresses and pathogen infections. This crucial phytohormone is synthesized from S-adenosylmethionine (SAM) through the intermediate 1-aminocyclopropane-1-carboxylic acid (ACC). In this chapter, we will delve into the techniques, materials, and methods for measuring internal and external ethylene concentrations in fruits and vegetables, both at the time of harvest and during storage, using gas chromatography (GC). Furthermore, it is essential to recognize that fruits and vegetables can be categorized based on their response to ethylene and their respiration patterns, distinguishing them as climacteric or non-climacteric. These protocols hold significant importance, as they enable us to identify the optimal maturity stage for fruits and vegetables, whether for short-term or long-term storage. This, in turn, helps extend the longevity and shelf life of these perishable crops.

Key words Ethylene, Gas chromatography, Climacteric, Non-climacteric, Maturity, Ripening, Senescence

1 Introduction

Ethylene, the simplest olefin gas, is a vital compound in the world of plant biology [1]. Plants employ ethylene for various purposes, including their development, growth, response to stress, and ripening of fruits [2]. Historically, ethylene found one of its earliest uses in ancient China around 2000 BCE, when it was employed as a smoking agent to hasten the ripening of pears. The initial scientific exploration of ethylene dates back to 1896 when DmitriNeljubov conducted pioneering research in Petersburg [3]. This research revolved around the observation of a triple response involving reduced elongation in pea seeds during germination in the absence of light. One of the remarkable attributes of ethylene is its ability to readily traverse cell membranes and dissolve, allowing it to diffuse freely within plant tissues.

Devesh Shukla (ed.), *Ethylene Sensing and Signaling*, Methods in Molecular Biology, vol. 2945,
https://doi.org/10.1007/978-1-0716-4650-2_6,

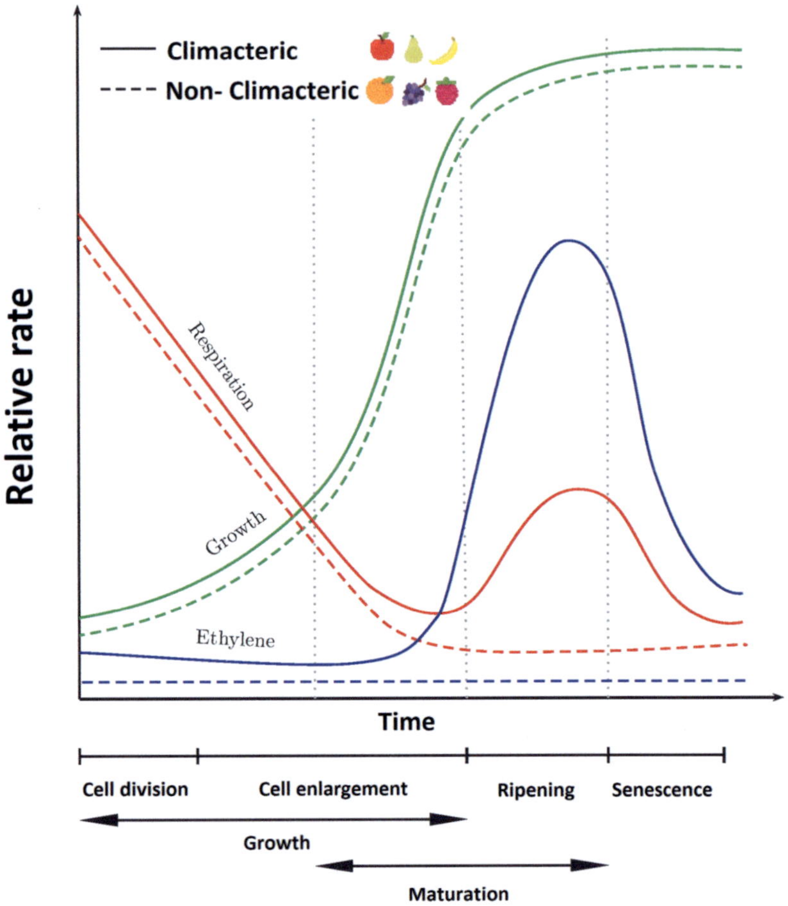

Fig. 1 Climacteric and non-climacteric pattern in horticultural crops. (Modified from Ref. [5])

Horticultural crops are classified into climacteric and non-climacteric crops based on their response to ethylene and their respiration pattern [4]. Climacteric fruits exhibit a notable surge in ethylene production and respiration rate as they begin to ripen. They possess the remarkable capability to further ripen post-harvest and during storage until reaching optimal ripeness. However, after reaching this peak, they become susceptible to fruit deterioration and senescence, as commonly seen in apples and pears. In contrast, non-climacteric fruits like grapes and citrus do not undergo additional ripening post-harvest, and their levels of ethylene and respiration remain low (Fig. 1) [5]. The term "climacteric" was first applied by Kidd and West (1927). Biale and Barcus (1970) published some selected respiration rates in different crops [6]. Wills et al. (1998) studied the effects of ethylene in climacteric and non-climacteric fruits. Paul et al. [7] discussed that some fruits show climacteric as well as non-climacteric behavior depending on the cultivar or genotype. Ethylene management has been involved in the precision of fruit and vegetable ripening, maintaining fruit quality, extending storage time, and prolonging shelf life [8–14].

Gas chromatography (GC), a groundbreaking innovation credited to the pioneering work of Martin and James in 1952, stands as one of the foremost and extensively employed analytical techniques within the realm of contemporary chemistry [15]. Gas chromatography techniques offer exceptional resolution, facilitating the precise differentiation of species, along with remarkable reliability and repeatability. Moreover, they seamlessly integrate with mass spectrometry (MS), ensuring a standardized method for compound identification. GC is ideally suited for analyzing volatile and semivolatile compounds, and its applicability can be expanded to encompass low or nonvolatile compounds through appropriate derivatization techniques. This process reduces molecular polarity and enhances thermal stability, streamlining the elution process in GC [16].

Bulens et al. [17] described an integrated methodology for the analysis of metabolites and enzyme activity of ethylene biosynthesis. Their article provides insights into the measurement of ethylene production rates in intact fruits. Furthermore, ongoing advancements in ethylene detection technologies have resulted in a thriving landscape of commercially available systems [18]. These innovative solutions excel at detecting ethylene even at remarkably low concentrations, making them invaluable for studying ethylene's role in non-climacteric fruits. Various methods have been employed, including electrochemical sensors, nondispersive infrared spectroscopy, metal oxide gas sensors [19, 20], and the continuous development of gas chromatography (GC). Notably, the miniaturization of GCs has been achieved through cutting-edge three-dimensional fabrication techniques for GC columns [21]. To enhance sensitivity, pre-concentrators made from materials like silicon or ceramics have been employed [22]. An extraordinary leap in ethylene detection technology is the laser-based photoacoustic ethylene detector (EDT-300, Sensor Sense), capable of quantifying ethylene with unparalleled sensitivity (0.3 nL L^{-1}) when compared to existing technologies. This remarkable device boasts a rapid response time of just 5 s [23]. In addition, optimization techniques for continuous flow, stop-and-flow, and sample collection methods have been detailed by experts in the field [23]. The applications of these detectors have proven versatile, enabling the measurement of ethylene production in fruits, assessment of ethylene removal rates, and investigation of ethylene diffusion properties in various polymeric materials. A recent breakthrough in ethylene detection involves the ingenious concept of fluorophore-tagged Grubbs catalysts, serving as turn-on fluorescent probes for the sensitive detection and quantification of ethylene [24]. In addition, wireless plant wearable ethylene sensors based on screen-printed MXene RF resonators, which can be flexibly deployed as smart tags on plant surfaces for in situ, continuous ethylene detection are poised to open new avenues for ethylene research and analysis [25]. However, it is important to note that chemical sensors can be influenced by environmental factors. They yield accurate results when properly calibrated to account for variations in humidity and temperature (Q. Liu, personal communication).

In this comprehensive chapter, we present a variety of refined techniques tailored to enable precise ethylene measurements both internally and externally within fruits and vegetables in the laboratory. These methods are designed for application at various stages, including development, harvest, storage, and shelf-life assessment.

2 Materials

2.1 GC Calibration

1. Gas chromatography (GC) is used to separate, detect, and quantify ethylene (*see* **Note 1**) (Fig. 2) [26].
2. Carrier gas (high-purity nitrogen or helium (based on the GC model, the type of detector used, the nature of the analytes, the column type, and the desired resolution and analysis speed), medical air, industrial hydrogen).
3. Disposable syringe (1 mL or larger based on the sample size).
4. Packed or capillary column (e.g., a 2 m × 2 mm i.d. stainless steel column packed with Alumina F-1 (60/80 mesh) lined with the appropriate stationary phase (e.g., nonpolar stationary phase).
5. Software and program for data presentation.
6. Ethylene standard (μL/L). The standard concentrations depend on the range of ethylene in the measured sample.

2.2 Internal Ethylene Measurement from Fruits and Vegetables

1. 18G needle.
2. Disposable syringe with 24G needle.

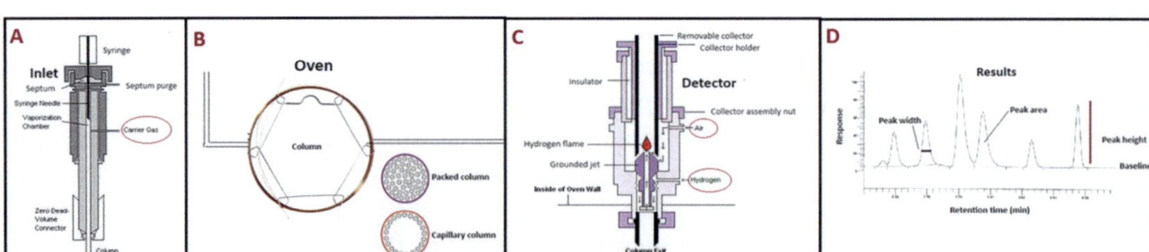

Fig. 2 A simplified diagram of a gas chromatograph showing (**a**) inlet, (**b**) oven and column, (**c**) detector, and (**d**) raw data. (Modified from Ref. [26])

3. 0.016-inch Jet cleaning wire.
4. Fruits or vegetables (samples).

2.3 External Ethylene Measurement from Sealed Containers

1. Plastic or glass container with a tight lid (the volume depends on the size of fruits and their ethylene production rate).
2. Rubber septa.
3. Disposable syringe (with volume based on the container size) with 24G needle.
4. Fruits or vegetables (samples).

2.4 External Ethylene Measurement from the Flow or Stop-Flow Method

1. Air source (air compressor or air cylinder).
2. Water trap (*see* **Note 8**).
3. Ethylene trap.
4. Containers with a known volume and tight lids.
5. Rubber septa.
6. Disposable syringe with 24G needle.
7. Fruits or vegetables (samples).

3 Methods

3.1 GC Calibration

1. Set the injector temperature (e.g., 230 °C).
2. Set the oven temperature (e.g., 160 °C).
3. Set the detector temperature (e.g., 245 °C).
4. Gas flow can be adjusted based on the GC type, used column, and the manufacturer recommendations (*see* **Note 1**).
5. Utilize the flame ionization detector (Fig. 2c).

3.2 Ethylene Standard

1. Use known concentrations of ethylene (μL/L), with the range of concentrations based on the measured samples, mixed with pure nitrogen (e.g., 10 μL/L).
2. Inject the standard to the GC and then compare the peak or the area under the curve with the results of the standard ethylene peak or the area under the curve (Fig. 2d).
3. Create an ethylene standard curve by diluting the known ethylene concentration (μL/L) with pure nitrogen and present the results of the peak or the area under the curve compared with the actual ethylene concentration in linear regression (e.g., y-axis represents different ethylene concentrations such as 1, 3, 5, 7, and 10 μL/L, while the x-axis represents the area under the curve for each concentration obtained from Fig. 2d).

3.3 Internal Ethylene Measurement from Fruits and Vegetables

1. The method is used on fruits and vegetables that accumulate ethylene in the core area (such as apple fruit).
2. Obtain 1 mL samples from core fruits or vegetables using a 1 mL disposable syringe.
3. Use 18G needle to penetrate to the core and use 0.016 Jet cleaning wire to prevent blockage.
4. Attach the 1 mL syringe to the 18G needle and pull more than 1 mL gas from the core.
5. Take the syringe out of the 18G needle and attach it to the 24G needle and adjust to 1 mL.
6. Inject your sample to the GC injector (*see* **Notes 2, 3,** and **4**).

3.4 External Ethylene Measurement from Sealed Containers

2. This method is suitable for all types of fruit and vegetables with different sizes.
3. Prepare a container with a known volume and a tightly sealing lid.
4. Use rubber septa to seal holes in the container lid completely, preventing gas diffusion and enabling the use of a syringe needle to collect the gas sample.
5. Use a 1 mL or larger volume disposable syringe, based on the container's volume, with a 24G needle.
6. Weigh your fruits before sealing them in the container.
7. Insert the needle through the container-sealed septa.
8. Flush the syringe plunger—three to four times in the sample headspace to mix the air in the container. Even we know that ethylene is lighter than air.
9. Pull more than 1 mL (or more depends on the syringe volume) of the gas samples.
10. Adjust the sample to 1 mL (in case you used 1 mL syringe).
11. Inject the sample into the GC injector.
12. The injection time between gas samples depends on the ethylene concentration in the sample. If ethylene is high in your samples, you should have more time between two injections to avoid column overload.
13. The measurement should be done based on the specific fruit or vegetable type (*see* **Note 5**).
14. Calculate the ethylene reading from the GC based on the peak or area under the curve, comparing it with the ethylene standard curve.
15. Calculate the results based on the sealing time, pressure inside the containers, fruit weight, and ethylene concentration from the GC. Present the results as (mL/kg/h) on a fresh weight basis (*see* **Note 9**).

3.5 External Ethylene Measurement from the Continuous or Stop-Flow Method

1. Use an air source from an air compressor or an air cylinder.
2. Ensure the airline contains a water trap and an ethylene trap.
3. The flow rate of the air from the main source should be higher than the total flow from the separate chambers (containers).
4. The chamber/container volume should be sufficient for the targeted fruit.
5. Drill a 5 mm diameter hole in the container lid and another one on the side of the container. Seal both holes with rubber septa. Use the one in the lid to collect gas samples using a 24G disposable 1 mL syringe. The other hole on the side of the container should have an 18G needle to allow air to flow out when the container is full.
6. Calibrate all containers with a flow meter and set them to the same flow rate.
7. Select the flow rate based on the fruit type and the level of ethylene production. If the fruit has a high ethylene and respiration rate, the flow rate should be high (e.g., 20 mL/min). If the fruit has a low ethylene and respiration rate, the flow rate should be low (e.g., 8 mL/min) (*see* **Note 7**).
8. Add the fruit with a known weight (e.g., units in kg).
9. Collect 1 mL of the gas samples every 24 h and inject the sample to the GC.
10. Calculate the readings based on the ethylene standard curve, as mentioned above.
11. Calculate the ethylene production rate based on the fruit weight, flow rate, and ethylene concentration from GC (*see* **Note 8**).
12. Present the results as mL/kg/h on a fresh weight basis (*see* **Note 9**).

4 Notes

1. **Example:** An example of the equipment used is Hewlett-Packard 5890 series II gas chromatograph (Wilmington, DE) with a Hewlett-Packard 3394A integrator and a 2 m × 2 mm i.d. stainless steel column packed with Alumina F-1 (60/80 mesh). The oven temperature is 160 °C, and injector and detector temperatures are 230 °C and 245 °C, respectively. The carrier gas was nitrogen (flow rate, 30 mL/min), while the detector was supplied with hydrogen and compressed air (flow rates 30 and 230 mL/min) [8, 27].
2. When collecting gas samples from the core of fruits or vegetables, ensure that your sample is free from juice (juice drops or

liquid vapor may be observed on the walls of the syringe; tap the syringe two to three times on a paper towel to ensure that no liquid remains inside the syringe before injection), as it can affect the results and column efficiency. If juice accidentally enters the GC, heat the column to a temperature higher than the oven temperature but lower than the column's maximum temperature tolerance.

3. When you are loading your samples into the GC injector, be sure to secure the syringe plunger to prevent it from being ejected due to GC pressure.
4. Replace the inlet septa in the GC injector when they show signs of wear, and use the gas pressure as an indicator for this replacement.
5. In crops with high respiration rates, the presence of CO_2 can inhibit ethylene production in fruits, impacting reading accuracy. To avoid CO_2 buildup, add NaOH to the sealed container. For fruits with a high ethylene production rate, sampling should be conducted more frequently to account for accelerated ripening.
6. For measuring ethylene from apple fruits, the recommended airflow rate ranges from 6 to 12 mL/min, for instance. Adjust the airflow rate up or down to prevent CO_2 accumulation in the container based on the fruit species, cultivar, and ripening stage. Use a continuous flow method when ethylene concentrations are moderate to high and the stop-flow method when ethylene concentrations are very low in the samples.
7. Silica gel is used in the water trap in the air line.
8. Potassium permanganate ($KMnO_4$) is used in the ethylene trap in the air line [28].
9. Ethylene calculation in a flow-through air system or in sealed container can be done as follows:
   (a) Ethylene readings from the GC are in μL/L.
   (b) Multiply by the flow rate in L/h if you are using the flow board.
   (c) If the flow rate is 20 mL/min, it is 1200 mL/h or 1.2 L/h.
   (d) If your GC reading is 0.001 mL/L $\times$ 1.2 L/h $=$ 0.0012 mL/h.
   (e) Adjust for the weight of the produce.
   (f) If the fruit weighs 200 g, that is equal to 0.200 kg.
   (g) Divide by the weight in kg ((0.0012 mL/h) / 0.200 kg $=$ 0.006 mL/kg/h).

Acknowledgments

I would like to express my sincere gratitude to Professor Chris B. Watkins of Cornell University for his invaluable editing of the text.

References

1. Bakshi A, Shemansky JM, Chang C, Binder BM (2015) History of research on the plant hormone ethylene. J Plant Growth Regul 34: 809–827. https://doi.org/10.1007/s00344-015-9522-9
2. Abeles FB, Morgan PW, Saltveit ME Jr (2012) Ethylene in plant biology. Academic Press
3. Vandenbussche F, Vriezen WH, Van Der Straeten D (2008) Ethylene biosynthesis and signaling: a puzzle yet to be completed. In: Plant hormone signaling, vol 9600. Blackwell, pp 125–145
4. Chen Y, Grimplet J, David K, Castellarin SD, Terol J, Wong DC, Luo Z, Schaffer R, Celton J-M, Talon M (2018) Ethylene receptors and related proteins in climacteric and non-climacteric fruits. Plant Sci 276:63–72. https://doi.org/10.1016/j.plantsci.2018.07.012
5. Valente J, Almeida R, Kooistra L (2019) A comprehensive study of the potential application of flying ethylene-sensitive sensors for ripeness detection in apple orchards. Sensors 19(2):372. https://doi.org/10.3390/s19020372
6. Laties GG (1995) Franklin Kidd, Charles west and FF Blackman: the start of modern postharvest physiology. Postharvest Biol Technol 5(1–2):1–10. https://doi.org/10.1016/0925-5214(94)00009-H
7. Paul V, Pandey R, Srivastava GC (2012) The fading distinctions between classical patterns of ripening in climacteric and non-climacteric fruit and the ubiquity of ethylene – an overview. J Food Sci Technol 49:1–21. https://doi.org/10.1007/s13197-011-0293-4
8. Cai W, Al Shoffe Y, Park D, Watkins CB (2023) Harvest maturity and preharvest aminoethoxyvinylglycine treatment effects on cold-induced ethylene production of 'Gala' apples. HortScience 58(5):532–538. https://doi.org/10.21273/HORTSCI17074-23
9. Al Shoffe Y, Park D, Algul BE, Watkins CB (2023) Short-term storage of 'Honeycrisp'apples at 0.5 oC does not improve control of bitter pit without increasing low temperature-induced physiological disorders. Sci Hortic 321:112364. https://doi.org/10.1016/j.scienta.2023.112364
10. Fu Y, Yang D, Chen Y, Shi J, Zhang X, Hao Y, Zhang Z, Sun Y, Zhang J (2023) MOF-based active packaging materials for extending post-harvest shelf-life of fruits and vegetables. Materials 16(9):3406. https://doi.org/10.3390/ma16093406
11. Barry CS, Giovannoni JJ (2007) Ethylene and fruit ripening. J Plant Growth Regul 26:143–159. https://doi.org/10.1007/s00344-007-9002-y
12. Tucker G, Yin X, Zhang A, Wang M, Zhu Q, Liu X, Xie X, Chen K, Grierson D (2017) Ethylene and fruit softening. Food Qual Saf 1(4):253–267. https://doi.org/10.1093/fqsafe/fyx024
13. Ebrahimi A, Zabihzadeh Khajavi M, Ahmadi S, Mortazavian A, Abdolshahi A, Rafiee S, Farhoodi M (2022) Novel strategies to control ethylene in fruit and vegetables for extending their shelf life: a review. Int J Environ Sci Technol 19(5):4599–4610. https://doi.org/10.1007/s13762-021-03485-x
14. Chen B, Zhang M, Chen H, Mujumdar AS, Guo Z (2023) Progress in smart labels for rapid quality detection of fruit and vegetables: a review. Postharvest Biol Technol 198:112261. https://doi.org/10.1016/j.postharvbio.2023.112261
15. Bartle KD, Myers P (2002) History of gas chromatography. TrAC Trends Anal Chem 21(9–10):547–557. https://doi.org/10.1016/S0165-9936(02)00806-3
16. Nolvachai Y, Amaral MS, Marriott PJ (2023) Foods and contaminants analysis using multidimensional gas chromatography: an update of recent studies, technology, and applications. Anal Chem 95(1):238–263. https://doi.org/10.1021/acs.analchem.2c04680
17. Bulens I, Van de Poel B, Hertog ML, De Proft MP, Geeraerd AH, Nicolaï BM (2011) Protocol: an updated integrated methodology for analysis of metabolites and enzyme activities of ethylene biosynthesis. Plant Methods 7:1–10. http://www.plantmethods.com/content/7/1/17

18. Hu B, Sun D-W, Pu H, Wei Q (2019) Recent advances in detecting and regulating ethylene concentrations for shelf-life extension and maturity control of fruit: a review. Trends Food Sci Technol 91:66–82. https://doi.org/10.1016/j.tifs.2019.06.010
19. Cristescu SM, Mandon J, Arslanov D, De Pessemier J, Hermans C, Harren FJ (2013) Current methods for detecting ethylene in plants. Ann Bot 111(3):347–360. https://doi.org/10.1093/aob/mcs259
20. Janssen S, Schmitt K, Blanke M, Bauersfeld M, Wöllenstein J, Lang W (2017) Ethylene detection in fruit supply chains. Philos Trans R Soc A Math Phys Eng Sci 2014(372):20130311. https://doi.org/10.1098/rsta.2013.0311
21. Sklorz A, Janßen S, Lang W (2013) Application of a miniaturised packed gas chromatography column and a SnO2 gas detector for analysis of low molecular weight hydrocarbons with focus on ethylene detection. Sensors Actuators B Chem 180:43–49. https://doi.org/10.1016/j.snb.2011.12.110
22. Zaidi NA, Tahir MW, Vellekoop MJ, Lang W (2018) Design of Novel Ceramic Preconcentrator and Integration in gas chromatographic system for detection of ethylene gas from ripening bananas. Sensors 18(8):2589. https://doi.org/10.3390/s18082589
23. Gwanpua SG, Jabbar A, Tongonya J, Nicholson S, East AR (2018) Measuring ethylene in postharvest biology research using the laser-based ETD-300 ethylene detector. Plant Methods 14:1–17. https://doi.org/10.1186/s13007-018-0372-x
24. Yang X, Leong JLK, Sun M, Jing L, Zhang Y, Wang T, Wang S, Huang D (2022) Quantitative determination of ethylene using a smartphone-based optical fiber sensor (SOFS) coupled with pyrene-tagged grubbs catalyst. Biosensors 12(5):316. https://doi.org/10.3390/bios12050316
25. Li X, Sun R, Pan J, Shi Z, Lv J, An Z, He Y, Chen Q, Han RP, Zhang F (2023) All-MXene-printed RF resonators as wireless plant wearable sensors for in situ ethylene detection. Small:2207889. https://doi.org/10.1002/smll.202207889
26. Thet K, Woo N Gas chromatography. https://chem.libretexts.org/Bookshelves/Analytical_Chemistry/Supplemental_Modules_(Analytical_Chemistry)/Instrumentation_and_Analysis/Chromatography/Gas_Chromatography
27. Al Shoffe Y, Nock JF, Zhang Y, Watkins CB (2021) Physiological disorder development of 'Honeycrisp'apples after pre-and post-harvest 1-methycyclopropene (1-MCP) treatments. Postharvest Biol Technol 182:111703. https://doi.org/10.1016/j.postharvbio.2021.111703
28. Kumar S, Kumar R, Bibwe BR, Nath P, Singh RK, Mandhania S, Pal A, Soni R, Kumar A (2023) Postharvest handling of ethylene with oxidative and absorptive means. J Food Sci Technol:1–20. https://doi.org/10.1007/s13197-023-05777-1

Chapter 7

Ethylene-Mediated Regulation of a Ripening-Specific N-glycan-processing enzyme β-D-N-Acetylhexosaminidase

Mohammad Irfan, Pankaj Kumar, and Asis Datta

Abstract

N-glycoproteins in plants play a critical role in various biological processes such as plant stress response, seed development, and fruit ripening. Further, N-glycans and N-glycan-processing enzymes have been identified as critical regulators of fruit ripening in various fruit crop including tomato. Fruit ripening, a multifaceted process influenced by diverse factors is controlled by the upregulation of cell wall degradation enzymes. Among these, N-glycan-processing enzymes, specifically β-D-N-acetylhexosaminidase (β-Hex), are involved in ripening-associated fruit softening, and its suppression enhances fruit shelf life and firmness in both climacteric and non-climacteric fruits. Ethylene is among the various factors that regulates the ripening-specific expression of *β-Hex*. Notably, *ethylene response factor 6* (*SlERF6*), also known as *SlERF. E4*, has been identified as a potential transcriptional regulator of *β-Hex* expression. However, in order to get deeper insights, a thorough understanding of the specific ethylene-mediated control of *β-Hex* during ripening is crucial. Thus, we present a protocol for investigating the ethylene-mediated regulation of this crucial ripening-related enzyme β-D-N-acetylhexosaminidase.

Key words Tomato, Fruit ripening, Ethylene, N-glycoproteins, N-glycans, β-D-N-acetylhexosaminidase, ACC, GUS assay

1 Introduction

Fruit ripening is a complex process that involves genetic and epigenetic regulators, transcription factors, and endogenous molecules that orchestrate a variety of metabolic alterations [1–3]. During fruit ripening, cell wall dissolution is a crucial process that contributes to the softening and overall textural changes in the fruit [1–3]. As a result, fruits become softer, sweeter, and more palatable; however, excessive softening during ripening results in huge postharvest losses across the globe [4–6]. The activity of various cell wall–modifying enzymes increases during fruit ripening, leading to

Authors Mohammad Irfan and Pankaj Kumar have equally contributed to this chapter.

Devesh Shukla (ed.), *Ethylene Sensing and Signaling*, Methods in Molecular Biology, vol. 2945,
https://doi.org/10.1007/978-1-0716-4650-2_7,

the depolymerization and solubilization of the cell wall components [3, 7]. Among these key factors, N-glycoproteins in the plant cell wall play a critical role in cell wall loosening during ripening [7, 8]. N-glycans and N-glycan-processing enzymes have been identified as critical regulators of ripening in various fruit crops including tomato (*Solanum lycopersicum*) [7–10].

The cell wall–located N-glycan processing enzymes, α-mannosidase (α-Man) and β-D-N-acetylhexosaminidase (β-Hex), have been identified as critical players during ripening, and their activity increases during ripening [7, 8, 11–13]. Down-regulation of *α-Man* and *β-Hex* through RNAi technology in tomato and capsicum leads to increased fruit firmness and extended shelf life, which suggests a novel approach to minimize post-harvest losses of fruits and vegetables [7, 8]. Recent studies have also indicated the involvement of N-glycans and β-Hex enzymes in the ripening-associated softening of strawberry and peach fruits [14, 15].

β-Hex (EC 3.2.1.52), a member of the glycosyl hydrolase family 20, acts on N-acetyl-β-D-hexosaminides and produces paucimannosidic N-glycans during ripening [7, 8, 16, 17]. The transcriptional regulation of *β-Hex* in tomato has been thoroughly investigated, revealing its regulation by the master regulator of fruit ripening, RIPENING INHIBITOR (RIN), and the ripening hormone ethylene [18, 19]. Ethylene induces the expression of *β-Hex* and its promoter activity [7, 18]. Moreover, *ABSCISIC ACID STRESS RIPENING 1 (SlASR1)* also influences the expression of *β-Hex* by acting as a positive transcriptional regulator. *RIN* and *SlASR1* are both ethylene-responsive genes, and *RIN* also regulates the expression of *SlASR1* [18].

Ethylene, a key ripening hormone, is known to play a pivotal role in this process, and many ethylene response factors (ERFs) have been identified, which play a critical role in coordinating the genetic regulation of fruit ripening by integrating signals from ethylene and other hormonal and environmental cues [1, 2, 19]. Among the various ethylene response factors, *SlERF6* acts as a key regulator of fruit ripening, and its expression particularly increases during ripening [20]. Recent studies have shed light on the transcriptional regulation of *β-Hex* by *SlERF.E4* (other name of *SlERF6, Solyc01g065980.2.1*) during the ripening of tomato fruits [21]. Various experimental approaches, including deletion mapping, site-directed mutagenesis, and DNA–protein interaction assays, were employed to investigate the regulatory mechanisms [21]. The findings confirmed that *RIN*, *SlASR1*, and *SlERF.E4* directly bind to the 200 bp region upstream to the start codon (D-200) of *β-Hex* gene, positively regulating the expression of *β-Hex* during fruit ripening [21]. Therefore, in this book chapter, we describe a procedure for ethylene-mediated regulation of *β-Hex (Solyc01g081610.2.1)*. The protocol encompasses several key experimental techniques, including the treatment of fruits with the ethylene precursor aminocyclopropane carboxylic acid (ACC), β-glucuronidase (GUS) histochemical and fluorometric assays, purification of SlERF.E4, and electrophoretic mobility shift assay (EMSA) for showing the binding of SlERF.E4 to the D-200 *β-Hex* promoter. This protocol can be useful in studying the ethylene-mediated gene expression in fruits.

2 Materials

2.1 Plant Materials

1. Tomato (*Solanum lycopersicum* cv. pusa ruby) WT seeds.
2. Transgenic tomato plants seeds containing plasmids D-200::GUS, mCArG/mASR1_D-200::GUS, as well as wild-type plants. (*see* **Note 1**).

2.2 Strains and Plasmids

1. *E. coli* BL21 with the following genotypes, *E. coli* B F- *dcm ompT hsdS*(r_B- m_B-) gal [malB$^+$]$_{K\text{-}12}$(λ^S).
2. Recombinant pGEX-4 T-2 vector expressing *SlERF.E4.*

2.3 Media and Chemicals

1. 2X YTA media: 1% Yeast extract; 1.6% Bacto tryptone; and 0.5% NaCl.
2. Antibiotics: Kanamycin—100 mg/mL in water; ampicillin—50 mg/mL in water.
3. Isopropylthio-β-galactoside (IPTG): 100 mM stock in H_2O, filter sterilized, and stored at −20 °C.
4. 200 mL 0.5 M sodium phosphate buffer (pH 7.2): 68.4 mL of 1 M Na_2HPO_4, 31.6 mL of 1 M NaH_2PO_4, and 100 mL water.
5. 1 L 10× PBS: 2.0 g KCl, 80 g NaCl, 17.8 g Na_2HPO_4, and 2.4 g KH_2PO_4.
6. Glutathione Sepharose 4B.
7. Elution buffer: 50 mM reduced glutathione in 125 mM Tris–HCl (pH 8.0) and 150 mM NaCl.
8. 1 L 10× TBE buffer (pH 8.3): dissolve 107.8 g Tris-base, 55 g boric acid, and 7.44 g Na_2EDTA.2H2O in 800 mL water and adjust the pH with boric acid.
9. X-Gluc: 100 mM stock in dimethyl formamide (DMF) and store at −20 °C.
10. GUS staining solution: 100 mM sodium phosphate buffer (pH 7.0), 10 mM EDTA, 0.5 mM $K_3Fe(CN)_6$, 0.5 mM $K_4Fe(CN)_6.3H_2O$, 0.1% Triton-X, 20% methanol, and 1 mM X-Gluc.

11. GUS extraction buffer: 50 mM sodium phosphate buffer (pH 7.0), 10 mM DTT, 10 mM EDTA, 0.1% sodium lauryl sarcosine, and 0.1% Triton X-100.
12. MUG assay buffer: 1 mM 4-methylumbelliferyl ß-D-glucuronide (4-MUG) in GUS extraction buffer.
13. Stop buffer: 0.2 M Na_2CO_3.
14. 4-MU: 10 mM stock to make a standard curve of 4-MU.
15. 5× gel shift assay binding buffer: 50 mM Tris–HCl (pH 7.5), 20% glycerol; 10 mM $MgCl_2$, 2.5 mM EDTA, 2.5 mM DTT, and 150 mM NaCl.
16. 10× gel loading buffer: 250 mM Tris–HCl (pH 7.5), 0.2% bromophenol blue, and 40% glycerol.
17. 6% non-denaturing (native) PAGE: 6 mL of 30% acrylamide: bisacrylamide (30: 0.8), 2 mL of 10× TBE buffer (pH 8.3), 28.4 mL water, 300 μL of 10% APS, 50 μL TEMED; and 1.25 mL of 80% glycerol.
18. 10 mM MES (pH 5.6).
19. poly(dI-dC).poly(dI-dC): 1 μg/uL.
20. 10× NEB buffer 2.
21. *Xba*I/*Hind*III digested ds D_4–200 *β-Hex* promoter.
22. dATP, dGTP, and dTTP: 0.5 mM of each.
23. [α-P^{32}] CTP: 3000 Ci/mmol, 50 μCi.
24. DNA polymerase I (Klenow) fragment.
25. Sorbitol.
26. ACC.

3 Methods

3.1 Seed Germination and Tagging for Fruit-Ripening Stages

1. Germinate the tomato seeds by placing them in presterilized soil.
2. Maintain greenhouse environment conditions with a temperature of 25 °C, 70% humidity, and a light cycle of 14 h of light (~250 μmoL m^{-2} s^{-1}), followed by 10 h of darkness.
3. Monitor the germinated seeds in the greenhouse for growth and development.
4. Tag the fruits when they reach the anthesis stage (*see* **Note 2**).
5. Define the ripening stages of the tagged fruits as follows:
   (a) Mature green stage: 40 days after anthesis.
   (b) Breaker stage: mature green stage +3 days.

(c) Pink stage: breaker stage +2 days.

(d) Red ripe stage: pink stage +3 days.

6. Observe and analyze the fruits at each ripening stage for further research or evaluation purposes.

3.2 Treatment of Fruits with Ethylene Precursor Aminocyclopropane Carboxylic Acid (ACC)

1. Harvest mature green fruits from transgenic plants containing mutated plasmids D-200::GUS, mCArG/mASR1_D-200::GUS, as well as wild-type plants.
2. Prepare a solution of 10 mM MES (pH 5.6), sorbitol (3% w/v), and 100 μM ACC.
3. Take six fruits from each group for the treatment.
4. Using a 1.0 mL syringe, infiltrate the fruits with the solution prepared in **step 2**. Inject through stylar apex to ensure precise application.
5. For the control group, infiltrate the fruits with the solution from **step 2** but without ACC.
6. Incubate the treated fruits at 25 °C under a 14 h light/10 h dark cycle until they reach the desired stage of ripening, which takes approximately 3–6 days.
7. Cut the fruits into transverse sections and perform the GUS histochemical assay immediately on 50% of total slices, and the remaining 50% slices can be flash-freezed in liquid nitrogen, followed by long-term storage at − 80 °C until further analysis.

3.3 GUS Histochemical and Fluorometric Assay

1. For GUS histochemical staining, completely dip the fruit slices in GUS staining solution in 140 mM petri dishes and perform vacuum infiltration (*see* **Notes 3–6**).
2. Incubate the petri dishes with the GUS staining solution in the dark at 37 °C for 10–12 h.
3. After incubation, rinse the sections with MQ water two to three times, and then destain with 75% ethanol for 1–2 days (*see* **Note 7**).
4. The fruit sections can be photographed for observation using any digital camera and microscope (e.g., Nikon AZ100 5× microscope). Figure 1a shows the GUS-stained fruits of D-200::GUS and mCArG/mASR1_D-200::GUS transgenic tomato with and without ACC treatment.
5. The quantitative analysis of GUS activity is determined by a fluorometric assay by measuring the production of 4-methylumbelliferone (4-MU) and expressed as pmol 4-MU/mg/min.
6. For GUS fluorometric assay, homogenize the fruit samples (100 mg) with liquid nitrogen and add 0.4 mL of GUS extraction buffer in a microcentrifuge tube.

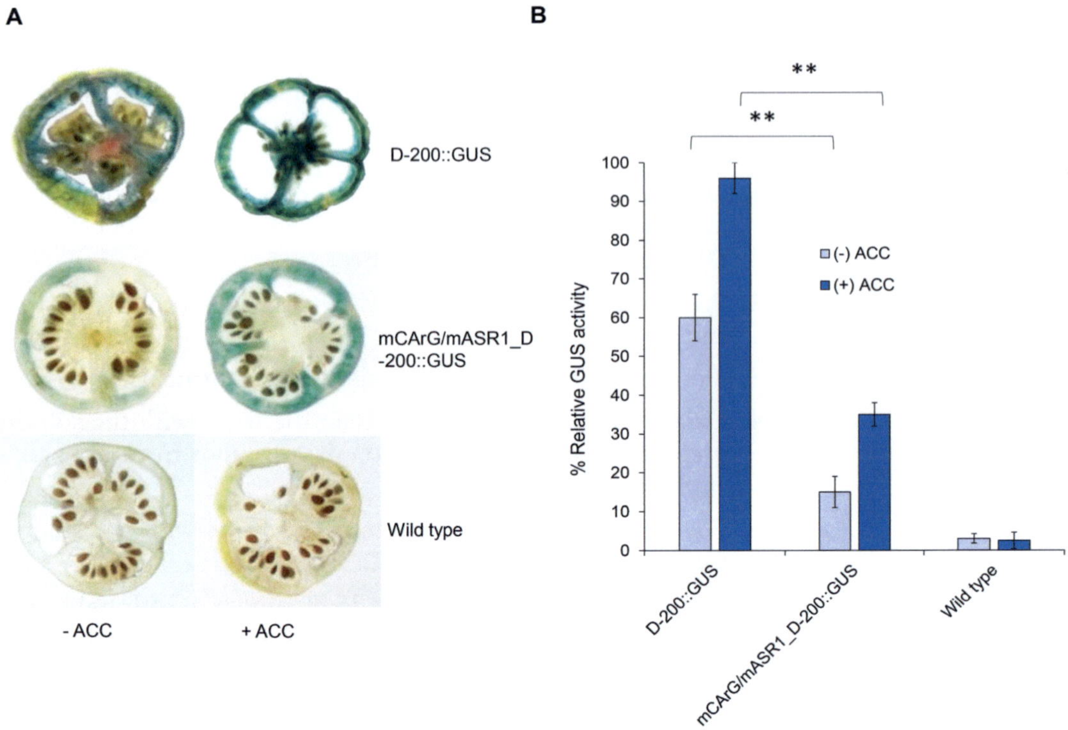

Fig. 1 The ethylene precursor aminocyclopropane carboxylic acid (ACC) treatment enhances GUS activity in D-200::GUS and mCArG/mASR1_D-200::GUS transgenic fruits, suggesting the regulation of *β-Hex* by ethylene and is independent of *SlASR1* and *RIN*. mCArG and mASR1 represent mutation in the binding sites of RIN and ASR1 located on the D-200 fragment of *β-Hex* promoter. (**a**) GUS histochemical assay. (**b**) GUS fluorometric quantitative assay. $^{**}P < 0.01$ displays significant differences (Student's *t* test). (Reproduced with permission, from Irfan et al. Ref. [21])

7. Centrifuge the homogenized mixture at 13,000 g at 4 °C for 15 min and collect the supernatant for protein quantification and a fluorometric assay (*see* **Note 8**).
8. For GUS enzyme assay, transfer 50 μL of the supernatant into a microcentrifuge tube containing 450 μL of prewarmed MUG assay buffer (protein extraction buffer containing 10 mM MUG) at 37 °C and mix them properly using a pipette.
9. Take 100 μL of this mixture and immediately transfer it into a separate microcentrifuge tube containing 900 μL of the reaction stop buffer (0.2 M Na_2CO_3) to serve as a control for 0 min time point.
10. Incubate the remaining sample reaction mixture at 37 °C.
11. At 15, 30, 60, and 120 min intervals, remove 100 μL aliquots from the reaction mixture and mix them with 900 μL of the stop buffer.

12. Measure the GUS activity using a fluorometer (Cary Eclipse, Varian) with excitation at 380 nm and emission at 454 nM.
13. Determine the amount of 4-MU from a standard curve of 4-MU. For making the standard curve, serially dilute 4-MU (0.05–1.5 μM) from the 10 mM stock and measure the fluorescence with excitation at 380 nM and emission at 454 nM. Plot the concentrations against the absorbance. Figure 1b represents the GUS fluorometric quantitative activity in D-200::GUS and mCArG/mASR1_D-200::GUS transgenic tomato fruits with and without ACC treatment.

3.4 Purification of *SlERF.E4*

1. For protein purification, *E. coli* BL21 cells harboring full-length *SlERF.E4* with N-terminal GST tag in pGEX4T-2-expression vector were used (*see* **Note 9**).
2. Grow a single colony in 5 mL of LB medium with 5 μL of ampicillin overnight at 37 °C in a shaking incubator.
3. Inoculate 1% of the overnight-grown culture into 50 mL of 2× YTA with ampicillin in a 250 mL conical flask and incubate at 28 °C with shaking until the optical density (OD) reaches mid-log phase (A_{600} 0.5–1.0).
4. Induce the culture with 0.6 mM concentration of IPTG for 4 h at 28 °C.
5. Harvest cells by centrifugation at 6000 rpm for 15 min at 4 °C and wash the pellets with prechilled 1× PBS twice.
6. Resuspend the pellets in 2.5 mL of 1× PBS and add 0.1 volume of lysozyme (10 mg/mL in 25 mM Tris–HCl, pH 8.0) and 5 mM DTT to the cell suspension.
7. Perform sonication of the cells with the following conditions: 0.5 cycle/70% amplitude, burst (20 s)–gap (2–3 min)–burst (20 s) 5–6 times.
8. Add Triton X-100 to a final concentration of 1% to the suspension and incubate for 30 min with gentle mixing.
9. Collect the supernatant by centrifugation at 10000 rpm for 10 min and purify the induced proteins from this crude extract by affinity chromatography using glutathione sepharose (*see* **Note 10**).
10. For making 200 μL bed volume of Glutathione Sepharose 4B, 275 μL slurry of original stock was added to 5 mL of ice cold 1× PBS and centrifuged at 500× g for 3 min at 4 °C.
11. Remove the 1× PBS and add crude cell lysate and incubate overnight at 4 °C with gentle agitation such as end-over-end rotation. (*see* **Notes 11** and **12**).
12. Centrifuge at 500× g for 3 min, discard the supernatant, and wash with 10 mL of 1× PBS.

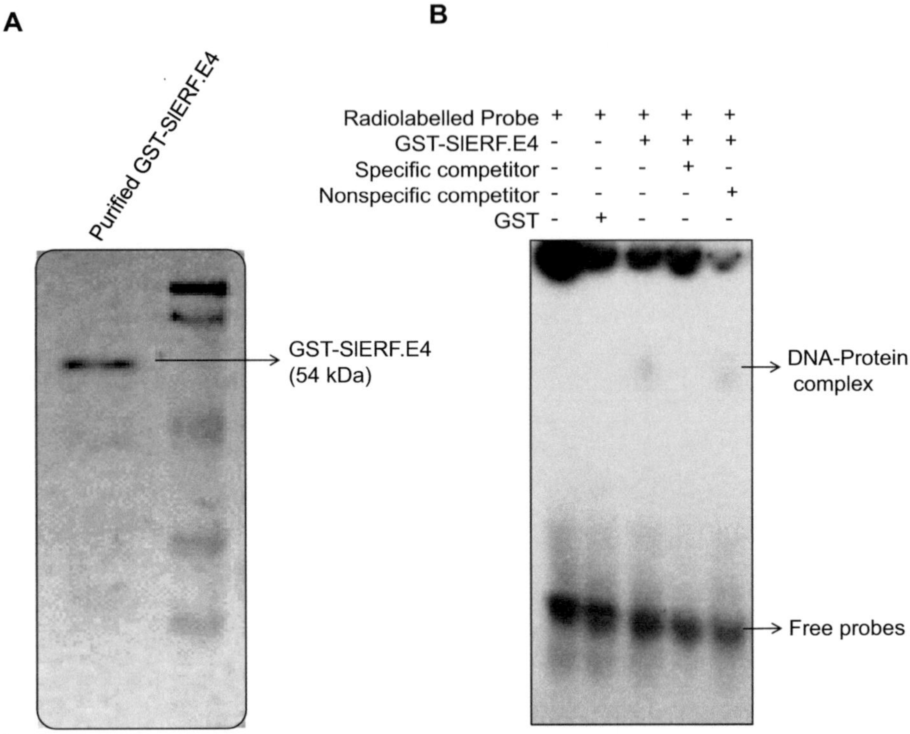

Fig. 2 (**a**) SDS-PAGE profile of purified SlERF.E4. (**b**) Electrophoretic mobility shift assay (EMSA) showing the binding of purified SlERF.E4 with the D-200 promoter of *β-Hex* gene. EMSA was carried out with 1 μg of SlERF.E4 protein and radioactively labeled D-200 *β-Hex* promoter. The reactions were resolved on 6% native polyacrylamide gels using 0.5× TBE buffer and visualized by autoradiography. The shift was competed out by 100-fold excess of specific and nonspecific competitors. (Reproduced with permission, from Irfan et al. Ref. [21])

13. Repeat **step 12** two or more times.
14. Elute the fusion protein with 100–200 μL of glutathione elution buffer by incubating 5–10 min, followed by centrifugation at 500 × g for 3 min.
15. Repeat the **step 14** three to four times and store all elutes separately.
16. For purity assessment of the purified proteins, perform the SDS-PAGE analysis using the method described by He et al. [22]. Figure 2a is a representative of purified SlERF.E4 protein analyzed on SDS-PAGE.
17. The purified samples can be pooled together after SDS-PAGE analysis.
18. To determine the concentration of purified proteins, perform the Bradford assay [23] and make 20% glycerol stocks of purified proteins and store them at −20 °C until further use.

3.5 Electrophoretic Mobility Shift Assay (EMSA)

1. To prepare the probe for EMSA, digest the pD-200 plasmid with *Hind*III/*Xba*I which releases the D-200 region of *β-Hex* promoter, followed by PCR purification of the fragment (*see* **Note 13**).
2. Perform the end filling of fragment with [α-P^{32}]CTP (3000 Ci/mmol, 50 μCi), using the DNA polymerase I (Klenow) fragment (New England Biolabs). For a 30 μL reaction, add the following components in a sterile microcentrifuge tube and incubate at room temperature for 30 min, followed by adding 1 μL of 0.5 M EDTA for stopping the reaction. Make the volume of the reaction mix upto 100 μL with nuclease-free water before purification by a Sephadex G-50 column to separate the labeled nucleotides (*see* **Note 14**).

10× NEB buffer 2	= 3 μL
*Xba*I/*Hind*III digested ds D_4–200 *β-Hex* promoter	= 3 μL (150 ng)
dATP (0.5 mM)	= 1 μL
dGTP (0.5 mM)	= 1 μL
dTTP (0.5 mM)	= 1 μL
[α-P^{32}]CTP (3000 Ci/mmol, 50 μCi)	= 1 μL
DNA polymerase I (Klenow) fragment	= 1 μL (5 units)
Nuclease-free water	= 19 μL

3. For the binding of SlERF.E4 with the D-200 β-Hex promoter, in 20 μL reaction, add the following components in a sterile microcentrifuge tube and incubate at room temperature for 30 min (*see* **Note 15**).

5× gel shift binding buffer	= 4 μL
[α-P^{32}]CTP labeled 200 bp dsDNA	= 2 μL (10 ng)
Purified SlERF.E4 (1 μg/μL)	= 1 μL
Poly(dI-dC).Poly(dI-dC) (1 μg/μL)	= 1 μL
Nuclease-free water	= 12 μL

4. For competition assay, unlabeled *XbaI/HindIII* digested ds D4-200 *β-Hex* promoter as a specific competitive inhibitor and an unrelated DNA as a nonspecific competitive inhibitor can be used. For negative control, use 1 μg of GST protein (without SlERF.E4) in the above gel shift assay.
5. Run the reaction on 6% native PAGE with a constant current (10 mA) using 0.5× TBE as a running buffer. Add 2 μL of 10×

loading buffer to only probe the sample for tracking the gel run.

6. Dry the gel using a gel dryer and visualize the protein–DNA complexes as well as free probes by autoradiography. Figure 2b shows the binding of SlERF.E4 protein with the promoter of *β-Hex,* suggesting SlERF.E4-mediated regulation of *β-Hex* gene.

4 Notes

1. The mutated plasmid used in this experiment is mCArG/ mASR1_D-200::GUS, while the control group consists of fruits without any mutations (D-200::GUS). The mCArG and mASR1 mutated binding sites of RIN and ASR1 located on the D-200 fragment of *β-Hex* promoter. The stable transgenic plants of tomato were generated using *Agrobacterium tumifaciens* EHA 105 as a standard protocol reported earlier [18, 21]. Irfan et al. developed transgenic tomato plants using cotyledons from 2-week-old seedlings. The tomato seeds undergo sterilization with 4% commercial bleach and germinate on the Murashige and Skoog (MS) medium. After 2 weeks, the cotyledons are cut and co-cultivated with Agrobacterium in the MS medium containing acetosyringone (0.1 μM) for 30 min. Subsequently, the cotyledons are collected for selection on MS plates supplemented with kanamycin (50 mg L^{-1}), cefotaxime (250 mg L^{-1}), and zeatine (1 ng L^{-1}). Upon regeneration of plantlets, they are transferred to the rooting medium consisting of the MS medium with kanamycin (50 mg L–1), cefotaxime (250 mg L^{-1}), and IAA (1 ng L^{-1}). Transgenic seeds are germinated in kanamycin-containing MS medium (50 mg L^{-1}) to obtain progeny plants [21].
2. This method outlines the steps to germinate tomato seeds and monitor the ripening stages of the resulting fruits in a greenhouse setting. Ripening stages can vary slightly depending on the tomato variety and environmental conditions. By observing the color, texture, and other indicators, you can determine the ideal ripeness stage for your experiment.
3. To ensure even staining, incubate samples in a flat position and make sure no fruit slices are on one another and GUS staining solution is infiltrated uniformly.
4. In order to reduce the high background GUS activity, all reagents should be prepared freshly, and aseptic conditions should be maintained during the assay setup. Increasing the washing steps to remove unbound substrate and optimizing

the staining duration and concentration of staining solution could also be helpful to minimize background staining.

5. In case of low GUS activity, check the purity, freshness, concentration, and light protection of the substrates (X-Gluc and 4-MUG).
6. For inconsistent GUS assay results, both a negative control and a positive control, such as p35S::GUS can be included in the assay.
7. GUS-stained tissue can be stored in 75% ethanol at 4 °C for 7–10 days.
8. The protein concentration of the samples can be determined using the standard Bradford assay, with BSA as the standard [23].
9. More detail on the construct preparation can be found in the work by Irfan et al. [21].
10. Immobilized glutathione affinity chromatography is used to separate GST fusion proteins from bacterial lysates. The GST fusion proteins are captured by the affinity medium, and contaminants are washed away. Fusion proteins are eluted under mild, non-denaturing conditions with low glutathione levels. On the pGEX plasmids, a site-specific protease is present upstream of the multiple cloning site and can, if needed, cleave the protein from GST. The best expression conditions are assessed after choosing clones that express the fusion proteins, taking into account elements like medium, growth temperature, culture density, and induction conditions. It is essential to make sure there is enough aeration, keep each growth stage brief, and use antibiotic resistance to positively select the plasmid. The expression optimization procedure may be impacted by the presence of inclusion bodies.
11. This is called batch purification. The column purification can also be done using commercially available columns which involves passing the mixture through a column packed with glutathione sepharose.
12. Incubation time can be as low as 2 h with some yield penalty.
13. The detail of the pD-200 plasmid can be found in the article by Irfan et al. [21].
14. While working with radiolabeled probes and in potentially radioactivity-contaminated environments, personal protective equipment (PPE) must be worn correctly and remain in place.
15. EMSA-binding conditions are very sensitive toward the binding buffer composition, incubation temperature, and time of binding reaction. Lowering the reaction incubation

temperature to 4 °C and running the PAGE at 4 °C may increase the stability of the DNA–protein complex.

Acknowledgments

The financial support for this study was provided by a core research grant from the National Institute of Plant Genome Research, New Delhi, and a research grant from the Department of Biotechnology, Government of India. MI acknowledges the Council of Scientific and Industrial Research (CSIR), India, for the Senior Research Associateship (CSIR Pool Number: 8862-A).

References

1. Giovannoni J, Nguyen C, Ampofo B, Zhong S, Fei Z (2017) The epigenome and transcriptional dynamics of fruit ripening. Annu Rev Plant Biol 68:61–84
2. Wang R, Angenent GC, Seymour G, de Maagd RA (2020) Revisiting the role of master regulators in tomato ripening. Trends Plant Sci 25: 291–301
3. Sharma M, Negi S, Kumar P, Srivastava DK, Choudhary MK, Irfan M (2023) Fruit ripening under heat stress: the intriguing role of ethylene-mediated signaling. Plant Sci 335: 111820
4. Kumar V, Irfan M, Ghosh S, Chakraborty N, Chakraborty S, Datta A (2016) Fruit ripening mutants reveal cell metabolism and redox state during ripening. Protoplasma 253:581–594
5. Kumari C, Sharma M, Kumar V, Sharma R, Kumar V, Sharma P, Kumar P, Irfan M (2022) Genome editing technology for genetic amelioration of fruits and vegetables for alleviating post-harvest loss. Bioengineering 9(4):176
6. Irfan M, Kumar P, Ahmad MF, Siddiqui MW (2023) Biotechnological interventions in reducing losses of tropical fruit and vegetables. Curr Opin Biotechnol 79:102850
7. Meli VS, Ghosh S, Prabha TN, Chakraborty N, Chakraborty S, Datta A (2010) Enhancement of fruit shelf life by suppressing N-glycan processing enzymes. Proc Natl Acad Sci 107: 2413–2418
8. Ghosh S, Meli VS, Kumar A, Thakur A, Chakraborty N, Chakraborty S, Datta A (2011) The N-glycan processing enzymes α-mannosidase and β-D-N acetylhexosaminidase are involved in ripening-associated softening in the non-climacteric fruits of capsicum. J Exp Bot 62:571–582
9. Priem B, Gross KC (1992) Mannosyl- and xylosyl-containing glycans promote tomato (*Lycopersicon esculentum* Mill.) fruit ripening. Plant Physiol 98:399–401
10. Priem B, Gitti R, Bush CA, Gross KC (1993) Structure of ten free N-glycans in ripening tomato fruit (arabinose is a constituent of a plant N-glycan). Plant Physiol 102:445–458
11. Hossain MA, Nakamura K, Kimura Y (2009) A-Mannosidase involved in turnover of plant complex type N -glycans in tomato (*Lycopersicum esculentum*) fruits. Biosci Biotechnol Biochem 73:140–146
12. Hossain MA, Roslan HA (2014) Molecular phylogeny and predicted 3d structure of plant beta -D- N -acetylhexosaminidase. Sci World J 2014:1–14
13. Irfan M, Ghosh S, Meli VS, Kumar A, Kumar V, Chakraborty N, Chakraborty S, Datta A (2016) Fruit ripening regulation of α-Mannosidase expression by the MADS box transcription factor ripening inhibitor and ethylene. Front Plant Sci 7:10
14. Bose SK, He Y, Howlader P, Wang W, Yin H (2021) The N-glycan processing enzymes β-D-Nacetylhexosaminidase are involved in ripening-associated softening in strawberry fruit. J Food Sci Technol 58:621–631
15. Wang T, Jia X, Liu L, Voglmeir J (2021) Changes in protein N-glycosylation during the fruit development and ripening in melting-type peach. Food Mater Res 1:2
16. Strasser R, Bondili JS, Schoberer J, Svoboda B, Liebminger E, Glössl J, Altmann F, Steinkellner H, Mach L (2007) Enzymatic properties and subcellular localization of 18rabidopsis β – n -acetyl hexosaminidase's. Plant Physiol 145:5–16

17. Liebminger E, Veit C, Pabst M, Batoux M, Zipfel C, Altmann F, Mach L, Strasser R (2011) Β-N-acetylhexosaminidases HEXO1 and HEXO3 are responsible for the formation of paucimannosidic N-glycans in *Arabidopsis thaliana*. J Biol Chem 286(12):10793–10802
18. Irfan M, Ghosh S, Kumar V, Chakraborty N, Chakraborty S, Datta A (2014) Insights into transcriptional regulation of β-D-N-acetylhexosaminidase, an N-glycan processing enzyme involved in ripening-associated fruit softening. J Exp Bot 65:5835–5848
19. Gao J, Zhang Y, Li Z, Liu M (2020) Role of ethylene response factors (ERFs) in fruit ripening. Food Qual Saf 4(1):15–20
20. Lee JM, Joung JG, McQuinn R, Chung MY, Fei Z, Tieman D, Klee H, Giovannoni J (2012) Combined transcriptome, genetic diversity and metabolite profiling in tomato fruit reveals that the ethylene response factor SlERF6 plays an important role in ripening and carotenoid accumulation. Plant J 70:191–204
21. Irfan M, Kumar P, Kumar V, Datta A (2022) Fruit ripening specific expression of β-D-N-acetylhexosaminidase (β-Hex) gene in tomato is transcriptionally regulated by ethylene response factor SlERF. E4. Plant Sci 323: 111380
22. He F (2011) Laemmli-SDS-PAGE. Bio 101: e80
23. Kielkopf CL, Bauer W, Urbatsch IL (2020) Bradford assay for determining protein concentration. Cold Spring Harb Protoc 4:102269

Chapter 8

The Role of Ethylene and Its Antagonists in Advancing Postharvest Management of Fruits and Flowers

Shweta Gupta and Devesh Shukla

Abstract

Ethylene, a key plant hormone that regulates fruit ripening, senescence, and abscission. Due to these functions, ethylene is often referred as "aging hormone." These roles help plant development and reproduction, yet complicate postharvest management for perishable fruits, vegetables, and cut flowers. Unchecked ethylene production and action accelerate spoilage, reduce shelf life, and diminish product quality during storage and transport. This chapter explores the synthesis, signaling pathways, and physiological effects of ethylene, providing an understanding for its impact on postharvest biology. It also examines control measures, such as temperature and humidity regulation, modified atmosphere storage, and ethylene inhibitors like 1-methylcyclopropene (1-MCP). Recent advances in ethylene antagonists and packaging systems infused with ethylene blockers are also discussed. This chapter offers strategies to manage ethylene and extend storage life, marketability, and acceptability of perishable goods. By integrating ethylene control into postharvest technologies, it balances optimal ripening with longer shelf life, ensuring that high-quality produce reaches consumers.

Key words Ethylene regulation, Postharvest management, Ripening and senescence, 1-Methylcyclopropene (1-MCP), Modified atmosphere packaging (MAP), Ethylene biosynthesis, Climacteric fruits, Ethylene antagonists, Nanotechnology in packaging, Food waste reduction

1 Introduction

Postharvest management encompasses essential practices performed after producing horticultural crops. These include cleaning, washing, selection, grading, disinfection, drying, packing, and storage, to remove undesirable elements, improving appearance, and meet quality standards [1–4]. Such practices maintain aesthetic appeal and ensure products meet regulatory and consumer expectations, thereby enhancing marketability [5].

Fresh horticultural produce continues to undergo physiological and physical changes postharvest [1, 3]. These changes, known as postharvest transformations, directly impact product quality and consumer satisfaction. Effective postharvest management reduces

Devesh Shukla (ed.), *Ethylene Sensing and Signaling*, Methods in Molecular Biology, vol. 2945,
https://doi.org/10.1007/978-1-0716-4650-2_8,

microbial contamination, minimizes water loss, prevents ethylene damage, decay, and insect infestation [6]. Losses during this stage can be qualitative, including nutritional and structural degradation, or quantitative, such as reduction in size, weight, or number. The overarching goal of is to curtail these losses, extend shelf life, and optimize resource utilization [1]. When implemented well, these practices prevent waste support global food security by reducing supply chain inefficiencies [7]. The UN/FAO "Global Initiative on Food Loss and Waste" (2011) and the Sustainable Development Goals (SDGs) (2015), highlighted the scale of postharvest losses. Reports suggest that approximately one-third of food produced worldwide is wasted annually, worsened by inadequate transportation and storage facilities [8]. In horticulture, postharvest losses can reach up to 80%, depending on type, season, and region [9]. Effective strategies include improved transportation, commodity-specific storage conditions, and efficient distribution to mitigate these losses. Additionally, educational programs for farmers and supply chain stakeholders promote best practices in postharvest handling [10]. In India, the second-largest producer of fruits and vegetables, postharvest challenges persist. Tomato is a widely grown and consumed horticultural crop in India. Before reaching the consumers, tomatoes undergo harvesting, storage, packaging, loading, unloading, and transportation. Inefficiencies in these processes cause postharvest losses (PHL), and affect the entire food supply chain (FSC). Research has focused on increasing productivity, while reducing PHL has less attention. Although technologies like cold storage and evaporative cooling exist, their adoption in Indian FSC is minimal. Using a case study approach, the study found key reasons for PHL, such as poor storage practices and long transport distances. It recommends temperature-controlled storage and better collaboration among FSC stakeholders as effective solutions for sustainability. India contributes 10.9% and 8.6% to the fruits and vegetables of the world production. Yet India faces severe food losses, estimated at 74 million tonnes annually (FAO, https://www.fao.org/india/fao-in-india/india-at-a-glance/en/?utm). These losses fuel food insecurity and demand advanced postharvest research and effective management to reduce waste and prolong shelf life. A strong focus on infrastructure, such as cold storage and robust transportation networks, is crucial to address these issues. Policies that incentivize private sector participation and technological innovation can improve postharvest systems [11].

Ripening and senescence processes are critical to the perishability of horticultural produce, particularly fruits and flowers [1]. Over-ripening, discoloration, and decay result in significant quality loss [2]. Ethylene, a pivotal plant hormone, has a dual postharvest role. It promotes ripening and aging, yet also creates management challenges. This chapter explores ethylene in

Table 1
Classification of select fruits and vegetables by their ripening behavior, ethylene production levels, and the corresponding sensitivity to ethylene exposure

Commodity	Type	Ethylene production	Ethylene sensitivity
Apple, kiwi, pear	Climacteric fruit	High	High
Avocado, cantaloupe melon, passion fruit	Climacteric fruit	High	Medium
Apricot, banana, mango	Climacteric fruit	Medium	High
Papaya, peach, plum, tomato, nectarine	Climacteric fruit	Medium	Medium
Broccoli, cabbage, carrot, cauliflower, cucumber, lettuce, potato, spinach	Vegetable	Low	High
Strawberry	Non-climacteric fruit	Low	High
Citrus		Low	Medium
Asparagus, bean, celery, eggplant	Vegetable	Low	Medium
Berries, cherry, grape, pineapple	Non-climacteric fruit	Low	Low
Artichoke	Vegetable	Low	Low
Pepper	Vegetable	Low	Low

ripening, senescence, and abscission, and strategies to mitigate negative impacts. A thorough understanding of mechanisms of ethylene enables tailored solutions for different crops and storage conditions. Ethylene is a gaseous plant growth regulator synthesized from methionine [12]. It affects key physiological and biochemical changes, particularly in climacteric fruits where it drives ripening [1, 3, 13]. Conversely, non-climacteric fruits ripen independently of ethylene [2, 14]. The dual nature of ethylene demands precise management to balance benefits and drawbacks. Table 1 illustrates ethylene production and sensitivity across climacteric and non-climacteric fruits [2, 15]. Ethylene's acceleration of ripening can be useful for uniform maturity when required. However, mismanagement risks over-ripening and premature spoilage [2]. Ethylene also impacts processes beyond ripening, including stress responses and pathogen resistance [2]. For instance, under certain environmental conditions, ethylene synthesis can be upregulated, worsening chilling injury in sensitive crops. As such, strategies to modulate ethylene activity are integral to postharvest management. Effects of ethylene have been noted since ancient practices of wounding fruits to induce ripening. In 1864, Girardin observed premature leaf senescence in trees near leaking gas pipelines. Subsequent research confirmed ethylene's role in plant development and senescence [1–3, 16]. Over the years, advances in analytical techniques have provided deeper insights into ethylene biosynthesis and signaling. These discoveries enabled innovative approaches to regulate ethylene activity in agricultural practices. Historical practices, such as exposing fruits to smoke for ripening, unknowingly used ethylene properties before formal discovery. Later identification of the ethylene molecular structure and physiological roles transformed its agricultural application, enabling more targeted and efficient interventions [1–3].

Applications of ethylene range from enhancing fruit ripening to altering flowering patterns. For example, ethylene promotes flowering in pineapple and modifies sex expression in cucumbers [17]. Postharvest applications include accelerating ripening and improving pigment development in fruits like tomatoes and bananas [1–3] [2]. Ethylene has broad agricultural applications, such as reducing bitterness in citrus juices by degrading limonin precursors. Despite these benefits, ethylene also poses challenges. Overexposure can cause premature aging, tissue softening, and decay, reducing shelf life. Thus, strategies like controlled atmosphere storage and ethylene antagonists mitigate these detrimental effects. For example, 1-methylcyclopropene (1-MCP) delays ripening by blocking ethylene receptors, extending the storage life of climacteric fruits [2]. Research into alternative inhibitors and their compatibility with various crops is ongoing, promising further advances in postharvest technology. The impact of ethylene includes ornamental flowers, where it can cause petal abscission and discoloration. Managing ethylene levels in storage for floriculture products is critical to preserve aesthetic and commercial value.

This chapter examines the dual role of ethylene in postharvest technology, its synthesis, impacts, and management strategies. It assesses how ethylene affects postharvest losses, progress in ethylene regulation, and related economic and environmental effects. It also reviews emerging technologies, such as genetic modification and nanotechnology, in fine-tuning ethylene responses for specific use. Taken together these topics offer a clear view of ethylene's role and potential in improving postharvest outcomes.

2 Ethylene Biosynthesis and Mechanism of Action

Ethylene, a vital plant hormone, shapes fruit ripening, abscission, and senescence [2]. Its synthesis and action are well defined and tightly linked to diverse physiological outcomes. Methionine is the biological precursor of ethylene in plants (Fig. 1). The pathway starts when methionine converts to SAM (S-adenosyl-methionine) via SAM synthetase, utilizing ATP. Subsequently, SAM is converted into ACC (1-amino-cyclopropane-1-carboxylic acid) and MTA (5′--methylthioadenosine) by enzyme ACC synthase (ACS) [2]. MTA recycles to re-form methionine, as the pathway conserves sulfur and maintains flux. Meanwhile, ACC oxidizes to ethylene via ACC

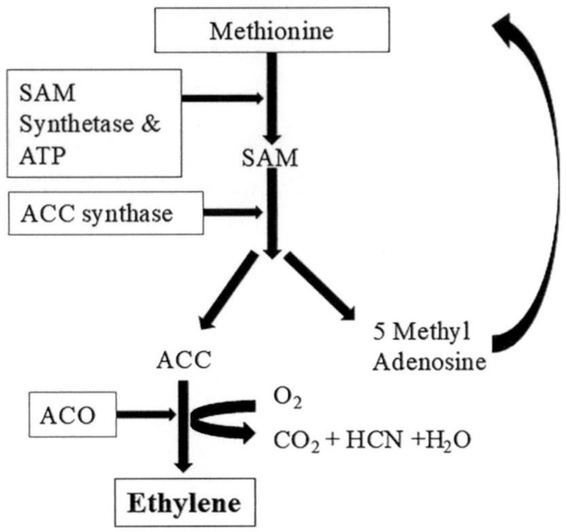

Fig. 1 Biosynthetic pathway of ethylene, starting from methionine and proceeding through S-adenosylmethionine (SAM) and 1-aminocyclopropane-1-carboxylic acid (ACC), resulting into ethylene formation and associated by-products

oxidase (ACO) in the presence of oxygen, completing biosynthesis (Fig. 1) [2, 3]. The regulation of ACS and ACO responds to environmental and developmental cues, showing pathway flexibility to stress and growth. This ethylene biosynthesis pathway underpins many physiological processes [16, 18]. Several factors modulate ethylene biosynthesis. Wounding, hypoxia, and hormonal interactions can either amplify or suppress ACS and ACO activities. For instance, auxin and cytokinin often act synergistically with ethylene, increasing biosynthesis and broadening growth and stress effects.

2.1 Ethylene Signaling Mechanism

Without ethylene, the receptor ETR1 regulates CTR1 (Constitutive Triple Response 1), a negative regulator of the ethylene signaling pathway. CTR1 suppresses EIN2, a key positive regulator, inhibiting downstream signaling [19]. When ethylene is present, it binds to ETR1, inhibits CTR1 [20]. Consequently EIN2 becomes active, leading to cleavage and translocation of its C-terminal to the nucleus. There, it promotes accumulation of EIN3 and EIL1, transcription factors which activate Ethylene Response Factors (ERFs) [20, 21]. The regulation of EIN2, EIN3, and ERFs allows plants to fine-tune ethylene responses. This cascade affects primary processes and secondary metabolism, including defense-related phytoalexins.

2.2 Effects of Ethylene on Fruit Ripening

Ethylene is pivotal in climacteric fruit ripening. In tomatoes, ethylene receptors (seven identified) interact with GR (Green-Ripe) and homologues like RAN1 and SlTPRI. RAN1 supplies copper to ethylene receptors. SlTPRI may lead to receptor degradation.

Among tomato-specific CTR1 homologues (SICTR1 to SICTR4), SICTR2 expression increases during ripening.

Ethylene also influences cell wall modification enzymes like:

I. **Expansins:** Loosen cell walls aiding ripening [22].
II. **Pectin Methyl Esterase (PME):** De-esterifies pectin, regulated by ethylene.
III. **Polygalacturonase:** Induced by ethylene, breaks down hydrogen bonds in pectin post-PME action.
IV. **Beta-Galactosidase:** Cleaves beta-(1-4)-galactan bonds in cell walls.
V. **Pectate Lyase:** Facilitates pectin degradation, with increased activity following ethylene treatment.

These modifications enhance fruit texture and flavor during ripening. For example, PME steers softening, crucial for consumer acceptability. 1-MCP, an ethylene inhibitor, suppresses PME, beta-galactosidase, and pectate lyase activities, delaying ripening [2, 3, 23].

2.3 Effect of Ethylene on Abscission and Leaf Senescence

Ethylene triggers flower senescence and regulates molecular events, including aquaporin modulation. Its effects vary across species. Silencing Ethylene Response Factor (ERF) genes can delay senescence, thereby extending flower longevity [24]. During leaf development, ethylene synthesis is initially high, decreases as the leaf matures, and spikes again at the senescence onset. Ethylene promotes chlorophyll degradation, contributing to abscission and necrosis. Additionally, it stimulates the expression of hydrolytic enzymes that weaken cell walls at abscission zones, facilitating organ separation. Its exposure accelerates senescence and abscission processes [23, 24].

Ethylene also regulates protease and lipase activities during senescence, ensuring nutrient remobilization to other parts of the plant. This dual role in decay and nutrient recycling supports plant survival and reproductive success.

2.4 Ethylene Production Under Storage Conditions

Ethylene synthesis is affected by factors such as temperature, mechanical stress, and microbial activity.

1. *Temperature Stress:* Studies on "Hayward" kiwifruit show ethylene production peaks at 30 °C – 34 °C, but drops above 38 °C [25]. Such temperature sensitivities reflect enzyme activity in the ethylene biosynthetic pathway. Consequently, controlled storage temperatures can mitigate excessive ethylene and extend shelf life.
2. *Mechanical Stress:* Handling injuries elevate ethylene synthesis, reducing fruit and vegetable shelf life and marketability. Thus,

use cushioned containers and reduce handling to minimize damage-induced ethylene bursts.

3. *Microbial Influence:* Microorganisms, including bacteria and fungi, produce ethylene. This production can enhance ethylene synthesis in infected climacteric fruits. Conversely, ethylene promotes fungal and bacterial proliferation, creating a feedback loop that accelerates decay.

The integration of ethylene inhibitors, such as 1-MCP, in storage can mitigate these effects. Packaging materials with ethylene scavengers are being developed to reduce accumulation, and preserve quality.

3 Physiological Roles of Ethylene in Postharvest Biology

Ethylene is termed as "ageing hormone" as it regulates fruit ripening, flower abscission, and senescence in plants. These effects make it pivotal in postharvest biology, particularly for climacteric fruits and ornamental flowers. Below is a comprehensive exploration of physiological roles of ethylene in these domains.

3.1 Fruit Ripening

The role of ethylene in fruit ripening is a well-established and extensively studied. Climacteric fruits, such as bananas, apples, tomatoes, and mangoes, show an ethylene surge during ripening. A respiratory burst accompanies this surge (Fig. 2). The surge is driven by upregulation of ACS (1-aminocyclopropane-1-carboxylic acid synthase) and ACO (1-aminocyclopropane-1-carboxylic acid oxidase) genes [2, 3, 22]. This event triggers changes in texture, color, firmness, flavor, aroma, and taste (Fig. 2). For example, cell wall breakdown softens texture, while starch conversion to sugars increases sweetness (Fig. 2). The autocatalytic nature of ethylene production is described through the two-system model proposed

Fig. 2 Demonstration of progressive ripening of climacteric fruit (bananas) over 6 days, displaying changes in peel color and fruit quality

by McMurchie et al. [26]. System 1 operates during normal vegetative growth, maintains basal ethylene levels. System 2 activates during ripening, driving exponential ethylene production. This dual system highlights sophisticated control of ethylene biosynthesis. In non-climacteric fruits like strawberries and citrus, role of ethylene is less dramatic, yet relevant. Ethylene facilitates chlorophyll breakdown, leading to peel degreening in citrus fruits. Limited ethylene production in non-climacteric fruits shows crosstalk with other hormones, subtly influencing ripening events [1–3]. Ethylene also induces biochemical changes that enhance fruit palatability. For example, it accelerates conversion of organic acids to sugars and promotes volatile synthesis responsible for fruit aroma. In addition, the exogenous ethylene application can synchronize ripening in commercially harvested fruits. This practice provides a consistent product for markets.

3.2 Flower Abscission and Senescence

The ornamental and cut-flower industries rely on technologies to extend the shelf life of flowers (Fig. 3). Ethylene accelerates flower senescence and abscission. Effects are pronounced in sensitive species such as carnations, orchids, Hibiscus, and petunias (Table 2) (Fig. 4). Abscission is the detachment of plant parts. It depends on the balance of ethylene and auxin at the abscission zone [24]. High ethylene sensitivity, coupled with reduced auxin, triggers cell wall modifications and enzymatic activities leading to separation. This process allows plants to conserve resources, but in horticulture, it challenges product quality. The role of ethylene in flower senescence is clarified by molecular studies. Downregulation of EIN2, a key component of the ethylene signaling pathway, delays senescence in petunia and dianthus [27]. Additionally, abscisic acid (ABA) has been associated with abscission. However, ethylene is now considered the primary hormone affecting this process. Environmental factors such as ozone exposure, wounding, and pollination enhance ethylene biosynthesis, worsening senescence and abscission [28]. Advancements in genetic studies have elucidated the molecular mechanisms of ethylene. For example, mutations in

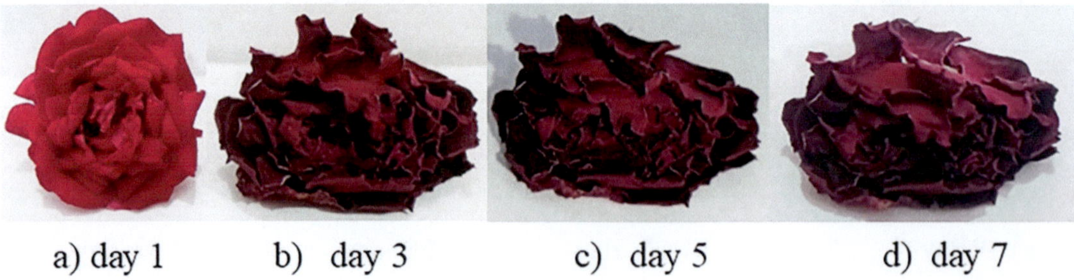

Fig. 3 Demonstration of progressive senescence of cut rose flowers over 7 days, displaying the influence of ethylene on floral longevity

Table 2
Classification of ornamentals by their ethylene sensitivity, including representative species and recorded responses, as detailed by [85] and discussed by Scariot [56]

Ethylene-sensitive flower	Ethylene effect	References
Orchids	Wilting	Celikel and Reid [82]
Hibiscus	Wilting	Celikel and Reid [82]
Wax flower	Petal/flower abscission	Macnish et al. [83]
Rose	Petal/flower abscission	Macnish et al. [83]
Clove pink	Wilting	Serek et al. [84]
Snapdragon	Petal/flower abscission	Serek et al. [84]

Fig. 4 Demonstration of progressive senescence (wilting and discoloration) of cut hibiscus flowers over 7 days

the ETR1 gene in petunias have delayed flower senescence. Such genetic insights have enabled innovative postharvest interventions. Furthermore, inhibitors like silver thiosulfate (STS) and 1-methylcyclopropene (1-MCP) block ethylene receptors, prolonging the vase life of cut flowers.

3.3 Environmental and External Sources of Ethylene

Ethylene production is not solely intrinsic; external factors also contribute. Environmental pollution, microbial activity, and handling damage can raise ethylene levels. These external sources demand stringent postharvest management. For example, bruising and cuts during harvest trigger localized ethylene, accelerating spoilage. Technologies such as modified atmosphere packaging (MAP) and ethylene scrubbers mitigate these effects. These solutions preserve produce quality during storage and transport. Nearby ripening fruits are natural ethylene sources that influencing neighboring produce. Therefore, segregating ethylene-sensitive commodities during storage and transport to prevent premature ripening or spoilage.

3.4 Interaction with Auxin

Interplay of auxin with ethylene in ripening and senescence is intricate. Auxin can diminish ethylene impact on abscission and senescence in some contexts. It also induces ethylene production in tissues like roots and young leaves. Auxin influence on ACS gene expression exemplifies this dual role. This interaction appears in etiolated mung beans. There, auxin promoted ethylene production by inducing ACS gene expression. Such findings highlight the complex hormonal crosstalk influencing postharvest physiology. The interplay extends to other developmental processes. For instance, auxin delays abscission by maintaining the integrity of the abscission zone. Ethylene overrides this effect as senescence progresses. This dynamic balance allows plants to adapt to environmental cues and internal signals, coordinating growth and stress responses.

4 Postharvest Challenges Associated with Ethylene

The term "ripe" comes from the Saxon word "ripi," meaning to gather or reap. It describes the stage when fruit reaches peak edible quality after harvest. Ripening encompasses early changes during fruit senescence, modifying unripe fruit structure and composition to make it palatable. These changes include softening, coloring, increased sweetness, and enhanced aroma [4]. Ripening enhances consumer appeal and marks the transition from growth to senescence.

4.1 Over-Ripening and Autocatalytic Nature of Ethylene

The beneficial effects of ethylene, such as reducing bitterness and initiating ripening, are well documented. However, excess ethylene accelerates over-ripening, leading to decay and reduced shelf life, particularly in climacteric fruits. This results from the autocatalytic nature of ethylene, so its production amplifies itself, and hastens ripening and senescence. Over-ripening driven by ethylene causes substantial postharvest losses, often compounded by environmental and handling factors [29]. Beyond economic losses, over-ripened produce complicates waste management and resource optimization in the supply chain.

Mechanical damage during handling stimulates ethylene synthesis, further promoting over-ripening and deterioration. Temperature-dependent processes, such as respiration and transpiration, compound these effects during cold storage [30]. Controlled storage with ethylene removal systems can mitigate losses and extend shelf life [31]. Additionally, gases like CO_2 and O_2 can accelerate over-ripening, so controlled atmospheres must manage ethylene levels. Packaging innovations, such as modified atmosphere packaging (MAP), help regulate gas concentrations and reduce ethylene exposure.

4.2 Effects of Ethylene on Quality Parameters

4.2.1 Taste

Ethylene enhances taste and flavor in many fruits. However, natural ripening often yields better sensory qualities than artificial ethylene-induced ripening [18]. For instance, naturally ripened mangoes may have a more complex flavor profile than mangoes ripened with exogenous ethylene applications.

4.2.2 Appearance

Ethylene shapes visual attributes such as peel color and texture, which strongly affect consumer acceptance. Because it alters pigments, ethylene-induced color change can improve citrus appeal by turning green peels orange or yellow. Conversely, it may cause undesirable yellowing in leafy vegetables or russet spotting in lettuce (Saltveit [18]). Over-ripening can lead to unappealing traits like banana browning or tomato skin shrinkage. In some cases, excess ethylene exposure can also result in uneven coloration, further diminishing the aesthetic value of produce.

4.2.3 Texture

Ethylene softens fruits for consumption, but prolonged exposure results in undesirable texture, such as mushiness in bananas or peppers. Over-ripening also causes tough texture in asparagus or sweet potatoes [18]. In climacteric fruits, such as kiwis, ethylene-induced softening can occur rapidly. Therefore, storage requires precise monitoring.

4.2.4 Nutritional Quality

The role of ethylene in nutritional composition varies. For instance, ethylene-treated papayas exhibit higher vitamin C content compared to naturally ripened ones. However, certain applications may elevate ethylene residues, influencing the overall nutritional value. Additionally, prolonged ethylene exposure can result in nutrient degradation, as seen in vitamin losses during extended storage periods.

4.3 Cross Contamination During Storage

Ethylene-sensitive products stored alongside climacteric fruits risk accelerated ripening and spoilage. Segregating climacteric and non-climacteric fruits, alongside storing unripe and ripe produce separately, minimizes cross contamination. Maintaining ethylene-free environments further reduces exogenous ethylene impacts. Ethylene scrubbers or absorbers in storage facilities are particularly effective at preserving mixed produce quality.

4.4 Broader Implications of Ethylene Management

Effective ethylene management extends beyond quality preservation. By reducing spoilage, producers and retailers can lower their carbon footprint, aligning with sustainability goals. Mitigating postharvest losses also supports global food security by delivering more produce in optimal condition. Research explores advanced technologies, such as biosensors, for real-time ethylene monitoring, offering promising avenues for innovation. In storage, use ethylene removal technologies and controlled atmospheres (manage O_2 and CO_2) to reduce ethylene levels. Techniques include activated carbon filters or potassium permanganate-based scrubbers. Avoid mechanical damage by using cushioning materials and gentle handling to minimize ethylene synthesis. Store climacteric and non-climacteric produce separately, with dedicated zones for high ethylene emitters. Optimize cold storage to balance respiration rates and mitigate ethylene effects, maintain uniform temperatures with advanced cooling systems. Educating farmers, handlers, and retailers about effective practices to ensure consistency across the supply chain. Role of ethylene and adopting targeted interventions, can reduce postharvest losses, improve economic returns and cut waste. Technology and education together will drive long-term success in ethylene management strategies.

5 Mechanisms and Applications of Ethylene Antagonists

5.1 1-MCP

In the mid-1990s, Professor E.C. Sisler of North Carolina State University discovered 1-MCP (1-methylcyclopropene). 1-MCP is a cyclopropane derivative and specific inhibitor of ethylene action [32]. It blocks ethylene production, action, and synthesis by occupying irreversible binding sites on ethylene receptors (Fig. 5) [33]. Consequently, 1-MCP halts signal transduction processes necessary for ripening, making it a widely used commercial ethylene antagonist. 1-MCP is primarily applied in gaseous form to fruits and flowers. It delays ripening in climacteric fruits such as apples and pears (marketed as SmartFresh by AgroFresh). It extends the vase life of cut flowers (EthylBloc by Rohn & Haas) [16]. Its applications reduce ethylene production and respiration rates, prolonging shelf life and reducing postharvest waste. It works at low concentrations, and short exposure times. Low phytotoxicity, and uniform delivery to bulky produce are additional advantages [34, 35]. For instance, bananas exposed to high concentrations of 1-MCP may show abnormal skin degreening. Bananas treated in pre-climacteric stages may also exhibit this effect. Such effects are mitigated by treating fruits after ripening initiation with moderate antagonist levels [36, 37]. However, for apples (*Malus sylvestris* L. var. domestica), 1-MCP maintains firm texture when applied pre-ripening [38]. 1-MCP suppresses softening, and reduces electrolyte leakage in papaya (*Carica papaya* L.) and "Galia" melon (*Cucumis melo* var. reticulatus L.) [39]. Multiple applications sustain its effects, as demonstrated in apple softening studies [35]. Moreover, it influences non-climacteric produce. Effects include delayed yellowing in broccoli, maintained firmness in watermelons, and preserving chlorophyll in parsley and coriander [35]. The integration of 1-MCP with modified atmosphere packaging (MAP) enhances its effectiveness. For example, combine 0.5 μL/L of 1-MCP with micro-perforated film packaging

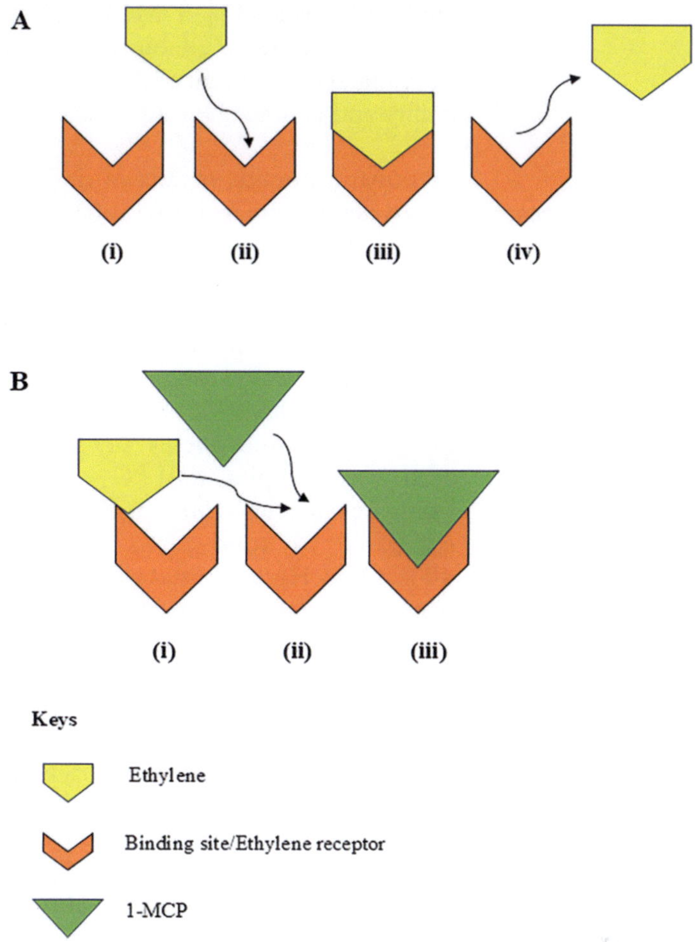

Fig. 5 Mechanism of 1-MCP action. 1-MCP prevents ethylene from binding to its receptor by competing for the same binding site (Sitbon et al. [81]). (**a**) In the absence of 1-MCP, an ethylene molecule in the air (i), binds to the receptor (ii), triggering the signaling cascade (iii). When ethylene concentration decreases, the molecule eventually dissociates from the receptor (iv). (**b**) In the presence of 1-MCP, both ethylene and 1-MCP compete for the receptor-binding site. 1-MCP binds more stably, blocking ethylene access and thus inhibiting the ethylene signaling pathway

(MFP). This treatment delays senescence, inhibits ethylene production, and reduces respiration in pears [40]. Similarly, 1-MCP combined with MAP preserved basil samples at 11 °C for up to 18 days [41]. Temperature sensitivity is a critical parameter for 1-MCP applications. It effectively inhibits ethylene-mediated softening, yellowing, and internal breakdown at −0.5 °C. However, it may exacerbate chilling injuries at lower temperatures. A study on Korla Xiang pears examined storage at −1.5 °C with 1-MCP. Under these conditions, TSS (total soluble solids) increased without inducing chilling injuries [42]. 1-MCP has become a pivotal tool for postharvest researchers, enabling the exploration of ethylene roles in ripening and senescence [38, 43]. Its applications enhance marketability and extend shelf life of horticultural commodities.

5.2 Impact of Hypobaric Conditions on Ethylene Production in Different Commodities

Hypobaric storage, means storing produce under low-pressure. It removes ethylene from the internal atmosphere of fruits and vegetables, improving storage quality. In apples, Thompson [44] reported a drop in ethylene concentration from 0.1 ppm to 0.01 ppm at 76 mm Hg. Bangerth [45] found 51 mm Hg for 11 months prevented autocatalytic ethylene production, though firmness decreased slightly. Similarly, bamboo shoots showed reduced ethylene at 50 kPa [46] and bananas did so in low-oxygen atmospheres [47]. Leafy vegetables like cabbages, stored with apples at 0 °C and 60 mm Hg, avoided chlorosis [48]. Cranberries stored at 80 mm Hg showed reduced ethylene compared to atmospheric pressure [49]. Loquat fruit under 40–50 mm Hg exhibited diminished ethylene production. Jujube fruits experienced slowed ethylene production under hypobaric storage [44]. Hypobaric conditions can extend storage life and maintain quality across many horticultural commodities.

5.3 Ethylene Scavengers/Absorbers

Ethylene scavengers can be used as part of packaging materials or placed them as sachets inside packaging. Since ethylene is a highly reactive, substances such as potassium permanganate, activated carbon, and different minerals absorb it [50].

5.3.1 Potassium Permanganate

Potassium permanganate ($KMnO_4$) is a strong oxidizing agent that converts ethylene into carbon dioxide (CO_2) and water (H_2O). It is one of the simplest methods for removing exogenous ethylene [51]. This compound is widely used in countries like Japan and the USA. Users embed it in inert, non-reactive carriers that absorb ethylene, such as silica, celite, alumina, or zeolite [51]. Embedded potassium permanganate oxidizes ethylene into acetaldehyde and acetate. Products produced using potassium permanganate are toxic, it cannot contact with food directly [31], so sachet use is essential [52]. Several commercial products use potassium permanganate embedded in alumina beads, such as Purafil, Ethysorb, Bloomfresh, and Circul-Aire [51]. Consequently, studies show that $KMnO_4$ sachets delay ripening, and extend the shelf life of perishable commodities.

5.3.2 Activated Carbon

Ethylene scavengers also use activated carbon integrated with metal catalysts. First, activated carbon absorbs ethylene, which then undergoes catalytic decomposition [50]. This method provides an effective alternative to traditional oxidation processes.

5.3.3 Nano- and Microparticles in Films as Ethylene Absorbers

Several minerals such as clays, nanoclays, pumice, and zeolite can be incorporated into films as ethylene absorbers. These films are opaque and absorb less ethylene than $KMnO_4$. Even though, this approach is gaining popularity due to integration into packaging systems is easy and materials are nontoxic.

5.4 Silver Thiosulfate (STS)

Silver ions, particularly silver thiosulfate (STS), are widely recognized for inhibiting ethylene action [15]. Silver blocks ethylene receptors, and thereby impedes ethylene signaling [18]. Despite effectiveness, STS use raises environmental concerns, as silver is a heavy metal and potential pollutant [15, 16]. Therefore, alternatives like 1-MCP are often preferred for food-related applications. Even with limitations, STS retains valuable applications. In bananas and tomatoes, silver treatment inhibits ethylene, delaying ripening and extending postharvest life [53]. In the floral industry, STS extends the shelf life and quality of cut flowers by inhibiting ethylene synthesis. The shoot apex, or growing tip, is a critical site, where ethylene signaling negatively affects flowering. Hyde et al. [54] reported benefits in cassava, mitigating detrimental effects of ethylene on floral development, and improving flowering. STS is recognized as a cost-effective tool for maintaining ornamental plant quality and longevity [55]. Nanotechnology now enables incorporation of nano-sized silver ions into packaging films. Such films retard senescence and reduce microbial growth, providing an ethylene-inhibitory effect at the nanoscale [56]. While not suitable for direct food contact, these innovations show STS versatility for nonfood uses, including ornamentals and postharvest technologies.

5.5 Amino-Ethoxy-Vinyl-Glycine (AVG)

Amino-ethoxy-vinyl-glycine (AVG) is a well-established irreversible inhibitor of ACC synthase [15]. ACC synthase is pivotal in ethylene biosynthesis. It catalyzes conversion of S-adenosylmethionine (SAM) into 1-aminocyclopropane-1-carboxylic acid (ACC). By inhibiting ACC synthase, AVG delays ripening of climacteric fruits [57]. Commercially marketed as ReTain, AVG is applied as a spray to fruits such as peaches shortly before harvest [51]. However, registration is not permitted in the European Union due to toxicity concern. Beyond ACC synthase, AVG inhibits other pyridoxal phosphate (PLP)-dependent enzymes. This includes enzymes involved in auxin biosynthesis and nitrogen metabolism [16]. AVG reduces endogenous ethylene synthesis, but does not mitigate effects of exogenous ethylene. Therefore, postharvest management use focuses on endogenous ethylene control to delay ripening and extend shelf life.

5.6 Modified Atmosphere Packaging (MAP)

Modified atmosphere packaging (MAP) is extensively used to extend the shelf life by altering the gaseous environment of the package. This method reduces oxygen and may raise carbon dioxide slowing respiration and ethylene production in perishable goods [50, 58]. Active MAP removes the initial atmosphere by evacuation, then introduces a desired gas mixture. When combined with ethylene absorbers, this approach is particularly effective for climacteric fruits. Reduced oxygen and increased carbon dioxide inhibit ethylene synthesis and mitigate its effects, improving quality and longevity of stored commodities.

6 Advances in Ethylene Research

The expected doubling of the global population by 2050 demands innovations in food production and postharvest management. Genetically engineered plants with delayed ripening traits offer promising solutions to extend shelf life and reduce food waste. Tomato, a climacteric ethylene dependent fruit, has been studied extensively for delayed ripening through genetic engineering. Zhou et al. [59] examined ERF069 an ethylene response factor gene, in transgenic tomatoes. Overexpression suppressed downstream ethylene signaling, and significantly delayed fruit ripening. Similarly, Reed et al. [60] introduced the ACCd (1-aminocyclopropane-1-carboxylic acid deaminase) gene, via *Agrobacterium tumefaciens.* ACCd reduced ethylene synthesis in tomatoes therefore extending the ripening phase and improving shelf life. As *ACCd* metabolizes ACC, the immediate ethylene precursor, it remains nontoxic and viable for commercial use. Beyond direct genetic interventions, sound waves can modulate ethylene production. Kim et al. [61] found that 1 kHz sound waves delayed tomato ripening by reducing ethylene synthesis. This suggests that physical stimuli can influence ethylene-dependent pathways and provide noninvasive methods for managing postharvest physiology. Hormonal interac tions also shape ripening. Park and Malaka et al. [62] highlighted antagonism between gibberellic acid and ethylene. Ethylene promotes ripening, whereas gibberellic acid delays it, providing a sophisticated postharvest management approach.

The CRISPR-Cas9 system has emerged as a transformative tool for genetic editing in plant systems. Pizarro et al. [63] targeted the RIPENING-INHIBITOR (RIN) gene. RIN is a transcription factor essential for regulating ethylene-dependent and independent pathways. CRISPR-mediated mutations in the RIN locus confirmed its role. These results demonstrated the potential for precise genome editing to control fruit ripening. Another avenue of research involves TALEN-induced mutations in the pro gene. The pro gene encodes DELLA proteins, negative regulators of gibberellin signaling. Lor et al. [64] revealed that enhanced gibberellin responses through TALEN-induced pro mutations could alter ripening timelines. These findings showcase synergy between hormonal regulation and genetic engineering. Collectively, these advancements underline the multifaceted approaches in ethylene research. These include direct genetic modifications, external physical stimuli, hormonal interplay, and genome editing technologies. By leveraging these insights, researchers and practitioners can develop comprehensive strategies to address postharvest losses.

Omics technologies, including transcriptomics and proteomics, reveal mechanisms driving fruit development and ripening [1–3, 22, 65]. Recent advancements have identified many genes and proteins linked to ethylene regulation in various fruits [1–3, 22]. Quantitative omic analyses revealed 2754 proteins in mango, 1810 in pear, and 4648 in citrus, illustrating the depth [66].

Transcriptomics examines all RNA transcripts in an organism, tissue, or cell at specific developmental stages or conditions [65]. It reveals genetic and regulatory networks by identifying ethylene-regulated genes important for ripening. Yu et al. [66] used RNA sequencing (RNA-seq) and quantitative techniques to analyze watermelon development across four stages. The study quantified 18,856 genes and 6226 proteins and validated results with sequencing and real-time PCR. Xu et al. [67] conducted RNA-seq on apricot fruit. They identified genes for ethylene metabolism (four genes), pectin metabolism (seven genes), and ethylene signaling (six genes).

Proteomics complements transcriptomics by identifying and quantifying proteins critical for developmental processes. Lyu et al. [68] used the ITRAQ technique to investigate ethylene-induced adventitious root development in cucumber seedlings. The study identified 115 proteins involved in root development. It showed that exogenous ethylene promotes endogenous ethylene synthesis, photosynthesis, and energy metabolism. This finding highlights broad role of ethylene in developmental biology. Together, these omics approaches reveal regulatory networks modulated by ethylene during fruit ripening and development. By integrating transcriptomic and proteomic data, researchers can decipher interactions between ethylene signaling and other metabolic pathways.

7 Innovative Delivery Systems

Nanotechnology has become a key tool in the twenty-first century. Its applications span pharmaceuticals and agricultural sciences. In postharvest handling, it improves packaging and preservation of climacteric fruits. These systems integrate ethylene scavengers in packaging. Consequently, nanotechnology has revolutionized postharvest technology.

7.1 Nanotechnology-Based Ethylene Scavengers

Traditional ethylene scavengers, such as potassium permanganate sachets, show uneven and localized action. Nanotechnology addresses these drawbacks by increasing scavenger surface area and efficiency. Ebrahimi et al. [51] developed nanostructured films with potassium permanganate. These films extended banana shelf life by up to 15 days. Joung et al. [69] loaded potassium permanganate into halloysite nanotubes within low-density polyethylene (LDPE). The P-HLNF system adsorbed ethylene effectively. Silver nanoparticles also inhibit ethylene. Scariot et al. [56] reported nanosized silver ions reduced ethylene-induced senescence and microbial growth. Sangtawesin et al. [70] used gamma irradiation to impregnate carbon-based materials with nano-silver particles. This approach significantly increased ethylene adsorption capacity.

7.2 1-MCP Delivery Innovations

1-Methylcyclopropene (1-MCP) is a widely used ethylene inhibitor delivered as a gas. Traditional fumigation uses enclosed treatments for 1–24 h. New delivery methods now complement fumigation. Pongprasert et al. [71] demonstrated microbubbles can apply 1-MCP effectively. Chen et al. [72] tested encapsulation for controlled 1-MCP release. These approaches improve use across diverse storage conditions.

7.3 Advanced Ethylene Scavenger Systems

Ethylene scavengers include sachets with potassium permanganate and clay-based systems. These materials remove ethylene by physical adsorption and catalytic reactions. Zeolites, activated carbon, and palladium-based systems show significant ethylene absorption. Consequently, polymer films infused with these scavengers extend shelf life of perishable commodities.

7.4 Monitoring Ethylene Levels

Precise ethylene monitoring is crucial for optimizing storage environments. Techniques range from gas chromatography and electrochemical sensors to advanced chemical sensors, including photoluminescence, CATA luminescence, and colorimetric sensors. Organic field-effect transistors have gained attention for sensitivity and specificity. In practice, the food industry widely use gas chromatography, optical methods, and electrochemical sensors for reliability and scale [73]. Integrating nanotechnology, advanced ethylene scavenger systems, and precise monitoring tools enables innovative delivery systems. These systems improve postharvest management of climacteric fruits. These advances improve shelf life and reduce food waste, supporting global food security efforts.

8 Environmental and Economic Considerations

Silver, widely used as silver thiosulfate (STS) as an ethylene antagonist, is one of the most toxic heavy metals. Exposure harms soil microorganisms, birds, and mammals. It reduces growth, reproduction, and increases mortality. STS is less toxic than silver ions, but its soluble forms are more hazardous. Safety of STS remains under investigation. Consequently, STS is avoided in food applications but continues in the floral industry.

1-MCP (1-methylcyclopropene), by contrast, is deemed environmentally safe by the US Environmental Protection Agency. It poses no toxicity risks to ecosystems or living organisms. Therefore, 1-MCP is preferred in postharvest technology to extend the shelf life of fruits and flowers.

Aminoethoxyvinylglycine (AVG), another ethylene inhibitor, shows no significant dietary risks and minimal aquatic toxicity. However, AVG shows moderate toxicity to specific birds like the northern bobwhite quail. These properties position AVG as a moderately safe antagonist.

8.1 Economic and Environmental Benefits of Ethylene Management

Managing ethylene is essential to reduce postharvest losses in fruits and vegetables. These products account for 50% of global food waste. In India, 74 million tonnes are wasted annually due to perishability and short shelf life. Ethylene accelerates ripening and senescence, causing over-ripening and waste. Effective control ensures optimal ripeness, extends shelf life, and reduces waste. These steps also cut pollution from discarded produce, delivering economic and environmental benefits. Practical strategies include using 1-MCP and ripening agents like ethephon into postharvest practices. Training smallholders on safe postharvest management and the use of ethylene inhibitors is also important. As growth regulators can shape development, leverage options such as auxins and gibberellic acids. Additionally, adopting transgenic plants with delayed ripening traits to improve quality and reduce losses.

8.2 Enhancing Awareness and Implementation

Train farmers on ethylene management, including antagonists such as STS, to improve profitability and sustainability. Providing low-cost solutions like 1-MCP and resources are accessible, smallholders can adopt these technologies effectively. At the same time, edible coatings and controlled ripening improve appearance, texture, and taste. These measures reduce postharvest losses and maintain the quality of perishables. Therefore, applying such practices across the supply chain yields economic gains, protects the environment, and supports food security.

9 Conclusions

Ethylene plays a multifaceted role in plant biology, influencing processes beneficial or detrimental to plants and human food systems. As a key plant hormone, ethylene regulates ripening, senescence, and abscission. Accelerated ripening and senescence help plant reproduction but reduce postharvest longevity and increase food wastage. As global hunger is rising, India is significantly impacted due to its growing population. To mitigate the challenges posed by ethylene, antagonists such as 1-MCP (1-methylcyclopropene), SLS (silver-based compounds), and AVG (aminoethoxyvinylglycine) extend shelf life. These treatments slow ripening and senescence. Postharvest practices such as edible coatings, modified atmosphere packaging, and ethylene absorbents complement these approaches. Looking ahead, advances in ethylene management through more efficient antagonists or innovative packaging, are critical.

10 Future Directions

Ethylene-dependent processes such as softening of flesh, yellowing, and climacteric respiration (ripening related respiratory burst) are well documented. Studies show fruit softening involves both ethylene-dependent and independent pathways [74]. Sugar accumulation predominantly follows an ethylene-independent mechanism. Compounds such as salicylic acid [75, 76] and melatonin [77] show potential to mitigate ethylene effects. Expanding the catalog of such natural inhibitors could offer sustainable, environmental friendly options for agricultural ethylene management. Systematic studies should define modes of action, cross crop efficacy, and real-world scalability. Digitalization and artificial intelligence are reshaping agriculture. Drones, 3D printing, virtual reality, and advanced biosensors improve efficiency and precision. Notable advances in ethylene management include carbon nanotube-based devices for ethylene gas detection [78]. Smartphone-integrated ethylene quantification methods [79] enable precise monitoring in storage environments. Additionally, enzymatic biosensors such as the ArM ethylene probe [80] detect endogenous and exogenous ethylene biosynthesis. These tools provide real-time insights into ethylene dynamics.

Developing cost-effective, and user-friendly devices to detect ethylene and associated enzymes will enhance postharvest management. Integrating these tools with AI and IoT (Internet of Things) platforms can enable predictive analytics. Consequently, stakeholders can make informed decisions for ethylene-sensitive commodities.

Acknowledgments

The Director, CSIR Central Institute of Medicinal and Aromatic Plants, Lucknow, India, is acknowledged for providing the facilities and support. The Institutional Communication No.: CIMAP/ PUB/ 2025/17.

References

1. Kesari R, Trivedi PK, Nath P (2007) Ethylene-induced ripening in banana evokes expression of defense and stress related genes in fruit tissue. Postharvest Biol Technol 46:136–143. https://doi.org/10.1016/j.postharvbio.2007.04.010
2. Nath P, Trivedi PK, Sane VA, Sane AP (2006) Role of ethylene in fruit ripening. In: Khan NA (ed) Ethylene action in plants. Springer, Berlin/Heidelberg, pp 151–184
3. Bapat VA, Trivedi PK, Ghosh A, Sane VA, Ganapathi TR, Nath P (2010) Ripening of fleshy fruit: molecular insight and the role of ethylene. Biotechnol Adv 28:94–107. https://doi.org/10.1016/j.biotechadv.2009.10.002
4. El-Ramady HR, Domokos-Szabolcsy É, Abdalla NA, Taha HS, Fári M (2015) Postharvest management of fruits and vegetables storage. In: Lichtfouse E (ed) Sustainable agriculture reviews. Springer, Cham, pp 65–152
5. Adhikari B, Aarati GC (2021) Post-harvest practices of horticultural crops in Nepal: issues and management. Arch Agric Environ Sci 6: 227–233. https://doi.org/10.26832/24566632.2021.0602015
6. FAO (2018) Guidelines on the measurement of harvest and post-harvest losses. Food and Agriculture Organization of the United Nations, Rome
7. Bisht A, Singh SP (2024) Postharvest losses and management of horticultural produce: a review. JSRR 30:305–320. https://doi.org/10.9734/jsrr/2024/v30i31881
8. FAO (2011) Global food losses and food waste: extent, causes and prevention. Study conducted for the International Congress "SAVE FOOD!" at Interpack 2011, Du¨sseldorf, Germany. Food and Agriculture Organization of the United Nations, Rome
9. Yahia EM, Fonseca JM, Kitinoja L (2019) Postharvest losses and waste. In: Postharvest technology of perishable horticultural commodities. Elsevier, pp 43–69
10. Mathushika JM (2021) Emerging concepts and practices in post-harvest management of horticultural crops revisited. IJCSRR 4. https://doi.org/10.47191/ijcsrr/V4-i8-04
11. Mohan A, Krishnan R, Arshinder K, Vandore J, Ramanathan U (2023) Management of post-harvest losses and wastages in the Indian tomato supply chain – a temperature-controlled storage perspective. Sustain For 15: 1331. https://doi.org/10.3390/su15021331
12. Alexander L (2002) Ethylene biosynthesis and action in tomato: a model for climacteric fruit ripening. J Exp Bot 53:2039–2055. https://doi.org/10.1093/jxb/erf072
13. Barry CS, Giovannoni JJ (2007) Ethylene and fruit ripening. J Plant Growth Regul 26:143. https://doi.org/10.1007/s00344-007-9002-y
14. Tipu MMH, Sherif SM (2024) Ethylene and its crosstalk with hormonal pathways in fruit ripening: mechanisms, modulation, and commercial exploitation. Front Plant Sci 15:1475496. https://doi.org/10.3389/fpls.2024.1475496
15. Martínez-Romero D, Bailén G, Serrano M, Guillén F, Valverde JM, Zapata P, Castillo S, Valero D (2007) Tools to maintain postharvest fruit and vegetable quality through the inhibition of ethylene action: a review. Crit Rev Food Sci Nutr 47:543–560. https://doi.org/10.1080/10408390600846390
16. Depaepe T, Van Der Straeten D (2017) Ethylene. In: Encyclopedia of applied plant sciences. Elsevier, pp 403–410
17. Huang W, Tan C, Guo H (2024) Ethylene in fruits: beyond ripening control. Hortic Res 11: uhae229. https://doi.org/10.1093/hr/uhae229
18. Saltveit ME (1999) Effect of ethylene on quality of fresh fruits and vegetables. Postharvest Biol Technol 15:279–292. https://doi.org/10.1016/S0925-5214(98)00091-X
19. Schaller GE, Binder BM (2017) Inhibitors of ethylene biosynthesis and signaling. In: Binder BM, Eric Schaller G (eds) Ethylene signaling. Springer, New York, pp 223–235
20. Binder BM (2020) Ethylene signaling in plants. J Biol Chem 295:7710–7725. https://doi.org/10.1074/jbc.REV120.010854

21. Liu M, Pirrello J, Chervin C, Roustan J-P, Bouzayen M (2015) Ethylene control of fruit ripening: revisiting the complex network of transcriptional regulation. Plant Physiol:01361.2015. https://doi.org/10.1104/pp.15.01361
22. Trivedi PK, Nath P (2004) MaExp1, an ethylene-induced expansin from ripening banana fruit. Plant Sci 167:1351–1358. https://doi.org/10.1016/j.plantsci.2004.07.005
23. Iqbal N, Khan NA, Ferrante A, Trivellini A, Francini A, Khan MIR (2017) Ethylene role in plant growth, development and senescence: interaction with other phytohormones. Front Plant Sci 08. https://doi.org/10.3389/fpls.2017.00475
24. Luo J, Chen S, Cao S, Zhang T, Li R, Chan ZL, Wang C (2021) Rose (*Rosa hybrida*) ethylene responsive factor 3 promotes rose flower senescence via direct activation of the abscisic acid synthesis–related *9-CIS-EPOXYCAROTENOID DIOXYGENASE* gene. Plant Cell Physiol 62:1030–1043. https://doi.org/10.1093/pcp/pcab085
25. Antunes MDC, Sfakiotakis EM (2002) Ethylene biosynthesis and ripening behaviour of 'Hayward' kiwifruit subjected to some controlled atmospheres. Postharvest Biol Technol 26:167–179. https://doi.org/10.1016/S0925-5214(02)00040-6
26. Mcmurchie EJ, Mcglasson WB, Eaks IL (1972) Treatment of fruit with propylene gives information about the biogenesis of ethylene. Nature 237:235–236. https://doi.org/10.1038/237235a0
27. Shibuya K, Barry KG, Ciardi JA, Loucas HM, Underwood BA, Nourizadeh S, Ecker JR, Klee HJ, Clark DG (2004) The central role of PhEIN2 in ethylene responses throughout plant development in petunia. Plant Physiol 136:2900–2912. https://doi.org/10.1104/pp.104.046979
28. Llop-Tous I, Barry CS, Grierson D (2000) Regulation of ethylene biosynthesis in response to pollination in tomato flowers. Plant Physiol 123:971–978. https://doi.org/10.1104/pp.123.3.971
29. Lang C, Hübert T (2012) A colour ripeness indicator for apples. Food Bioprocess Technol 5:3244–3249. https://doi.org/10.1007/s11947-011-0694-4
30. Schudel S, Shoji K, Shrivastava C, Onwude DI, Defraeye T (2023) Solution roadmap to reduce food loss along your postharvest supply chain from farm to retail. Food Packaging and Shelf Life 36: 101057. https://doi.org/10.1016/j.fpsl.2023.101057
31. Prasad P, Kochhar A (2014) Active packaging in food industry: a review. IOSRJESTFT 8:01–07. https://doi.org/10.9790/2402-08530107
32. Paul V, Pandey R (2017) 1-Methylcyclopropene (1-MCP) treatments. In: Pareek S (ed) Novel postharvest treatments of fresh produce, 1st edn. CRC Press, Boca Raton, pp 149–216
33. Serek M, Sisler EC, Reid MS (1994) Novel gaseous ethylene binding inhibitor prevents ethylene effects in potted flowering plants. JASHS 119:1230–1233. https://doi.org/10.21273/JASHS.119.6.1230
34. Sisler EC (2006) The discovery and development of compounds counteracting ethylene at the receptor level. Biotechnol Adv 24:357–367. https://doi.org/10.1016/j.biotechadv.2006.01.002
35. Huber DJ (2008) Suppression of ethylene responses through application of 1-methylcyclopropene: a powerful tool for elucidating ripening and senescence mechanisms in climacteric and nonclimacteric fruits and vegetables. HortScience 43:106–111. https://doi.org/10.21273/HORTSCI.43.1.106
36. Bagnato N, Barrett R, Sedgley M, Klieber A (2003) The effects on the quality of Cavendish bananas, which have been treated with ethylene, of exposure to 1-methylcyclopropene. Int J Food Sci Technol 38:745–750. https://doi.org/10.1046/j.1365-2621.2003.00726.x
37. Golding JB, Shearer D, Wyllie SG, McGlasson WB (1998) Application of 1-MCP and propylene to identify ethylene-dependent ripening processes in mature banana fruit. Postharvest Biol Technol 14:87–98. https://doi.org/10.1016/S0925-5214(98)00032-5
38. Watkins CB (2006) The use of 1-methylcyclopropene (1-MCP) on fruits and vegetables. Biotechnol Adv 24:389–409. https://doi.org/10.1016/j.biotechadv.2006.01.005
39. Ergun M, Jeong J, Huber DJ, Cantliffe DJ (2005) Suppression of ripening and softening of 'Galia' melons by 1-methylcyclopropene applied at preripe or ripe stages of development. HortScience 40:170–175. https://doi.org/10.21273/HORTSCI.40.1.170
40. Li F, Zhang X, Song B, Li J, Shang Z, Guan J (2013) Combined effects of 1-MCP and MAP on the fruit quality of pear (Pyrus bretschneideri Reld cv. Laiyang) during cold storage. Sci Hortic 164:544–551. https://doi.org/10.1016/j.scienta.2013.10.018
41. Patiño LS, Castellanos DA, Herrera AO (2018) Influence of 1-MCP and modified

atmosphere packaging in the quality and preservation of fresh basil. Postharvest Biol Technol 136:57–65. https://doi.org/10.1016/j.postharvbio.2017.10.010

42. Jia X, Wang W, Du Y, Tong W, Wang Z, Gul H (2018) Optimal storage temperature and 1-MCP treatment combinations for different marketing times of Korla Xiang pears. J Integr Agric 17:693–703. https://doi.org/10.1016/S2095-3119(17)61872-0
43. Blankenship SM, Dole JM (2003) 1-Methylcyclopropene: a review. Postharvest Biol Technol 28:1–25. https://doi.org/10.1016/S0925-5214(02)00246-6
44. Thompson AK (2016) Hypobaric storage. In: Fruit and vegetable storage. Springer, Cham, pp 37–92
45. Bangerth F, Bufler G, Halder-Doll H (1984) Experiments to prevent ethylene biosynthesis and/or action and effects of exogenous ethylene on ripening and storage of apple fruits. In: Fuchs Y, Chalutz E (eds) Ethylene. Springer, Dordrecht, pp 291–301
46. Thompson AK (2016) Fruit and vegetable storage: hypobaric, hyperbaric and controlled atmosphere. Springer, Cham
47. Hesselman CW, Freebairn HT (1969) Rate of ripening of initiated bananas as influenced by oxygen and ethylene1,2. J Am Soc Hortic Sci 94:635–637. https://doi.org/10.21273/JASHS.94.6.635
48. McKeown AW, Lougheed EC, Murr DP (1978) Compatability of cabbage, carrots, and apples in low pressure storage1. J Am Soc Hortic Sci 103:749–752. https://doi.org/10.21273/JASHS.103.6.749
49. Pelter GQ (1975) The effects of a subatmospheric storage pressure upon the breakdown, color, respiration, and ethylene evolution of cranberry fruit. Master thesis, Washington State University
50. Ohlsson T (2002) Minimal processing technologies in the food industries. CRC Press/Woodhead Publishing, Boca Raton/Boston/New York/Cambridge
51. Ebrahimi A, Zabihzadeh Khajavi M, Mortazavian AM, Asilian-Mahabadi H, Rafiee S, Farhoodi M, Ahmadi S (2021) Preparation of novel nano–based films impregnated by potassium permanganate as ethylene scavengers: an optimization study. Polym Test 93:106934. https://doi.org/10.1016/j.polymertesting.2020.106934
52. Abeles FB, Saltveit ME, Morgan PW (1992) Ethylene in plant biology, 2nd edn. Academic Press, San Diego
53. Zhang J, Cheng D, Wang B, Khan I, Ni Y (2017) Ethylene control technologies in extending postharvest shelf life of climacteric fruit. J Agric Food Chem 65:7308–7319. https://doi.org/10.1021/acs.jafc.7b02616
54. Hyde PT, Guan X, Abreu V, Setter TL (2020) The anti-ethylene growth regulator silver thiosulfate (STS) increases flower production and longevity in cassava (Manihot esculenta Crantz). Plant Growth Regul 90:441–453. https://doi.org/10.1007/s10725-019-00542-x
55. Veen H (1983) Silver thiosulphate: an experimental tool in plant science. Sci Hortic 20:211–224. https://doi.org/10.1016/0304-4238(83)90001-8
56. Scariot V, Paradiso R, Rogers H, De Pascale S (2014) Ethylene control in cut flowers: classical and innovative approaches. Postharvest Biol Technol 97:83–92. https://doi.org/10.1016/j.postharvbio.2014.06.010
57. Khan AS, Ali S (2018) Preharvest sprays affecting shelf life and storage potential of fruits. In: Preharvest modulation of postharvest fruit and vegetable quality. Elsevier, pp 209–255
58. Vakkalanka MSR, D'Souza T, Ray S, Yam KL, Mir N (2012) Emerging packaging technologies for fresh produce. In: Emerging food packaging technologies. Elsevier, pp 109–133
59. Zhou M, Guo S, Tian S, Zhang J, Ren Y, Gong G, Li C, Zhang H, Xu Y (2020) Overexpression of the watermelon ethylene response factor ClERF069 in transgenic tomato resulted in delayed fruit ripening. Hortic Plant J 6:247–256. https://doi.org/10.1016/j.hpj.2020.05.004
60. Reed AJ, Magin KM, Anderson JS, Austin GD, Rangwala T, Linde DC, Love JN, Rogers SG, Fuchs RL (1995) Delayed ripening tomato plants expressing the enzyme 1-aminocyclopropane-1-carboxylic acid deaminase. 1. Molecular characterization, enzyme expression, and fruit ripening traits. J Agric Food Chem 43:1954–1962. https://doi.org/10.1021/jf00055a036
61. Kim J-Y, Lee J-S, Kwon T-R, Lee S-I, Kim J-A, Lee G-M, Park S-C, Jeong M-J (2015) Sound waves delay tomato fruit ripening by negatively regulating ethylene biosynthesis and signaling genes. Postharvest Biol Technol 110:43–50. https://doi.org/10.1016/j.postharvbio.2015.07.015
62. Park M-H, Malka SK (2022) Gibberellin delays metabolic shift during tomato ripening by inducing auxin signaling. Front Plant Sci 13:1045761. https://doi.org/10.3389/fpls.2022.1045761

63. Martín-Pizarro C, Posé D (2018) Genome editing as a tool for fruit ripening manipulation. Front Plant Sci 9:1415. https://doi.org/10.3389/fpls.2018.01415
64. Lor VS, Starker CG, Voytas DF, Weiss D, Olszewski NE (2014) Targeted mutagenesis of the tomato *PROCERA* gene using transcription activator-like effector nucleases. Plant Physiol 166:1288–1291. https://doi.org/10.1104/pp.114.247593
65. Asif MH, Lakhwani D, Pathak S, Gupta P, Bag SK, Nath P, Trivedi PK (2014) Transcriptome analysis of ripe and unripe fruit tissue of banana identifies major metabolic networks involved in fruit ripening process. BMC Plant Biol 14:316. https://doi.org/10.1186/s12870-014-0316-1
66. Yu Y, Guo S, Ren Y, Zhang J, Li M, Tian S, Wang J, Sun H, Zuo Y, Chen Y, Gong G, Zhang H, Xu Y (2022) Quantitative transcriptomic and proteomic analysis of fruit development and ripening in watermelon (Citrullus lanatus). Front Plant Sci 13:818392. https://doi.org/10.3389/fpls.2022.818392
67. Xu M, Zhou W, Geng W, Zhao S, Pan Y, Fan G, Zhang S, Wang Y, Liao K (2021) Transcriptome analysis insight into ethylene metabolism and pectinase activity of apricot (Prunus armeniaca L.) development and ripening. Sci Rep 11:13569. https://doi.org/10.1038/s41598-021-92832-6
68. Lyu J, Wu Y, Jin X, Tang Z, Liao W, Dawuda MM, Hu L, Xie J, Yu J, Calderón-Urrea A (2021) Proteomic analysis reveals key proteins involved in ethylene-induced adventitious root development in cucumber (*Cucumis sativus* L.). PeerJ 9:e10887. https://doi.org/10.7717/peerj.10887
69. Joung J, Boonsiriwit A, Kim M, Lee YS (2021) Application of ethylene scavenging nanocomposite film prepared by loading potassium permanganate-impregnated halloysite nanotubes into low-density polyethylene as active packaging material for fresh produce. LWT 145:111309. https://doi.org/10.1016/j.lwt.2021.111309
70. Sangtawesin T, Weerasuk B, Piangsongchan K, Janjamrus N, Chutimasakul T (2024) Carbon-based materials impregnated with silver nanoparticle synthesized by gamma irradiation for ethylene gas scavenger application. J Agric Food Res 16:101204. https://doi.org/10.1016/j.jafr.2024.101204
71. Pongprasert N, Srilaong V (2014) A novel technique using 1-MCP microbubbles for delaying postharvest ripening of banana fruit. Postharvest Biol Technol 95:42–45. https://doi.org/10.1016/j.postharvbio.2014.04.003
72. Chen M, Chen X, Ray S, Yam K (2020) Stabilization and controlled release of gaseous/volatile active compounds to improve safety and quality of fresh produce. Trends Food Sci Technol 95:33–44. https://doi.org/10.1016/j.tifs.2019.11.005
73. Hu B, Sun D-W, Pu H, Wei Q (2019) Recent advances in detecting and regulating ethylene concentrations for shelf-life extension and maturity control of fruit: a review. Trends Food Sci Technol 91:66–82. https://doi.org/10.1016/j.tifs.2019.06.010
74. Pech JC, Bouzayen M, Latché A (2008) Climacteric fruit ripening: ethylene-dependent and independent regulation of ripening pathways in melon fruit. Plant Sci 175:114–120. https://doi.org/10.1016/j.plantsci.2008.01.003
75. Leslie CA, Romani RJ (1988) Inhibition of ethylene biosynthesis by salicylic acid. Plant Physiol 88:833–837. https://doi.org/10.1104/pp.88.3.833
76. Yuan R, Mao L, Min T, Zhao Y, Duan Y, Wang H, Lin Q (2023) Salicylic acid treatment inhibits ethylene synthesis and starch-sugar conversion to maintain apple fruit quality during shelf life. Sci Hortic 308:111586. https://doi.org/10.1016/j.scienta.2022.111586
77. Onik JC, Wai SC, Li A, Lin Q, Sun Q, Wang Z, Duan Y (2021) Melatonin treatment reduces ethylene production and maintains fruit quality in apple during postharvest storage. Food Chem 337:127753. https://doi.org/10.1016/j.foodchem.2020.127753
78. Esser B, Schnorr JM, Swager TM (2012) Selective detection of ethylene gas using carbon nanotube-based devices: utility in determination of fruit ripeness. Angew Chem Int Ed 51:5752–5756. https://doi.org/10.1002/anie.201201042
79. Yang X, Leong JLK, Sun M, Jing L, Zhang Y, Wang T, Wang S, Huang D (2022) Quantitative determination of ethylene using a Smartphone-Based Optical Fiber Sensor (SOFS) coupled with pyrene-tagged grubbs catalyst. Biosensors 12:316. https://doi.org/10.3390/bios12050316
80. Vong K, Eda S, Kadota Y, Nasibullin I, Wakatake T, Yokoshima S, Shirasu K, Tanaka K (2019) An artificial metalloenzyme biosensor can detect ethylene gas in fruits and Arabidopsis leaves. Nat Commun 10:5746. https://doi.org/10.1038/s41467-019-13758-2
81. Sitbon C, Paliyath G (2011) Pre- and postharvest treatments affecting nutritional quality. In: Comprehensive biotechnology. Elsevier, pp 349–357

82. Çelikel FG, Reid MS (2005) Temperature and postharvest performance of rose (Rosa Hybrida l. 'first red') and gypsophila (gypsophila Paniculata l. 'Bristol fairy') flowers. Acta Hortic:1789–1794. https://doi.org/10.17660/ActaHortic.2005.682.239
83. Macnish AJ, Hofman PJ, Joyce DC, Simons DH, Reid MS (2000) 1-Methylcyclopropene treatment efficacy in preventing ethylene perception in banana fruit and grevillea and waxflower flowers. Aust J Exp Agric 40:471. https://doi.org/10.1071/EA99120
84. Serek M, Sisler EC, Reid MS (1995) Effects of 1-MCP on the vase life and ethylene response of cut flowers. Plant Growth Regul 16:93–97. https://doi.org/10.1007/BF00040512
85. van Doorn WG (2001) Categories of petal senescence and abscission: a re-evaluation. Annals of Botany 87: 447–456. https://doi.org/10.1006/anbo.2000.1357

Chapter 9

Isolation and Characterization of Proximal Promoters Regulating Ethylene-Induced Ripening in Banana

Devesh Shukla, Ravi Kesari, Prabodh Kumar Trivedi, and Pravendra Nath

Abstract

Fruit ripening is a key developmental phase with biochemical and physiological changes. In climacteric fruits (ethylene responsive), such as bananas, ethylene predominantly regulates these changes. This study presents a protocol to isolate and functionally analyse proximal promoters of banana ripening genes. Using genome walking, we identify promoter fragments of key ripening genes. We then characterize their ethylene responsiveness. Promoter–GUS fusion constructs are prepared for functional testing. We deliver these constructs transiently into banana fruit tissues. Promoter activity is quantified by histochemical (color-based) and fluorometric (fluorescence-based) β-glucuronidase (GUS) assays. Findings clarify the regulation of ripening-associated pathways. Therefore, they suggest genetic engineering options to extend shelf life and improve banana nutritional quality. This protocol provides a practical guide for researchers to study and manipulate the molecular mechanisms underlying fruit ripening.

Key words Banana, Fruit ripening, Ethylene, Promoter isolation, Genome walking, Transient expression, Genetic engineering

1 Introduction

Fruit ripening is a key final phase in the ontogeny of many angiosperms. During ripening, fruits become nutritious and visually appealing, which optimizes seed dispersal [1, 2]. This developmental transformation increases plant reproductive success. It also supports ecosystems by feeding fruit-eating animals that spread seeds [3]. Ripening includes broad biochemical, physiological, and structural changes, including color, texture, flavor, and aroma. An intricate gene expression network coordinates these changes, modulated by developmental signals, environmental factors, and hormonal signaling. Ethylene controls this process and chain of events, and it is central in climacteric fruits [1, 2]. Climacteric fruits such as bananas, tomatoes, and apples show sharp increases in ethylene biosynthesis and respiration during ripening [2]. In contrast, non-climacteric fruits like grapes and strawberries mature

Devesh Shukla (ed.), *Ethylene Sensing and Signaling*, Methods in Molecular Biology, vol. 2945,
https://doi.org/10.1007/978-1-0716-4650-2_9,

independently of ethylene signaling [4]. In climacteric species, ethylene triggers and regulates ripening, affecting pigment biosynthesis, carbohydrate metabolism, and cell wall disassembly [5, 6]. The role of ethylene in these processes underlines major agronomic and commercial significance [1]. Therefore, many studies on ripening mechanisms focus on climacteric fruits like banana and tomatoes, given their agricultural relevance. These studies show ethylene coordinates ripening pathways through its biosynthesis, perception, and downstream signaling [1, 2, 5–7]. High-throughput methods, including transcriptomics, subtractive hybridization, and RNA sequencing identified genes in ethylene-regulated cell wall remodeling, secondary metabolite production, and stress response [6, 8, 9]. Despite advances in banana ripening biology, an economically vital and globally consumed climacteric fruit, many aspects need further study. Banana is highly susceptible to ethylene-induced postharvest ripening. Because low temperatures cause chilling injury, banana cannot be stored cold, complicating post-harvest management. Thus, urgent genetic interventions should extend banana shelf life and improve fruit quality [6, 10].

One promising strategy is identifying and functionally analyzing of fruit promoters that regulate ripening-related gene expression. Promoters, as regulatory DNA elements, control spatial and temporal expression of genes and can help control ripening [2, 11]. Engineering promoters to modulate gene expression offers potential solutions for improving fruit shelf life. Among available methods, genome walking has emerged as a robust approach, especially in species lacking a genome sequence. Genome walking extends from known DNA into adjacent unknown regions. This chapter demonstrates experimental steps for isolating and characterizing proximal promoters of ripening-associated genes in bananas. These findings support genetic engineering, such as extending shelf life or improving fruit nutritional quality [2, 3].

2 Materials

2.1 Plant Material

The local edible variety of banana (*Musa acuminata*; Dwarf Cavendish, genome AAA, var. Robusta (Harichhal)) is utilized. This banana variety requires ethylene exposure to initiate ripening, as it does not ripen naturally in a reasonable time. For experiments, mature unripe banana hands are procured from a local farm and treated with ethylene or other hormones/chemicals as per the experimental design. Additionally, other banana plant parts, such as leaves, flowers, and fruits at various developmental stages, are collected from the farm when required.

2.2 Chemicals and Molecular Biology Kits

- **Ethylene and Inhibitors:** Ethylene (0.5% v/v) and 1-methylcyclopropene (1-MCP) are used from recognized suppliers.
- **Ultrapure Chemicals:** These are sourced from general laboratory-grade suppliers for molecular biology applications.
- **Molecular Biology-Grade Reagents:** High-purity reagents are suitable for sensitive molecular biology techniques.
- **Enzymes and Modifying Agents:** Restriction enzymes and DNA-modifying enzymes are suitable for genome walking and cloning experiments.
- **Molecular Biology Kits:** GenomeWalker Universal Kit and other kits for DNA isolation, amplification, and purification are required.
- **Water:** Sterile, deionized water from a Milli-Q synthesis system is used throughout the study.

2.3 Buffers and Solutions

- **TE Buffer:** 10 mM Tris–HCl (pH 8.0), 0.1 mM EDTA (pH 7.5).
- **Phenol:Chloroform:** Prepared as a 1:1 solution for DNA extraction.
- **GUS Extraction Buffer:** 50 mM sodium phosphate buffer (pH 7.0), 10 mM DTT, 1 mM Na_2EDTA, 0.1% SDS, 0.1% Triton X-100.
- **Assay Buffer:** 4 mM 4-Methylumbelliferyl-β-D-glucuronide (MUG) in GUS extraction buffer.
- **Ethanol:** 70% absolute ethanol is used for washing and precipitation.
- **NaOAc Solution:** 3 M sodium acetate (pH 4.5).

2.4 Primers and Oligonucleotides

A list of primers and oligonucleotides used for isolation of promoters and preparation of constructs is detailed in Table 1 [12]. Primers are designed for various applications, including genome walking and promoter analysis [12] (*see* **Note 4**).

2.5 Specialized Equipment

- **Thermal Cycler:** For all PCR reactions, ensuring precision in temperature cycles.
- **Gel Electrophoresis System:** Agarose gel electrophoresis setup for DNA quality and size verification.
- **UV–Visible (UV–vis) Spectrophotometer:** For qualitative and quantitative analysis of DNA samples.
- **Particle Delivery System:** Equipment for particle bombardment experiments.
- **Spectrofluorimeter:** To measure GUS activity quantitatively.
- **Centrifuge:** Suitable for DNA precipitation and purification.

Table 1
List of primers and oligonucleotides

Oligo name	Sequence (5′ to 3′)	Gene/promoter	Used for
IFR Pro GSPR1	GCGCTCGCAAACACGATGAACTTGCC AAG CTT TTT	Isoflavone reductase	For genome walking
IFR Pro GSPR2	CGATGATCAGGATCTTGCTGTTCTCG	Isoflavone reductase	For genome walking
IFR 1.2 prom SalI	TCACAAATGGTCGACATTGGAATCC	Isoflavone reductase	For IFR 1.2 Prom:GUS Vector
IFR 1.4 prom SalI	AAAGAGTAGAGGTCGACCATTAT	Isoflavone reductase	For IFR 1.4 Prom:GUS Vector
IFR prom SmaI	CTTGCTGTTCCCGGGTCGCCATTCTCTCCTC	Isoflavone reductase	For IFR1.2/IFR1.4 Prom:GUS Vector
GenomeWalker Adapter	GTAATACGACTCACTATAGGGCACGCGTGG TCGACGGCCCGGGCTGGT	–	For making GenomeWalker library
AP1	GTAATACGACTCACTATAGGGC	–	For genome walking
AP2	ACTATAGGGCACGCGTGGT	–	For genome walking
101 Forward	TCG TAT GTT GTG TGG AAT TGT GAG	–	pBI101 vector specific primer
GUS Reverse	ATT CCA CAG TTT TCG CGA TCC AGA C	–	pBI101 vector specific primer
M13	TGTAAAACGACGGCCAG	–	pTZ57R vector specific primer
M13r	CAGGAAACAGCTATGACC	–	pTZ57R vector specific primer

- **Liquid Nitrogen:** For immediate preservation and grinding of tissue samples.
- **Milli-Q Water System:** For preparing sterile and deionized water used in all solutions and reagents.

2.6 Miscellaneous

- Agar plates (1% w/v) are used for tissue support during particle bombardment.
- Sterile razor blades are used for tissue slicing and DNA band extraction.
- X-Gluc staining solution is used for histochemical GUS activity assays.
- Petri dishes and sterile microcentrifuge tubes are used for sample handling and processing.

3 Methods

3.1 Isolation of Proximal Promoters of Ripening-Related Genes

3.1.1 Construction of GenomeWalker Library

1. **Genomic DNA Isolation:** Extract genomic DNA from banana leaves (*Musa acuminata*; var. Robusta Harichhal) using the Dellaporta et al. [13] protocol.
   - I. **Tissue Preparation:** Weigh 1 g of leaf tissue, quick-freeze in liquid nitrogen, and grind into a fine powder using a mortar and pestle. Make sure the tissue remains frozen through the grinding to prevent DNA degradation.
   - II. **Extraction:** Transfer the powder into a 30 mL tube and add 15 mL of Extraction Buffer 1 (EB1; 100 mM Tris–HCl pH 8.0, 50 mM EDTA pH 8.0, 500 mM NaCl, with freshly added 0.7 mL of β-mercaptoethanol per liter of EB1). Mix well.
   - III. **Cell Lysis:** Add 1 mL of 20% SDS to the tube. Mix thoroughly by shaking and incubate at 65 °C for 10 min with occasional mixing by inversion.
   - IV. **Optional RNase Treatment:** After lysis, cool the mixture to room temperature and add 20 μg/mL RNase A to the lysate. (*see* **Note 2**).
   - V. **Chloroform Extraction:** Add an equal volume of chloroform: isoamyl alcohol (24:1, v/v) to the lysate. Mix thoroughly by inversion for 10 min to ensure complete deproteinization.
   - VI. **Centrifugation:** Centrifuge at 13,000 rpm (approximately 25,000 × g) for 15 min at room temperature. Separate the aqueous phase carefully.

VII. **Potassium Acetate Treatment:** Separate the aqueous phase (top layer) to a fresh tube. Add 5 mL of 5 M potassium acetate (KOAc) to the aqueous phase. Mix well and incubate on ice for 20 min (*see* **Note 1**).

VIII. **Supernatant Collection:** Centrifuge the mixture at 13,000 rpm (approximately 25,000 × g) for 20 min at 4 °C. Filter the supernatant through a muslin cloth (cheesecloth) into a new tube containing an equal volume of isopropanol.

IX. **DNA Precipitation:** Mix by inverting 20 times and incubate at −20 °C for 30 min. Centrifuge at 12,000 rpm (approximately 20,000 × g) for 15 min at 4 °C (*see* **Note 1**).

X. **DNA Washing:** Discard the supernatant, wash the DNA pellet with 70% ethanol, and air-dry. Confirm that pellets are completely dried.

XI. **Dissolution:** Dissolve the DNA pellet in 0.7 mL of EB2 (50 mM Tris–HCl pH 8.0, 10 mM EDTA). Incubate overnight at 4 °C if necessary (*see* **Note 1**).

2. **Restriction Enzyme Digestion:**

I. Digest 2.5 μg of genomic DNA separately with blunt end producing restriction enzymes *Dra*I, *Eco*RV, *Pvu*II, and *Stu*I (Fig. 1).

II. Use 8 μL of each enzyme (10 units/μL), 10 μL 10× restriction enzyme buffer, and sterile water to adjust the total volume to 100 μL.

III. Incubate at 37 °C overnight. Confirm digestion by loading 6 μL of each reaction on a 0.6% agarose (w/v) gel.

3. **Adapter Ligation:**

I. After digestion, extract the reaction mixture using phenol:chloroform and precipitate DNA with 2 volumes of ice-cold 95% ethanol, 1/10 volume of 3 M sodium acetate (pH 4.5), and 20 μg glycogen (Fig. 1).

II. Pellet DNA via centrifugation at 15,000 rpm for 10 min. Wash with 80% ethanol and suspend in 20 μL TE buffer (pH 7.5).

III. Ligate GenomeWalker Adapter "GTAATACGACT-CACTATAGGGCACGCGTGGTC-GACGGCCCGGGCTGGT" (25 μM) to 4 μL of purified DNA with T4 DNA ligase at 16 °C overnight (Table 1).

IV. Store the resulting four libraries at −20 °C (*see* **Note 3**).

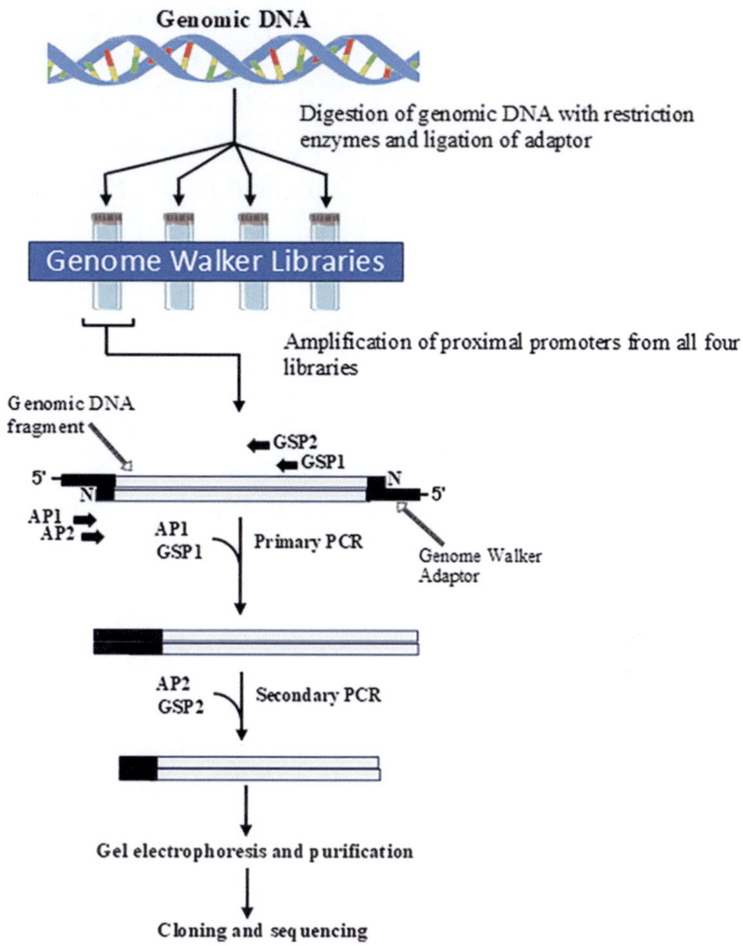

Fig. 1 Schematic representation of genome walking shown to isolate proximal promoters of ethylene-responsive banana genes. The procedure involves extracting genomic DNA, performing restriction digestion and adapter ligation to build GenomeWalker libraries. Next, two PCR rounds use adapter primers (AP1, AP2) and gene-specific primers (GSP1, GSP2). The resulting amplicons are resolved, purified, cloned, and sequenced to reveal upstream regulatory regions of ripening-related genes

3.1.2 Genome Walking for Promoter Isolation

1. **Primer Design:**
   I. Design gene-specific primers (GSP R1 and R2) based on the nucleotide sequence of the 5′ region of the target gene (Fig. 1 and Table 1) [12] (*see* **Note 4**).
2. **Primary PCR:**
   I. Set up a 20 μL reaction for each of four GenomeWalker libraries containing:
   - 13 μL sterile H_2O.
   - 1 μL GenomeWalker library.

- 2 μL 10× PCR buffer.
- 1 μL dNTP mix (10 mM each).
- 1 μL AP1 (Adapter Primer 1).
- 1 μL GSP R1 primer.
- 1 μL high-fidelity DNA polymerase (e.g., Phusion).

II. Thermocycling conditions:
- Initial denaturation: 94 °C for 2 min.
- 34 cycles of 94 °C for 20 s, 50–68 °C for 30 s (gradient PCR to optimize annealing temperature), and 72 °C for 3 min.
- Final extension: 72 °C for 7 min.

3. **Secondary PCR:**

I. **Template Preparation:** Dilute the primary PCR product 100-fold in sterile H_2O to reduce excess primers and non-specific amplification products (Fig. 1).

II. **Reaction Setup:** Prepare a 20 μL reaction mix containing:
- 13 μL sterile H_2O.
- 1 μL diluted primary PCR product.
- 2 μL 10× PCR buffer.
- 1 μL dNTP mix (10 mM each).
- 1 μL AP2 (Adapter Primer 2).
- 1 μL GSP R2 primer.
- 1 μL high-fidelity DNA polymerase (e.g., Phusion).

III. **Thermocycling Conditions:** Use the following conditions:
- Initial denaturation: 94 °C for 2 min.
- 34 cycles of 94 °C for 20 s, 50–68 °C for 30 s (gradient PCR for optimized annealing), and 72 °C for 2 min.
- Final extension: 72 °C for 7 min.

IV. **Analysis:** Load the amplified products on a 1% agarose (w/v) gel stained with ethidium bromide or a safer dye (e.g., SYBR Safe). Identify and excise the largest and most prominent band amplified among the four GenomeWalker libraries.

V. **Product Purification:** Purify the desired amplicon using a gel extraction kit.

4. **Cloning and Sequencing:**

I. Excise the largest amplicon, purify, and clone into the pTZ57R/T vector (Table 1).

Table 2
Proximal promoters isolated for putative fruit-specific and ripening-related genes

Gene name	Length of proximal promoters isolated
Isoflavone reductase, IFR 1.4	1359 bp
IFR 1.2	1215 bp

II. Sequence both strands for confirmation (Table 2).

III. The lengths of the isolated IFR 1.4 and IFR 1.2 proximal promoters are given in Table 2.

3.2 Preparation of Promoter–GUS Fusion Constructs

1. **Restriction Site Introduction:**
   I. Specific restriction sites (*Sal*I and *Sma*I) are introduced at the promoter ends by designing forward and reverse primers containing these sites (Table 1) [12].
2. **PCR Amplification of Promoters:**
   I. The amplification of promoters is performed using a recombinant pTZ57R plasmid as the template.
   II. PCR setup includes:
   - 13 μL sterile H_2O.
   - 1 μL recombinant plasmid DNA template.
   - 2 μL 10× PCR buffer.
   - 1 μL dNTP mix (10 mM each).
   - 1 μL forward primer (with *Sal*I site).
   - 1 μL reverse primer (with *Sma*I site).
   - 1 μL high-fidelity DNA polymerase (e.g., Phusion).

   III. Thermocycling conditions:
   - Initial denaturation: 94 °C for 2 min.
   - 34 cycles of 94 °C for 20 s, primer-specific annealing temperature for 30 s, and 72 °C for 1 min/kb of expected amplicon size.
   - Final extension: 72 °C for 10 min.

   IV. The PCR products are purified using a gel extraction kit and digested with *Sal*I and *Sma*I enzymes.
3. **Vector Preparation and Ligation:**
   I. The pBI101 vector is digested with *Sal*I and *Sma*I enzymes to create compatible ends for ligation.
   II. Ligation reaction setup includes:
   - 50 ng of digested promoter insert.
   - 200 ng of predigested pBI101 vector.

- 1.5 μL of 10× ligation buffer.
- 1 μL of T4 DNA ligase.
- Total reaction volume: 15 μL.

III. The ligation is carried out at 16 °C overnight.

4. ***E. coli* Transformation and Screening:**

I. After ligation, 15 μL of the ligation mixture is transformed into *E. coli* DH5α-competent cells using the heat shock method:

- Incubate the competent cells with the ligation mixture on ice for 30 min.
- Heat shock at 42 °C for 90 s.
- Immediately place on ice for 2 min.
- Add 500 μL of SOC medium and incubate at 37 °C with shaking (200 rpm) for 1 h.

II. Plate the transformed cells on LB agar plates containing kanamycin (50 μg/mL) and incubate overnight at 37 °C.

III. **Screening:**

- Perform colony PCR to identify transformants containing the desired promoter–GUS construct.
- Set up a PCR reaction using gene-specific forward and reverse primers.
- Analyze the PCR products on a 1% agarose (w/v) gel to confirm the presence of the insert.

IV. **Plasmid Isolation:**

- Isolate plasmid DNA from positive colonies using a Plasmid Mini Kit following the manufacturer's protocol.
- Verify the integrity and sequence of the insert using Sanger sequencing.

V. The confirmed plasmids are then used for transient expression analysis and plant transformation.

3.3 Transient Expression Analysis of Promoters

3.3.1 Preparation of Particles for Bombardment

1. **Preparation of Gold Particles:**

I. Weigh 30 mg of gold particles and suspend them in 500 μL of sterile water. Briefly vortex to homogenize the suspension.

II. Centrifuge the suspension at 10,000 rpm for 10 s and discard the supernatant. Wash the gold particles once with 70% ethanol, followed by two washes with 500 μL sterile water (*see* **Note 5**).

III. After the final wash, resuspend the particles in sterile water and aliquot into 50 μL volumes in microcentrifuge tubes. Store at 4 °C until use.

2. **DNA Coating on Gold Particles:**

   I. Resuspend the gold particles in 50 μL of fresh sterile water. While vortexing slowly and continuously, sequentially add:

   - 5–10 μL of plasmid DNA (1–2 mg/mL concentration).
   - 50 μL of 2.5 M $CaCl_2$.
   - 20 μL of 0.1 M spermidine free base.

   II. Continue vortexing for 3 min to ensure uniform coating of DNA onto the particles.

   III. Pellet the DNA-coated gold particles by centrifugation at 10,000 rpm for 10 s at 4 °C. Discard the supernatant and wash the particles with 250 μL of 100% ethanol (*see* **Note 5**).

   IV. Resuspend the coated particles in 60 μL of 100% ethanol and store on ice until used.

3.3.2 Particle Bombardment to Banana Fruit Slices

1. **Preparation of Tissue Samples:**

   I. Cut banana fruit pulp into thin, uniform slices (approximately 2–3 mm thick) and place them on 1% agar (w/v) plates to maintain moisture and prevent desiccation during bombardment.

2. **Particle Bombardment Procedure:**

   I. Use a particle delivery system (e.g., Bio-Rad PDS-1000/He) for bombardment.

   II. Load 10 μL of the prepared DNA-coated gold particle suspension (as described earlier) onto the macrocarrier and allow it to dry.

   III. Assemble the macrocarrier and rupture disk (1100 psi) into the particle delivery system according to the manufacturer's instructions.

   IV. Bombard the banana slices under vacuum at the specified pressure.

3. **Post-bombardment Treatment:**

   I. After bombardment, expose the tissue slices to ethylene gas to enhance promoter activity and keep them in the dark at room temperature overnight.

   II. Depending on the downstream analysis, either stain the slices with X-Gluc for histochemical GUS activity or freeze them in liquid nitrogen for fluorometric analysis (Fig. 2) (*see* **Note 6**).

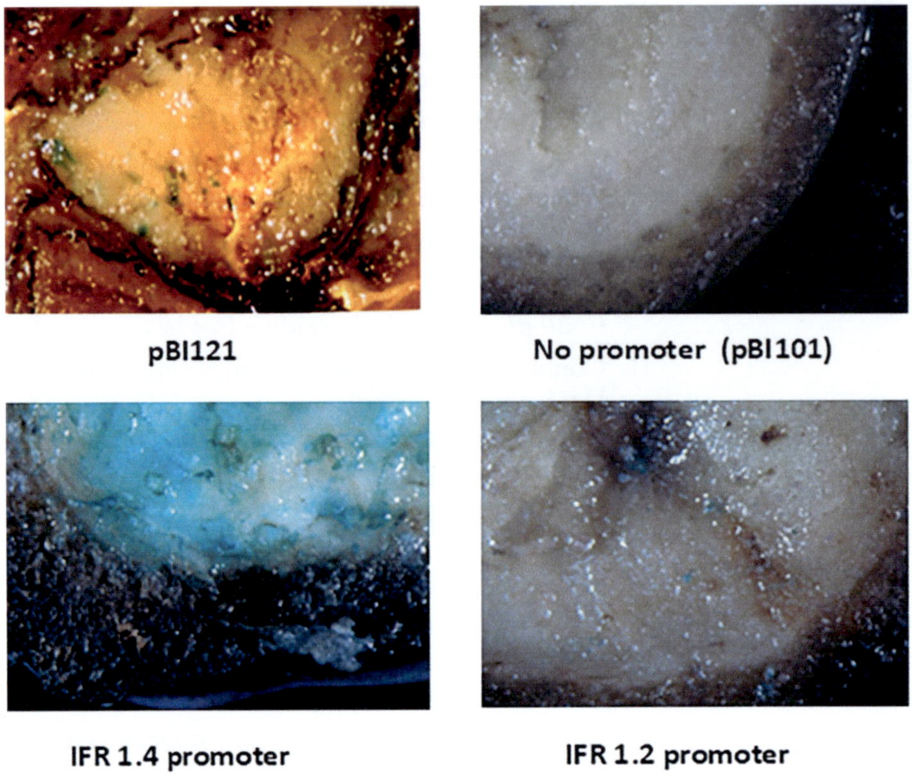

Fig. 2 Transient GUS expression in banana fruit slices using different promoters. The tested promoters effectively drive GUS activity in pulp tissue following ethylene treatment

3.3.3 Fluorogenic β-Glucuronidase (GUS) Assay

1. **Sample Preparation:**
   I. Take three bombarded banana slices from each experimental plate and grind them into a fine powder using liquid nitrogen.
   II. Weigh 0.1 g of the powdered tissue and mix it with 1 mL of GUS extraction buffer:
      - 50 mM sodium phosphate buffer (pH 7.0)
      - 10 mM DTT
      - 1 mM Na_2EDTA
      - 0.1% SDS
      - 0.1% Triton X-100
2. **Extraction:**
   I. Vortex the mixture briefly and centrifuge at 10,000 × g for 10 min to separate the supernatant containing soluble proteins.

3. **GUS Enzyme Reaction:**
   I. Transfer 75 μL of the supernatant to a fresh tube.
   II. Add 25 μL of prewarmed 4× assay buffer (4 mM MUG in GUS extraction buffer) and 25 μL of methanol.
   III. Incubate the reaction at 55 °C for 10 min to enhance enzyme activity.
   IV. Further, incubate at 37 °C for 3 h to allow the complete enzymatic reaction (*see* **Notes 6**, **7**, and **8**).
4. **Reaction Termination:**
   I. Stop the reaction by adding 0.9 mL of 0.2 M Na_2CO_3.
5. **Fluorometric Analysis:**
   I. Measure the fluorescence of the reaction using a spectrofluorometer calibrated with standard 4-methylumbelliferone (MU) solution.
   II. Use excitation at 365 nm and emission at 455 nm for the readings.
6. **Data Interpretation:**
   I. Compare the fluorescence intensities to evaluate the activity of the promoters in the bombarded tissues (*see* **Notes 8, 9, 10,** and **11**).
   II. For absolute quantification of GUS activity, prepare a standard curve using known 4-methylumbelliferone (MU) concentrations and convert fluorescence readings into pmol MU/min/mg protein (*see* **Notes 7, 8, 9, 10,** and **11**).

4 Notes

1. Make sure genomic DNA extracted from banana tissues is high purity. A typical A260/A280 ratio should be close to 1.8. The A260/A230 ratio should not go below 2.0. These measures prevent inhibition in downstream applications like genome walking. If values are lower or show deviations, reprecipitate with sodium acetate (3 M, pH 5.2) and ethanol to improve quality.
2. Calculate amount of RNase A from the total volume of the lysate. For example, for 15 mL lysate, add 30 μL of 10 mg/mL RNase A stock solution. Incubate at 37 °C for 15 min to remove residual RNA. This step ensures RNA-free DNA for downstream applications.
3. Ligation of the genome walking adapter to digested DNA is critical for efficient PCR amplification of proximal promoter

regions. Suboptimal conditions may yield weak or nonspecific PCR products. Confirm ligation by performing control PCR with known adapter primers before proceeding.

4. Specificity in genome walking largely depends on gene-specific primers (GSPs) design. Avoid repetitive sequences, and make sure high GC content (40–60%) to increase primer-binding specificity and consistent melting temperature. Adjust annealing temperatures during PCR cycles to prevent nonspecific amplifications.
5. During gold particle preparation for DNA coating, thorough ethanol washes are essential to confirm the absence of salt contamination. Residual salts can interfere with particle bombardment efficiency.
6. For promoter activity analysis, prepare banana slices with uniform thickness (approximately 3 mm) to enable consistent bombardment with DNA-coated gold particles.
7. Maintain strict incubation times and temperatures during the GUS assay for reliable results. Prewarm the substrate solution to 37 °C before adding it to tissue extracts to optimize enzymatic activity. Include technical triplicates for each sample to capture experimental variability.
8. Include vectors such as pBI121 (CaMV 35S:GUS as a positive control) and pBI101 (empty vector as a negative control) in every transient assay. This confirms the reliability of observed promoter activity.
9. Ethylene induced promoter activity depends on both concentration and exposure time. Overexposure can cause tissue degradation, affecting downstream assays.
10. Make sure complete chlorophyll removal during X-Gluc staining to minimize background color. Use freshly prepared staining solution and extend incubation for weak promoters to visualize expression clearly.
11. For reproducibility, conduct all experiments with at least three biological replicates. Statistical analyses, such as Student's t-test and if required ANOVA, should be applied to compare promoter activities across constructs.

Acknowledgments

The Institutional Communication No. CIMAP/PUB/ 2025/08.

References

1. Nath P, Trivedi PK, Sane VA, Sane AP (2006) Role of ethylene in fruit ripening. In: Khan NA (ed) Ethylene action in plants. Springer, Berlin/Heidelberg, pp 151–184
2. Bapat VA, Trivedi PK, Ghosh A, Sane VA, Ganapathi TR, Nath P (2010) Ripening of fleshy fruit: molecular insight and the role of ethylene. Biotechnol Adv 28:94–107. https://doi.org/10.1016/j.biotechadv.2009.10.002
3. Giovannoni J (2001) Molecular biology of fruit maturation and ripening. Annu Rev Plant Physiol Plant Mol Biol 52:725–749. https://doi.org/10.1146/annurev.arplant.52.1.725
4. Ledger SE, Gardner RC (1994) Cloning and characterization of five cDNAs for genes differentially expressed during fruit development of kiwifruit (Actinidia deliciosa var.deliciosa). Plant Mol Biol 25:877–886. https://doi.org/10.1007/BF00028882
5. Lohani S, Trivedi PK, Nath P (2004) Changes in activities of cell wall hydrolases during ethylene-induced ripening in banana: effect of 1-MCP, ABA and IAA. Postharvest Biol Technol 31:119–126. https://doi.org/10.1016/j.postharvbio.2003.08.001
6. Kesari R, Trivedi PK, Nath P (2007) Ethylene-induced ripening in banana evokes expression of defense and stress related genes in fruit tissue. Postharvest Biol Technol 46:136–143. https://doi.org/10.1016/j.postharvbio.2007.04.010
7. Alba R, Payton P, Fei Z, McQuinn R, Debbie P, Martin GB, Tanksley SD, Giovannoni JJ (2005) Transcriptome and selected metabolite analyses reveal multiple points of ethylene control during tomato fruit development. Plant Cell 17:2954–2965. https://doi.org/10.1105/tpc.105.036053
8. Diatchenko L, Lau YF, Campbell AP, Chenchik A, Moqadam F, Huang B, Lukyanov S, Lukyanov K, Gurskaya N, Sverdlov ED, Siebert PD (1996) Suppression subtractive hybridization: a method for generating differentially regulated or tissue-specific cDNA probes and libraries. Proc Natl Acad Sci USA 93:6025–6030. https://doi.org/10.1073/pnas.93.12.6025
9. Kesari R, Trivedi PK, Nath P (2010) Gene expression of pathogenesis-related protein during banana ripening and after treatment with 1-MCP. Postharvest Biol Technol 56:64–70. https://doi.org/10.1016/j.postharvbio.2009.11.012
10. Gupta SM, Srivastava S, Sane AP, Nath P (2006) Differential expression of genes during banana fruit development, ripening and 1-MCP treatment: presence of distinct fruit specific, ethylene induced and ethylene repressed expression. Postharvest Biol Technol 42:16–22. https://doi.org/10.1016/j.postharvbio.2006.05.002
11. Hart CM, Nagy F, Meins F (1993) A 61 bp enhancer element of the tobacco β-1,3-glucanase B gene interacts with one or more regulated nuclear proteins. Plant Mol Biol 21:121–131. https://doi.org/10.1007/BF00039623
12. Rozen S, Skaletsky H (1999) Primer3 on the WWW for general users and for biologist programmers. In: Bioinformatics methods and protocols. Humana Press, pp 365–386
13. Dellaporta SL, Wood J, Hicks JB (1983) A plant DNA minipreparation: version II. Plant Mol Biol Report 1:19–21. https://doi.org/10.1007/BF02712670

Part III

Tools and Methods for Investigating Ethylene Action

Chapter 10

Deciphering Ethylene Signaling via Protein–Protein Interaction in Tomato Fruit

Yusuke Kamiyoshihara, Yuki Achiha, Shin Ishikawa, Shinji Mizuno, Hitoshi Mori, Akira Tateishi, Donald J. Huber, and Harry J. Klee

Abstract

Plant ethylene receptors (ETRs) form complicated protein complexes at endoplasmic reticulum membranes. The receptors function as dimers, and they gather to form higher-order protein complexes, together with other proteins that modulate ethylene signaling. Co-immunoprecipitation is a useful biochemical tool for determining the composition of protein complexes. Here, we introduce a basic co-immunoprecipitation scheme that includes preparation of microsomal fraction, immunoprecipitation, and mass spectrometry (MS) analysis for protein identification. Although the ETR protein complex in tomato fruit is addressed here, the protocol may be adapted to other fruit species with some modifications.

Key words Antibody, Ethylene receptor, Immunoprecipitation, Mass spectrometry, Membrane-associated proteins, Protein complexes, Tomato fruit

1 Introduction

There are diverse in vivo and in vitro methods for addressing protein–protein interactions. Co-immunoprecipitation (IP) is a common and powerful strategy for this purpose (Fig. 1). IP itself is a classic technique in which a target antigen is recognized by a specific antibody (polyclonal or monoclonal), and the immune complex is precipitated from the protein extract using protein A/G-conjugated resins. Alternatively, an epitope-tagged (e.g., c-Myc) target antigen in a transgenic organism can be purified using an anti-tag antibody. Through this process, the binding partners that physically interact with the target antigen can be co-purified under the appropriate conditions. Following IP, the samples are analyzed by using immunoblotting (IB) with other antibodies against the proteins of interest to test the interactions. Alternatively, the IP fraction may be subjected to mass spectrometry (MS) analysis to identify novel candidate proteins that interact with the antigen. In

Devesh Shukla (ed.), *Ethylene Sensing and Signaling*, Methods in Molecular Biology, vol. 2945,
https://doi.org/10.1007/978-1-0716-4650-2_10,

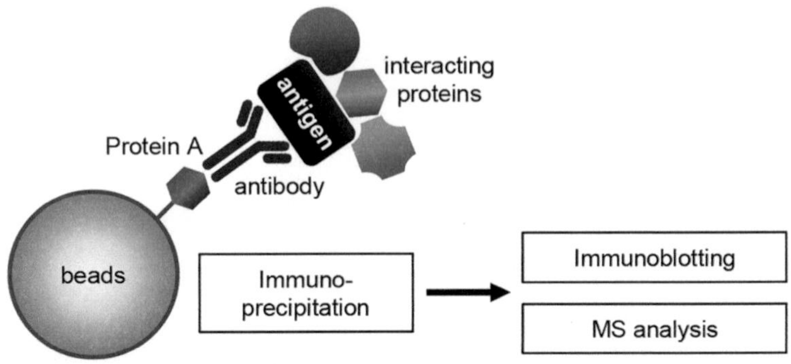

Fig. 1 Co-immunoprecipitation strategy to identify interacting proteins with an antigen protein. Following immunoprecipitation, the samples are subjected to either immunoblotting with other antibodies for possible interacting proteins to test the interactions or MS analysis to identify candidate interacting proteins

such cases, the interactions must be verified using different types of assays such as pull-down assays and yeast two-hybrid assays.

The plant ethylene receptors (ETR) are known to form complicated protein complexes. ETRs are related to bacterial two-component receptors [1, 2] and are associated with the endoplasmic reticulum membranes [3]. Biochemical analyses have indicated that ETR isoforms interact with each other as either dimers or higher-order clusters in homomeric and heteromeric combinations [2, 4–8]. ETRs also interact with other proteins that regulate ethylene signaling. Physical interactions with the receptors have been indicated with central signaling components such as CTR1, a Raf-like protein kinase [9–11], and EIN2, a transmembrane protein related to a family of metal transporters [12]. Copper delivery proteins [13], a Golgi-/ER-associated protein named RTE1 [14], and tetratricopeptide repeat proteins [15, 16] have been identified as receptor-interacting proteins that regulate the receptor functionality. While many of these factors have been identified using genetic approaches, biochemical experiments have verified specific interactions with the receptors.

Ethylene perception mechanisms and the signaling pathway described above have been mainly studied in the seedlings of the model plant *Arabidopsis thaliana* [17]. This mechanism is predicted to be highly conserved among higher plants. However, it would not be surprising that diverged regulations are present in different developmental processes or different plant species. Ethylene is intimately involved in fruit ripening. In climacteric fruits such as tomato, apple, and banana, ethylene triggers a rapid progression in ripening processes, including fruit softening, accumulation of sugars and pigments, and production of flavor volatiles [18]. Given that the ethylene-induced fruit ripening is observed in a limited number of species, unique regulations in the signaling might have been acquired through evolutionary processes. Thus, it is important to address the composition of receptor complexes and the regulation of signal transmission from the complexes to the downstream processes that induce the expression of ripening-related genes. Although the receptor and signaling-related genes have been identified in major fruit species [19], biochemical analyses addressing protein interactions among these factors tend to be insufficient compared with those in *A. thaliana*. Thus, it remains unknown how the complexity of these interactions regulates the signal for fruit ripening.

To address this question, we have analyzed receptor complexes in tomato (*Solanum lycopersicum*) fruit. Since tomato has been a model for climacteric fruit ripening, many ethylene-related proteins have been identified. Also, the tomato genome (protein) databases are well established, which is a big advantage for protein identification with database search in MS analysis. Using the IP–IB combination, we showed the interaction between SlETR3 and SlETR4, which are the two major ripening-related ETRs in tomato fruit [20]. Furthermore, the MS analysis following IP with an anti-SlETR4 antibody identified other ETR isoforms and CTR1-like proteins as candidate proteins that interact with SlETR4 (Table 1). Together with the results of the yeast two-hybrid assays, we predicted that different ETR isoforms form higher-order heteromeric clusters in tomato fruit. The interaction rules between these proteins appeared to be partially different from those found in *A. thaliana*. Here, we introduce the basic procedures for the preparation of microsomal fraction from tomato fruit, co-IP, and subsequent MS analysis. Modifications of these protocols may result in the identification of other interacting proteins. For example, the choice of detergents for preparing microsomal fractions would influence the recovery of protein complexes. These protocols would also be applied to the analyses of other fruit species.

2 Materials

Prepare all solutions using ultrapure water (prepared by purifying deionized water to attain a sensitivity of 18 MΩ- cm at 25 °C) and analytical grade reagents. While preparing reagents, wear lab gloves to prevent contamination with skin proteins. All the solutions should be filtered to remove possible contaminants.

2.1 Preparation of Microsomal Fractions

1. Pericarp tissues of tomato fruit stored at −80 °C (*see* **Note 1**).
2. Homogenization buffer: 100 mM Tris–HCl (pH 8.2), 300 mM NaCl, 20 mM ethylene diamine tetra acetic acid (EDTA), 20% (v/v) glycerol.

Table 1
Example of protein identification list of MS analysis

Accession	Description	Gene symbol	Sum PEP score	Sequence coverage [%]	Number of hit peptides	Number of unique peptides
52222394	Ethylene receptor homologue	SlETR4	195.09	32	20	20
1043261366	Dehydrin		193.88	63	19	19
K4BVG7	Uncharacterized protein		92.35	13	10	10
1163081	Ethylene receptor homologue	SlETR1	92.15	30	16	11
52222392	Ethylene receptor never-ripe	SlETR3	78.84	31	16	13
723734218	Predicted: Low-quality protein: Prohibitin-1, mitochondrial-like		69.66	40	8	8
2828252	Cyp1		55.17	14	5	5
1008806506	Actin-7-like		45.19	24	8	1
957571263	Actin-7		45.01	24	8	1
K4ASU4	Uncharacterized protein		38.65	16	7	6
460412932	Predicted: zeaxanthin epoxidase, chloroplastic-like		38.36	23	9	9
5669642	Ethylene-responsive protein kinase TCTR1	SlCTR1	35.06	12	7	7
460375108	Predicted: mitochondrial outer membrane protein porin of 34 kDa		35.01	34	7	5
350537551	UDP-glucose:protein transglucosylase-like protein SlUPTG1		31.41	29	10	3
460382474	Predicted: ATP synthase subunit beta, mitochondrial		30.21	13	5	4
991820329	Alpha-1,4-glucan-protein synthase [UDP-forming] 2		27.51	25	8	1
K4D7X4	Uncharacterized protein		27.13	52	8	8
460401690	Predicted: acetyl-coenzyme A carboxylase carboxyl transferase subunit alpha		25.78	7	4	4
1890116	ADP/ATP translocator		25.30	22	7	7
544163598	ATP synthase CF1 alpha subunit (chloroplast)		24.94	14	6	6

(continued)

Table 1 (continued)

Accession	Description	Gene symbol	Sum PEP score	Sequence coverage [%]	Number of hit peptides	Number of unique peptides
460364985	Predicted: mitochondrial outer membrane protein porin of 36 kDa		22.92	29	7	7
460374479	Predicted: low-temperature-induced cysteine proteinase-like		22.77	4	1	1
460369422	Predicted: uncharacterized protein LOC101249308		20.35	7	4	4
806776648	Glyceraldehyde 3-phosphate dehydrogenase		18.76	18	6	1
K4BKB2	Uncharacterized protein		18.47	4	2	2
2852993	Ethylene receptor homologue	SlETR2	18.43	7	5	1
960513859	Protein disulfide-isomerase-like 2–1 precursor		17.38	17	5	5

Anti-SlETR4 antibody reacted with microsomal proteins from tomato fruit (turning stage) to precipitate SlETR4 and its interacting proteins. Proteins with higher identification scores are shown in the list. The result indicates that SlETR1, SlETR2, SlETR3, and SlCTR1 are candidate proteins that form protein complexes with SlETR4. Many of the other proteins were identified in a negative control experiment (with pre-immune serum), indicating that they would be nonspecific proteins bound to the beads. The sum PEP score means the score calculated based on the posterior error probability (PEP) values of the peptide spectrum matches (PSM). This table is a part of exported data from Proteome Discoverer™ 2.2. This data was obtained from an independent experiment from the previous study [20]

3. 50× solution of cOmplete™, EDTA-free Protease Inhibitor Cocktail (Roche Diagnostics): Dissolve one tablet in 1 mL of ultrapure water to obtain 50× solution (*see* **Note 2**).
4. Crushed ice.
5. Regular centrifugation tubes (50 mL size), ultracentrifugation tubes (the tube size depends on the rotor used), and micro-centrifuge tube (1.5 mL size).
6. Polytron homogenizer.
7. Miracloth (Millipore).
8. Refrigerated high-speed centrifuge.
9. Glass funnel.
10. Refrigerated ultracentrifuge.
11. Solubilization buffer: Add Triton X-100 [final 1% (v/v)] and 50× solution of cOmplete™, EDTA-free Protease Inhibitor Cocktail (final 1× concentration) to homogenization buffer.

12. Shaking apparatus.
13. Refrigerated microcentrifuge.

2.2 Immunoprecipitation and SDS-Polyacrylamide Gel Electrophoresis (PAGE)

1. Specific antibody for immunoprecipitation (*see* **Note 3**).
2. Dynabeads™ Protein A (Thermo Fisher Scientific) (*see* **Note 4**).
3. Dynal MPC-S™ (Thermo Fisher Scientific) or regular magnet.
4. Vortex mixer.
5. Rotating tube mixer.
6. PBS: 10 mM Na_2HPO_4, 1.8 mM KH_2PO_4, 140 mM NaCl, 2.7 mM KCl, pH 7.3.
7. PBS-T: PBS containing 0.1% (v/v) Tween 20.
8. Conjugation buffer: 200 mM triethanolamine–HCl (pH 8.2).
9. BS^3 solution: Dissolve 3 mg of bis(sulfosuccinimidyl)suberate (BS^3) in 1 mL conjugation buffer. This solution must be freshly prepared immediately before use.
10. Quenching buffer: 200 mM monoethanolamine–HCl (pH 8.2).
11. SDS sample buffer: 100 mM Tris–HCl (pH 6.8), 2.5% (w/v) sodium dodecyl sulfate (SDS), 25% (v/v) glycerol, 0.002% (w/v) bromophenol blue (*see* **Note 5**).
12. 1 M dithiothreitol (DTT). This solution can be stored at −20 ° C.
13. 50% (w/v) acrylamide solution. This solution can be stored at −20 °C.
14. Conventional polyacrylamide mini gel [stacking gel, 5.5% (w/v) acrylamide, 125 mM Tris–HCl (pH 6.8), 0.1% (w/v) SDS; separating gel, 10% (w/v) acrylamide, 375 mM Tris–HCl (pH 8.8), 0.1% (w/v) SDS] (*see* **Note 6**). 30% (w/v) acrylamide solution is used [acrylamide/bis-acrylamide mix (37.5:1)].
15. Conventional apparatus for SDS-PAGE with a power supply.
16. SDS-PAGE running buffer: 25 mM Tris base, 192 mM glycine, 0.1% (w/v) SDS.
17. Protein molecular weight marker (*see* **Note 7**).
18. MS-compatible silver staining kit.

2.3 Sample Preparation for MS Analysis

1. Single-edge razor blade and tweezers.
2. Light panel (or white paper).
3. Protein-low-binding microcentrifuge tubes (1.5 mL size) (*see* **Note 8**).
4. Ultrapure water.

5. Microtube mixer.
6. De-staining solution: 15 mM potassium ferricyanide and 50 mM sodium thiosulfate. Prepare 2× stock solution for each and mix equal volumes immediately before use.
7. Dehydration solution: 25 mM ammonium bicarbonate and 50% (v/v) acetonitrile.
8. Centrifugal evaporator.
9. Trypsin solution [10 μg/mL, Trypsin Gold, mass spectrometry-grade (Promega)]: Add 50 mM acetic acid to lyophilized powder to make a stock solution (100 μg/mL). Dispense small volumes in microtubes and keep at −80 °C. For trypsin digestion, dilute the stock with 50 mM ammonium bicarbonate to obtain 10 μg/mL solution.
10. Peptide extraction solution: 5% (v/v) trifluoroacetic acid (TFA), 50% (v/v) acetonitrile.
11. 0.1% (v/v) formic acid.
12. Microcentrifuge.
13. Analytical vials adapted for an autosampler.

3 Methods

For all the procedures, wear lab gloves to prevent contamination with skin proteins.

3.1 Preparation of Microsomal Fraction from Tomato Fruit

During the preparation of microsomal fractions, the samples should be kept at a low temperature (on ice or at 4 °C) to prevent potential protein degradation even in the presence of protease inhibitors.

1. Dispense 20 mL of homogenization buffer to 50 mL-centrifugation tubes (*see* **Note 9**). Add 400 μL of 50× solution of cOmplete™ to each tube.
2. Add 10 g of frozen tomato tissues to each tube and homogenize the tissues using a Polytron homogenizer (at 10,000–15,000 rpm). Once the chunks of tissues disappear, keep the tube in crushed ice and continue homogenization for one more minute (*see* **Note 10**).
3. Centrifuge at 27,000 × *g* for 15 min at 4 °C (high-speed centrifuge) and filter the resultant supernatant through two layers of Miracloth set in a glass funnel.
4. Transfer the filtrate to ultracentrifugation tubes and centrifuge at 100,000 × *g* for 1 h at 4 °C (ultracentrifuge). Discard the supernatant.
5. Resuspend the pellet in 1 mL of solubilization buffer (*see* **Note 11**).

6. Incubate the tube on ice for 1 h with gentle agitation using a shaker to solubilize membrane-associated proteins.
7. Transfer the samples to microcentrifuge tubes and centrifuge at 13,000 rpm for 5 min at 4 °C (microcentrifuge).
8. Transfer the supernatant to new microcentrifuge tubes. Keep on ice until use (*see* **Note 12**).

3.2 Preparation of Antibody-Bound Beads

Dynabeads™ Protein A used in this protocol are magnetic beads covalently coupled with Protein A. Since antibodies (immunoglobulin G, IgG) strongly bind to Protein A, IgG contained in ascites, serum, or tissue culture supernatants can be efficiently purified by the beads. Once a tube containing the beads is placed on a magnet, the beads migrate to the side of the tube facing the magnet, which allows for easy removal of the supernatant.

1. Transfer the optimal amount of Dynabeads™ Protein A suspension to a microcentrifuge tube (*see* **Note 13**).
2. Place the tube on Dynal MPC-S™ to collect the beads and remove the supernatant (*see* **Note 14**).
3. Resuspend in 1 mL of PBS-T and briefly vortex.
4. Repeat **steps 2–3**.
5. Remove the supernatant and add 0.5 mL of PBS-T and an optimal amount of antibody (*see* **Note 15**).
6. For antibody binding to the beads, incubate the mixture for 1 h at 20–25 °C with rotation on a rotating tube mixer to prevent beads' sedimentation.
7. Wash the beads with 1 mL of PBS-T three times (as in **steps 2–3**).
8. Remove the supernatant and resuspend in 1 mL of conjugation buffer and briefly vortex.
9. Repeat **step 8**.
10. Remove the supernatant and resuspend in 1 mL of BS^3 solution. Incubate for 45 min at 20–25 °C with rotation to cross-link the antibodies and the beads (*see* **Note 16**).
11. Remove the supernatant and resuspend in 1 mL of quenching buffer and briefly vortex.
12. Repeat **step 11** twice.
13. Incubate the mixture for 1 h at room temperature with rotation to quench cross-linking.
14. Remove the supernatant and wash the beads with 1 mL of PBS three times. Resuspend the beads in 1 mL of PBS. It is now ready to use for immunoprecipitation (*see* **Note 17**).

3.3 Immunoprecipitation and Gel Electrophoresis

In this method, antibodies are attached to Protein A-beads in advance (direct IP). If the free antibody first reacts with the target antigen (indirect IP), the antibody cannot be covalently immobilized to the beads by cross-linking. It is important to perform a negative control of the experiment. Since MS analysis is highly sensitive, many proteins that bind nonspecifically to magnetic beads are expected to be detected. Thus, an experiment with pre-immune serum is recommended to be performed simultaneously to compare the identified proteins. Alternatively, a mutant or transgenic suppression line for an interest gene that lacks the corresponding protein is useful as a negative control.

1. Remove the supernatant from the antibody-bound beads using Dynal MPC-S™ (from Subheading 3.2).
2. Resuspend the beads in 1 mL of solubilization buffer (without protease inhibitors) and briefly vortex.
3. Repeat **steps 1–2**.
4. Remove the supernatant and add the microsomal fraction (from Subheading 3.1) to the beads.
5. Incubate the mixture with rotation overnight at 4 °C (*see* **Note 18**).
6. Wash the beads five times with 1 mL of solubilization buffer (with protease inhibitors) as in **steps 1–2**.
7. Completely remove the supernatant. Add 50 μL of SDS sample buffer containing 100 mM DTT and briefly vortex. The antigen proteins (and interacting proteins) are eluted at this point. Then, incubate samples for 1 h at 37 °C (*see* **Note 19**).
8. Spin down the samples briefly and set the tubes on Dynal MPC-S™. Transfer the supernatant to new tubes.
9. Add 5 μL of 50% (v/v) acrylamide solution for Cys alkylation and incubate for 30 min at 20–25 °C (*see* **Note 20**).
10. Apply the samples to a polyacrylamide gel and run electrophoresis (*see* **Note 21**). When the front line of bromophenol blue reaches the edge of the gel, stop electrophoresis.
11. Visualize protein bands using a silver staining kit compatible with MS, according to the manufacturer's instructions. Take an image of the stained gel.

3.4 In-gel Digestion and MS Analysis

There are several possible combinations of peptide preparation methods and MS instruments. In this subheading, we present an example of the basic scheme. Alternatively, at this point, the gel bands may be asked to be analyzed by companies.

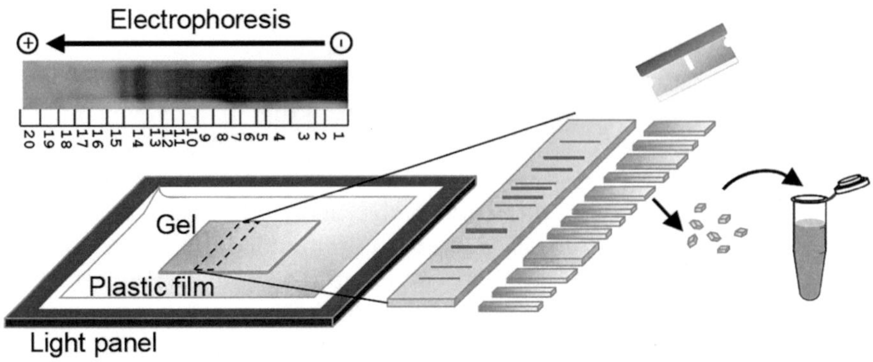

Fig. 2 Gel band preparation for in-gel digestion. Make a design for how to excise gel bands, as shown in the upper gel image. The sizes of pieces need not be equal. Peptides of minor proteins may be masked to be detected in MS analysis if those of major proteins coexist. Put a gel on a clean plastic film with a white background. Excise gel bands into small pieces using a razor blade and tweezers

1. Make a design for excising gel bands using a gel image (Fig. 2).
2. Prepare low-protein-binding microcentrifuge tubes and write numbers on the tubes for each gel band. Then, dispense 1 mL of ultrapure water to each tube.
3. Place the silver-stained gel on a transparent plastic sheet with white background (*see* **Note 22**). As planned in **step 1**, prepare gel bands using a razor blade and tweezers (Fig. 2, *see* **Note 23**). Excise each gel band into small pieces (approximately 1–2 mm dices). Transfer the gel pieces to the prepared tubes using tweezers.
4. Discard water using a micropipette.
5. Add 500 μL of de-staining solution and shake for 10 min at room temperature using a microtube mixer (1000–1300 rpm).
6. Discard the liquid and add 0.5 mL of ultrapure water. Shake for 15 min at room temperature using a microtube mixer (1000–1300 rpm).
7. Repeat **step 6** until the yellow color is diminished (normally three times in total).
8. Discard the liquid and add 200 μL of dehydration solution. Shake for 10 min at 20–25 °C using a microtube mixer (500–650 rpm).
9. Repeat **step 8**.
10. Discard the liquid. Dry gel pieces using a centrifugal evaporator until they become completely white (approximately 30 min).
11. Add 30–60 μL of trypsin solution to the gel pieces in each tube. Keep the tubes on ice until the gel pieces absorb the solution (*see* **Note 24**). Remove the excess trypsin solution.

12. Incubate them for 12–16 h at 37 °C.
13. Add 50 μL of peptide extraction solution to the gels. Shake for 30 min at room temperature using a microtube mixer (1000–1300 rpm).
14. Transfer the liquid to new low-protein-binding microcentrifuge tubes.
15. Repeat **step 13**.
16. Combine the liquid with the previous extract.
17. Completely dry the extracts using a centrifugal evaporator. It may take a few hours.
18. Add 20 μL of 0.1% (v/v) formic acid and vortex vigorously. Centrifuge at 13,000 rpm for 10 min.
19. Transfer 15 μL of the supernatant to analytical vials adapted for an autosampler. Store the samples at −20 °C until MS analysis.
20. Perform MS analysis (*see* **Note 25**). There are various instruments for liquid chromatography and MS systems. Since it is difficult to cover all of them here, our condition is shown below as an example. Adapt conditions for any specific instrument.
    - Instrument: Q Exactive mass spectrometer (Thermo Fisher Scientific) equipped with captive spray ionization and interfaced online using a high-performance liquid chromatography system [Advance LC (Bruker)].
    - Sample injection volume: 1 μL.
    - Trap column: L-column ODS, 0.3 × 5 mm, 5 μm (CERI Tokyo, Japan).
    - Analytical column: Zaplous column αPep-C18, 0.2 mM × 50 mM, 3 μm (AMR Tokyo, Japan).
    - Gradient elution: Elute peptides with a gradient mixture of 0.1% (v/v) formic acid (Sol. A) and 100% (v/v) acetonitrile (Sol. B) as follows: 5–65% Sol. B for 20 min, 65–95% Sol. B for 1 min, and 95% Sol. B for 3 min. Flow rate, 1500 nL/min.
    - Collected spectra: *m/z* range of 350–2000.
    - Analytical software: Proteome Discoverer™ (Thermo Fisher Scientific).
    - Database: *Solanum lycopersicum*.

4 Notes

1. Cut fresh tomato pericarp tissues into small pieces (0.5–1 cm dices) and freeze in liquid nitrogen. Then, store the samples at −80 °C.

2. The 50× protease inhibitor solution can be stored at −20 °C. Protease inhibitors provided by other suppliers could also be used.
3. Some antibodies may require antigen purification in advance. It should also be noted that some antibodies are compatible with immunoblotting but not immunoprecipitation. The antibody must recognize the surface of the protein.
4. Magnetic beads are convenient for the removal of supernatant. Magnetic beads provided by other suppliers can also be used. Protein G-conjugated beads can be used instead of Protein A-conjugated beads. The binding capacities of antibodies depend on the immune animals and immunoglobulin subclasses.
5. Since the SDS sample buffer is often contaminated with keratin from the human skin, the buffer should be carefully prepared and divided into small portions for storage.
6. The concentration of the acrylamide gels could vary depending on the target proteins. All the aqueous reagents should be filtered to remove any possible contamination. The glass plates should be completely cleaned in advance. Alternatively, precast gels provided by suppliers can be used.
7. Since silver staining is highly sensitive for protein detection, molecular weight markers may need to be diluted. When applying the marker on a gel, it is recommended to keep a distance from the experimental samples (keep one or two blank lanes) to prevent the detection of marker proteins in MS analysis.
8. Regular microcentrifuge tubes may absorb small amounts of proteins and peptides. The use of low-protein binding tubes is strongly recommended, especially during the peptide extraction step.
9. The total experimental volume may vary depending on the research goals. In our case, 60 g of tissues is generally processed in one experimental unit (six tubes, 10 g of fruit pericarp tissue, and 20 mL of buffer per tube).
10. The homogenization buffer does not need to be chilled in advance. Mixing the frozen tissues results in an appropriate cold temperature. However, for the last 1 min of homogenization, it is recommended to maintain a low temperature with crushed ice to prevent a temperature increase due to frictional heat. Through the following procedures, samples should be kept at 4 °C or on ice to prevent potential proteolysis.
11. For the efficient resuspension, pass the samples through a micropipette tip repeatedly. The use of different detergents or concentrations may result in the recovery of different protein complexes. We have constantly used 1% (v/v) Triton X-100

(nonionic detergent) to solubilize membrane proteins. Use of other nonionic detergents (i.e., Nonidet P-40, *n*-octyl-β-D-glucoside) or zwitterionic detergents (i.e., CHAPS) may influence the solubilization efficiency and maintenance of protein complexes, resulting in the identification of other interacting proteins. It is important that ionic detergents [i.e., sodium dodecyl sulfate (SDS) and cetyltrimethylammonium bromide (CTAB)] would not be suitable for this purpose because they generally disrupt protein interactions.

12. It is recommended that immunoprecipitation is performed quickly to avoid the dissociation of protein complexes and protein degradation.
13. The amount of beads suspension varies depending on the experimental scale and antibody properties. A certain amount for multiple batches may be prepared once because antibody-bound beads can be kept for at least 2 weeks at 4 °C in our condition. We use 125 μL of the bead suspension (original volume) for 60 g of tissues.
14. A regular magnet can be used for bead collection; however, Dynal MPC-S™ is far more efficient in a series of procedures.
15. We normally use 30 μL of serum for 125 μL of the bead suspension. Preliminary experiments are required to determine the optimal bead/antibody ratio.
16. It is strongly recommended to use a cross-linking reagent. If not cross-linked, antibodies are eluted with the SDS sample buffer, which results in the detection of IgG proteins by MS analysis. Although BS^3 is a common cross-linker, there are various cross-linking reagents with different chemical properties, which may influence the recovery of protein complexes.
17. Antibody-bound beads can be stored for at least 2 weeks at 4 ° C in our condition.
18. The incubation time and temperature vary depending on the antibodies used and the properties of the target protein complex. In our lab, overnight incubation at 4 °C is adopted. In some cases, a shorter incubation period at room temperature may be better for capturing target antigens.
19. Avoid heating at the boiling temperature. Membrane proteins tend to be aggregated by heating at higher temperatures.
20. Although alkylation of Cys residues could be performed later in the gels before trypsin digestion, we do alkylation at this point. In addition, other alkylation reagents, such as iodoacetamide, may be used. However, Cys residues in proteins may be partially decorated with unpolymerized acrylamide (propionamide) during gel electrophoresis. Therefore, we use acrylamide to ensure consistent Cys alkylation.

21. All the volume (50 μL) from one tube may be applied to multiple lanes. Alternatively, a part of the sample can be stored at −20 °C. For example, 20 μL out of 50 μL is used, and the rest is stored. Normally, we run with constant currents (20 mA for the stacking gel and 30 mA for the separation gel for one gel); however, a constant voltage is also fine.
22. To clearly visualize the gel bands, white paper or a light panel should be placed under a transparent plastic film.
23. Tools to prepare gel bands need to be cleaned well.
24. When gel pieces absorb the trypsin solution enough, they would become transparent. If not enough, they will partially remain white.
25. Since MS analysis is in general sensitive, a plenty of nonspecific proteins are expected to be identified. Nonspecific ones tend to be originally abundant in the prepared protein samples, and part of them would be recovered through nonspecific binding to the beads. Thus, it is important to perform a negative control experiment simultaneously (i.e., with pre-immune serum). Also, to reduce the amounts of nonspecific proteins, the washing buffer constituent (**step 6** in Subheading 3.3) may need to be considered. Repeated experiments are needed to get the reliability of identified candidate proteins. Even if the identification score is low for a certain protein in MS analysis, it might be a genuine one as long as it is detected in repeated experiments. Some interacting proteins might exist at low levels or exhibit weak interactions with an antigen. In that case, it might be worth validating with other interaction assays.

Acknowledgments

This work was supported by the Nihon University College of Bioresource Sciences Research Grant for 2020 and 2021 (To Y.K.).

References

1. Chang C, Kwok SF, Bleecker AB et al (1993) Arabidopsis ethylene-response gene ETR1: similarity of product to two-component regulators. Science 262:539–544
2. Schaller GE, Ladd AN, Lanahan MB et al (1995) The ethylene response mediator ETR1 from Arabidopsis forms a disulfide-linked dimer. J Biol Chem 270:12526–12530
3. Chen YF, Randlett MD, Findell JL et al (2002) Localization of the ethylene receptor ETR1 to the endoplasmic reticulum of Arabidopsis. J Biol Chem 277:19861–19866
4. Hall AE, Findell JL, Schaller GE et al (2000) Ethylene perception by the ERS1 protein in Arabidopsis. Plant Physiol 123:1449–1457
5. Gao Z, Wen CK, Binder BM et al (2008) Heteromeric interactions among ethylene receptors mediate signaling in Arabidopsis. J Biol Chem 283:23801–23810
6. Grefen C, Staedele K, Ruzicka K et al (2008) Subcellular localization and in vivo interactions of the Arabidopsis thaliana ethylene receptor family members. Mol Plant 1:308–320

7. Chen YF, Gao Z, Kerris RJ III et al (2010) Ethylene receptors function as components of high-molecular-mass protein complexes in Arabidopsis. PLoS One 5:e8640
8. Berleth M, Berleth N, Minges A et al (2019) Molecular analysis of protein-protein interactions in the ethylene pathway in the different ethylene receptor subfamilies. Front Plant Sci 10:726
9. Clark KL, Larsen PB, Wang X et al (1998) Association of the Arabidopsis CTR1 Raf-like kinase with the ETR1 and ERS ethylene receptors. Proc Natl Acad Sci USA 95:5401–5406
10. Huang YF, Li H, Hutchison CE et al (2003) Biochemical and functional analysis of CTR1, a protein kinase that negatively regulates ethylene signaling in Arabidopsis. Plant J 33:221–233
11. Zhong S, Lin Z, Grierson D (2008) Tomato ethylene receptor-CTR interactions: visualization of NEVER-RIPE interactions with multiple CTRs at the endoplasmic reticulum. J Exp Bot 59:965–972
12. Bisson MMA, Bleckmann A, Allekotte S et al (2009) EIN2, the central regulator of ethylene signalling, is localized at the ER membrane where it interacts with the ethylene receptor ETR1. Biochem J 424:1–6
13. Hoppen C, Müller L, Hänsch S et al (2019) Soluble and membrane bound protein carrier mediate direct copper transport to the ethylene receptor family. Sci Rep 9:10715
14. Dong H, Jang M, Scharein B et al (2010) Molecular association of the Arabidopsis ETR1 ethylene receptor and a regulator of ethylene signaling, RTE1. J Biol Chem 285: 40706–40713
15. Lin Z, Arciga-Reyes L, Zhong S et al (2008) SlTPR1, a tomato tetratricopeptide repeat protein, interacts with the ethylene receptors NR and LeETR1, modulating ethylene and auxin responses and development. J Exp Bot 59: 4271–4287
16. Lin Z, Ho CW, Grierson D (2009) AtTRP1 encodes a novel TPR protein that interacts with the ethylene receptor ERS1 and modulates development in Arabidopsis. J Exp Bot 60: 3697–3714
17. Binder BM (2020) Ethylene signaling in plants. J Biol Chem 295:7710–7725
18. Klee HJ, Giovannoni JJ (2011) Genetics and control of tomato fruit ripening and quality attributes. Annu Rev Gen 45:41–59
19. Chen Y, Grimplet J, David K et al (2018) Ethylene receptors and related proteins in climacteric and non-climacteric fruits. Plant Sci 276: 63–72
20. Kamiyoshihara Y, Achiha Y, Ishikawa S et al (2022) Heteromeric interactions of ripening-related ethylene receptors in tomato fruit. J Exp Bot 73:6773–6783

Chapter 11

Fundamental Techniques to Evaluate the Roles of Ethylene Receptors in Seed Germination Under Salt Stress in *Arabidopsis thaliana*

Arkadipta Bakshi

Abstract

Seed germination plays a critical role in the reproduction, growth, and survival of plants, enabling them to emerge and establish themselves in their environment. During the process of seed germination, plants experience several biotic and abiotic stresses, affecting their growth and development and leading to morphological, physiological, biochemical, and molecular changes. Therefore, studying seed germination is important to understand plant biology and genetics. This chapter elucidates the basic techniques employed to investigate the contrasting roles of two ethylene receptors, ETHYLENE RECEPTOR 1 and 2 (ETR1 and ETR2), in seed germination under salt stress in *Arabidopsis thaliana*. We begin with an overview of the ethylene receptor family in *Arabidopsis*, focusing on the overlapping and nonoverlapping roles of ETR1 and ETR2 in regulating physiological functions such as seed germination. The Subheading Methods details the experimental setup, starting with sterilizing the seeds, preparing the plates, and assessing the germination under both no salt and salt conditions, to understanding changes at the molecular level in the seedlings by isolating RNA and assessing the expression of salt-responsive genes by qPCR analysis. The detailed protocol offers strategies for researchers investigating the role of other hormone signaling pathways in seed germination under biotic and abiotic stresses.

Key words *Arabidopsis thaliana*, Seed germination, Abscisic acid, Ethylene receptors, ETR1, ETR2, Salt stress, RNA isolation, qPCR, Gene expression

1 Introduction

Ethylene is a simple, unsaturated, gaseous plant hormone that intricately regulates numerous developmental and physiological processes in higher plants, such as seed germination, senescence, abscission, fruit ripening, and response to various stressors [1]. The biosynthesis of ethylene in plants is tightly regulated, and the production is initiated by a diverse set of stimuli, such as attack by pathogens and various stress factors like high salinity, submergence, wounding, and mechanical stress. Due to its signaling activities and interactions with hormones like auxin, cytokinin, and ethylene

Devesh Shukla (ed.), *Ethylene Sensing and Signaling*, Methods in Molecular Biology, vol. 2945,
https://doi.org/10.1007/978-1-0716-4650-2_11,

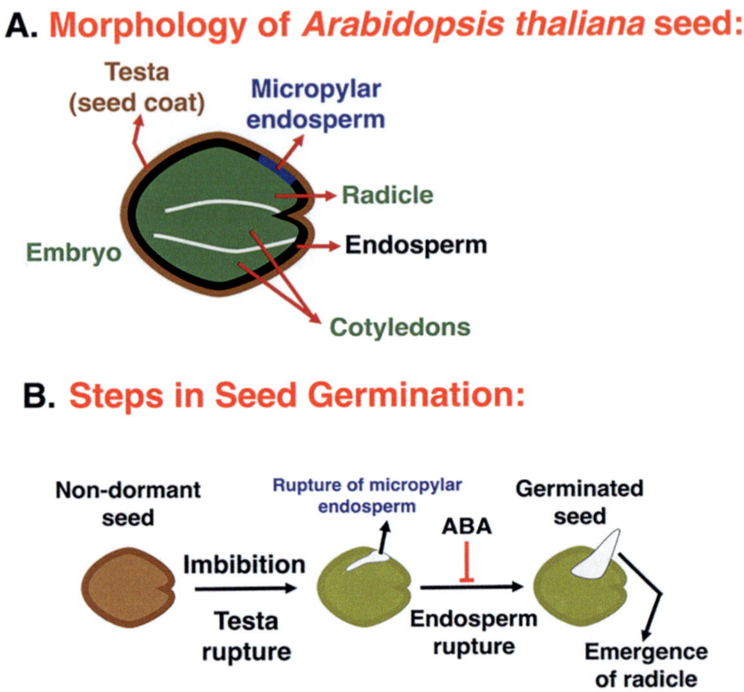

Fig. 1 Working model for seed germination in *Arabidopsis thaliana*. (**a**) Morphology of a mature seed of *Arabidopsis*. (**b**) In *Arabidopsis*, germination follows a two-step process involving the rupture of the seed coat (testa) upon imbibition, followed by the endosperm rupture. Abscisic acid (ABA) inhibits the rupture of the endosperm but does not affect the seed coat rupture in after-ripened seeds. Modified from 22.

itself, it influences various stages of the plant developmental cycle [2] (Figs. 1 and 2).

In *Arabidopsis*, responses to ethylene are mediated by a family of endoplasmic reticulum–localized ethylene receptors [3]. All these receptors bind ethylene with high affinity. In *Arabidopsis*, the family of ethylene receptors has been classified into two subfamilies based on phylogenetic analysis and structural similarities: Subfamily 1 consists of ETHYLENE RECEPTOR1 (ETR1) and ETHYLENE RESPONSE SENSOR1 (ERS1), and subfamily 2 consists of ETR2, ERS2, and ETHYLENE INSENSITIVE4 (EIN4) [4–7]. All of them are predicted to consist of a conserved transmembrane domain, also called the ethylene-binding domain (EBD), which also includes a copper cofactor at the N terminus. In addition, subfamily 2 contains an additional transmembrane signal sequence at the N terminus. The EBD is followed by a GAF domain allowing for protein–protein interactions and a kinase domain [8]. Subfamily 1 shows histidine kinase activity in vitro, whereas subfamily 2 contains degenerate histidine kinase domains and shows serine/threonine kinase activity in vitro. Out of the five

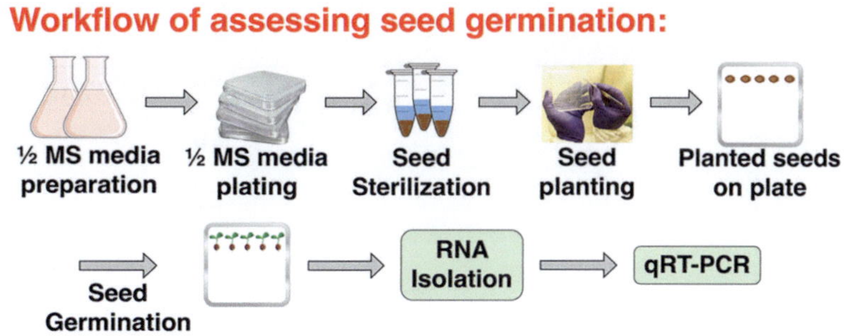

Fig. 2 Schematic representation of the workflow of assessing seed germination in *Arabidopsis*. The process begins with the preparation of growth media, followed by pouring the media onto plates. Next, seeds are sterilized and then planted onto the prepared plates. Assays are conducted to monitor the rate of seed germination. Subsequently, RNA is isolated from the germinated seeds. Finally, gene expression analysis is performed using quantitative PCR (qPCR) to evaluate the transcript levels of relevant genes

receptor isoforms, ERS1 has been shown to have both histidine and serine/threonine kinase activities in vitro. Three of the five-receptor isoforms (ETR1, ETR2, and EIN4) also contain a receiver domain, which contains the conserved aspartic acid residue required for the phosphorelay function [9, 10]. Previous reports show that the five-receptor isoforms in *Arabidopsis* have become subfunctionalized, showing both overlapping and nonoverlapping roles in regulating various physiological processes such as growth in air, growth recovery after the removal of ethylene, ethylene-stimulated nutational bending, and germination under salt stress [11–18]. The receiver domain containing receptors ETR1 and ETR2 has a contrasting role in seed germination under salt stress involving the phytohormone abscisic acid (ABA). Loss-of-function *etr1-6* mutants germinate better, and *etr2-3* loss-of-function mutants germinate worse than wild-type seedlings under salt stress (Fig. 3a, b) [17–20]. The results from these studies highlight the intricate balance between ethylene signaling and other hormonal pathways, such as ABA, in regulating seed germination under different abiotic stress conditions (Fig. 3c). These findings contribute to a deeper understanding of plant resilience mechanisms, which can be beneficial for improving crop tolerance toward salinity stress.

To delve deeper into these complex interactions, it is essential to understand the physiological and molecular mechanisms underlying seed germination, which is a critical developmental stage in plants that ensures their survival in the environment but also a highly delicate one. The model plant *Arabidopsis thaliana* has been extensively used to understand the physiological and molecular intricacies of seed germination in angiosperms [21]. *Arabidopsis* seeds consist of an embryo with two cotyledons (embryonic leaves)

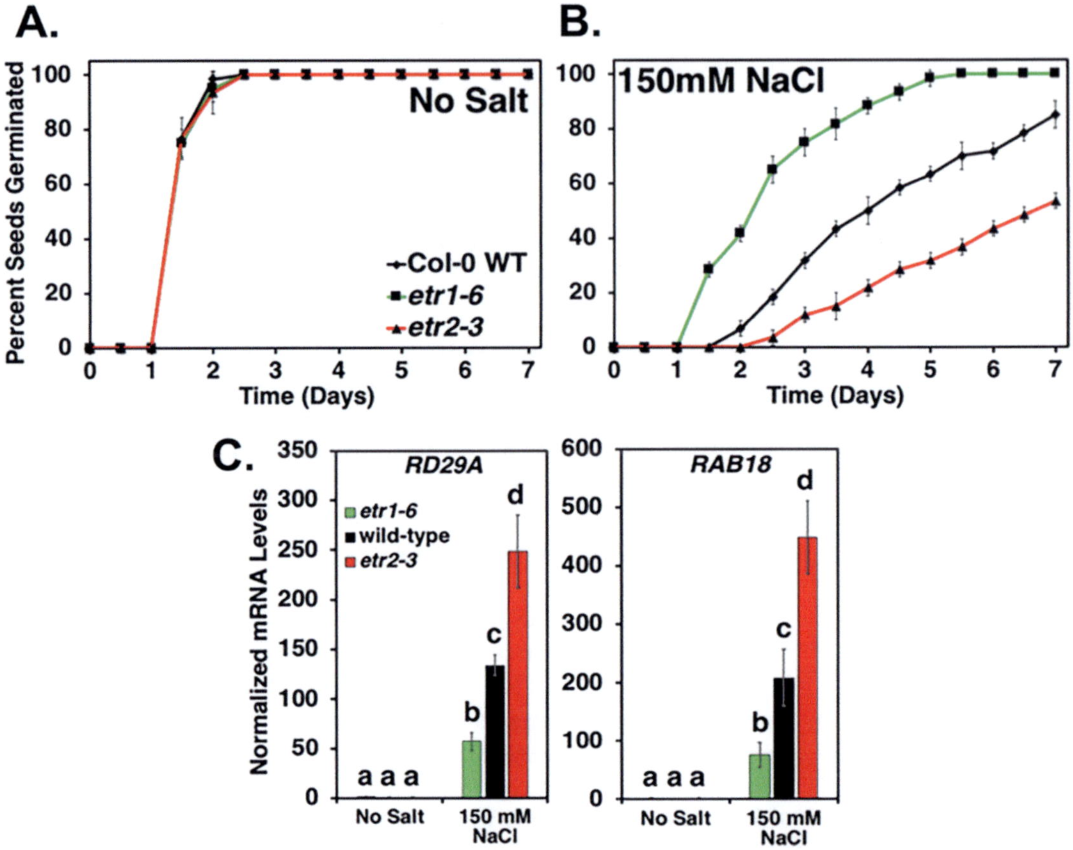

Fig. 3 Effect of salt stress on the germination of *etr1-6*, wild-type, and *etr2-3* seeds. (**a, b**) Germination time courses for wild-type, *etr1-6*, and *etr2-3* single mutants were examined both in the absence and in the presence of 150 mM NaCl. Seed germination percentages were recorded every 12 h. All germination experiments were conducted in triplicate. The average percentage of seed germination ± standard deviation (SD) at each time point is plotted for each seed line. (**c**) The transcript levels of *RD29A* and *RAB18* were analyzed with quantitative real-time RT-PCR. Data were normalized to the levels of *At3g12210* in each seed line to calculate the relative transcript levels for each gene and further normalized to the levels in untreated wild-type seeds. Data represent mean ± SE for three to four independent biological replicates, with three technical replicates. In each panel, bars sharing the same letters are not significantly different (one-way ANOVA and Tukey's multiple comparisons test, $P < 0.05$). *RESPONSIVE TO DESICCATION 29A (RD29A); RESPONSIVE TO ABA18 (RAB18)*

and a radicle (embryonic root), which emerges through the micropylar endosperm. The mature embryo is encased in a single layer of endosperm and an outer layer of the testa or seed coat (Fig. 1a). Germination starts with imbibition or water uptake, followed by the rupture of the seed coat, leading to the emergence of a radicle through the micropylar endosperm (Fig. 1b). Plant hormones are crucial in regulating seed germination and dormancy, thereby controlling the growth and development of plants as well as their responses to different environmental stimuli. Although ABA does not affect testa rupture, it is essential for inhibiting the rupture of endosperm and, thereby, promoting seed dormancy (Fig. 1b). In contrast, plant hormones, ethylene and gibberellic acid (GA), negate the inhibitory effects of ABA on seed germination [22].

In this chapter, we describe the detailed steps to analyze seed germination in *Arabidopsis* as part of experiments aimed at understanding the opposing roles of the two ethylene receptors, ETR1 and ETR2, in seed germination under salt stress. These studies use the respective single mutants of ETR1 and ETR2 (*etr1-6* and *etr2-3* loss-of-function mutants) and wild-type controls. We have induced salt stress using sodium chloride (NaCl) at 150 mM concentration to stimulate environmental stress conditions [18, 20]. To quantify seed germination rates, we have employed seed germination assays under controlled laboratory growth conditions (Fig. 2).

To adapt to stresses like salt stress, plants upregulate or downregulate the expression of numerous stress-responsive genes at the mRNA level. These changes in transcription trigger a cascade of downstream changes within the plants [23]. In order to investigate the changes at the molecular level, we have included the steps to isolate RNA from seeds germinating under salt stress and the subsequent use of quantitative real-time polymerase chain reaction (qRT-PCR) to measure the expression levels of stress-responsive genes (Fig. 2) such as *RESPONSIVE TO ABA18* (*RAB18*) encoding a dehydrin protein and *RESPONSIVE TO DESICCATION 29A* (*RD29A*) encoding a leucine zipper transcription factor in seeds (Fig. 3c). Both these genes have been previously shown to accumulate in plants in response to salt stress [24, 25]. Although the methods described here have been effectively utilized in *Arabidopsis*, these protocols can be adapted directly or with modifications for other plant species. With appropriate controls, these protocols can also be employed to uncover the crosstalk between different hormone-signaling pathways affecting seed germination under different environmental stress conditions.

2 Materials

2.1 Surface Sterilization of Arabidopsis Seeds

1. Sterile double-distilled water (ddH_2O).
2. Ethanol: 70% (v/v) solution in ddH_2O.
3. Bleach: 50% (v/v) solution in ddH_2O.
4. Biosafety cabinet or laminar flow hood.
5. Microcentrifuge tubes: 2 mL.
6. Autoclaved filter paper: ~5 cm diameter.
7. Sterile tips: 1000 μL pipette tips.
8. Single-channel pipette (range: 200–1000 μL).

9. Beaker: 250 mL.
10. Timer.
11. Clear square Petri dish: 10 cm by 10 cm.
12. Micropore surgical tape (3 M).

2.2 Preparation of ½ Strength Murashige and Skoog Basal Medium Plates With and Without NaCl

1. Murashige and Skoog basal medium (MS medium) with Gamborg's vitamins (*see* **Note 1**) [26].
2. Sodium chloride (NaCl): 150 mM.
3. Phytagel.
4. Sterile double-distilled water (ddH_2O).
5. Potassium hydroxide (KOH): 0.1 M.
6. Glass conical flask: 2 L.
7. Glass beaker: 1 L.
8. Graduated measuring cylinder: 1 L.
9. Magnetic stir bar.
10. Magnetic plate.
11. Spatula.
12. Weighing boats.
13. Weighing scale machine.
14. pH meter.
15. Aluminum foil.
16. Autoclave tape.
17. Autoclave tray.
18. Autoclave.
19. Pipette gun.
20. Glass pipette: 50 mL.
21. Clear square Petri dish: 10 cm by 10 cm.
22. Biosafety cabinet or laminar flow hood.

2.3 Plating and Assessing of Seed Germination Under No Salt and 150 mM NaCl Salt Stress Conditions

1. Sterile double-distilled water (ddH_2O).
2. Ethanol: 70% (v/v) solution in ddH_2O.
3. Biosafety cabinet or laminar flow hood.
4. Surface-sterilized *Arabidopsis* seeds: Columbia-0 wild-type; *etr1-6* single mutant; *etr2-3* single mutant.
5. Filter paper.
6. Tweezer.
7. ½ MS strength plates with and without 150 mM NaCl.
8. Micropore surgical tape (3 M).
9. Aluminum foil.

10. Plate racks.
11. Growth chamber.
12. Magnifying glass.
13. Camera.

2.4 Molecular Markers to Assess Germination Under Salt Stress

1. Extraction buffer: 100 mM Tris–HCl (pH 9.5), 150 mM NaCl, 1% Sarkosyl, 5 mM DTT, TRIzol reagent.
2. Chloroform.
3. Acid phenol.
4. Sodium acetate (pH 5.2): 3 M solution in ddH_2O.
5. Isopropanol: 100%.
6. Ethanol: 75% (v/v) solution in ddH_2O.
7. Spectrum plant total RNA kit: STRN10/STRN50/STRN250, Sigma Aldrich.
   (a) Lysis solution: Sodium dodecyl sulfate salt and sodium hydroxide (Catalog #L1912, Sigma Aldrich).
   (b) β-mercaptoethanol (Catalog #M3148, Sigma Aldrich).
   (c) Binding solution (Catalog #B9560, Sigma Aldrich).
   (d) Wash solution 1 (Catalog #W0263, Sigma Aldrich).
   (e) Wash solution 2 (Catalog #W4639, Sigma Aldrich).
   (f) Binding column (Catalog #C6991, Sigma Aldrich).
   (g) Collection tubes: 2 mL (Catalog #T5449, Sigma Aldrich).
8. Liquid nitrogen.
9. Nuclease-free double-distilled water.
10. DNase 1 Amp Grade (Catalog #AMPD1, Sigma Aldrich).
   (a) 10× DNase buffer (Catalog #R6273, Sigma Aldrich).
   (b) DNase 1 reaction buffer (Catalog #D5307, Sigma Aldrich).
11. Luna Universal One-Step RT-qPCR Kit (Catalog #E3005, New England Biolabs).
12. Beaker: 250 mL.
13. Clear square Petri dish.
14. Weighing scale.
15. Metal beads: 2.4 mm (Catalog # 10032–370, Omni International, VWR).
16. Microcentrifuge tubes: 1.5 mL.
17. Microcentrifuge tubes: 2 mL.
18. Mortar and pestle.
19. Single-channel pipette (range: 200–1000 μL) and sterile tips.

20. Single-channel pipette (range: 20–200 μL) and sterile tips.
21. Single-channel pipette (range: 2–20 μL) and sterile tips.
22. Centrifuge.
23. Heat block.
24. Nanodrop machine.
25. Real-time qPCR instrument.

3 Methods

3.1 Surface Sterilization of Arabidopsis Seeds

While performing a germination experiment, it is crucial to use seeds that are age-matched (seeds that are bulked from plants grown together under long-day conditions and harvested on the same day) and size-matched (seeds ranging between 250 and 300 μm in size) to avoid discrepancies in the rate of germination. After harvesting, the seeds should be stored in tubes inside a desiccator for at least 2–3 weeks at room temperature to allow after-ripening and then mechanically sorted by size using sieves [27].

1. Take your seeds in a 2 mL microcentrifuge tube (100–200 *Arabidopsis* seeds per tube) in a sterile biosafety cabinet or laminar flow hood.
2. Add 1.5 mL 70% (v/v) ethanol to the tube and suspend the seeds in ethanol by shaking the tube for 2 min (*see* **Note 2**).
3. Let the seeds settle down, and then pour ethanol into a beaker by discarding the liquid or using a pipette.
4. Add 1.5 mL of 50% bleach to the tube and suspend the seeds in the bleach by intermittent mixing from time to time for 10 min (*see* **Note 3**).
5. Let the seeds settle down and then discard the bleach into the beaker without disturbing the seeds at the bottom of the tube.
6. Add 1.5 mL ddH_2O to wash the seeds thoroughly by shaking for 2 min.
7. Discard the water in the beaker and repeat **step 6** five times (*see* **Note 4**).
8. Discard the water after the fifth time and add a final volume of 1.5 mL ddH_2O to the tube.
9. Place a piece of autoclaved filter paper in a clear square Petri dish. Pour water containing the sterile seeds onto the paper using a pipette (*see* **Note 5**).
10. Leave the Petri dish inside the biosafety cabinet to allow the seeds to dry for 1–2 h. Your seeds are now surface-sterilized and ready to be used for planting (*see* **Note 6**).

3.2 Preparation of ½ Strength Murashige and Skoog Basal Medium Plates With and Without NaCl

1. Add a magnetic stir bar in a 1 L glass beaker and pour 800 mL of double-distilled water into it.
2. Place the glass beaker containing water and the stir bar on a magnetic plate. Add 2.2 g of MS salt (½ strength) to it.
3. Stir the solution until the MS salt is completely dissolved. Then, using a pH meter, check the pH and adjust it to pH 5.7 with 0.1 M KOH (*see* **Note 7**).
4. Pour the media into a 1 L graduated cylinder and adjust the volume to 1 L with double-distilled water.
5. To prepare ½ MS with 150 mM NaCl, take another 1 L glass beaker, add 8.76 g of NaCl and 2.2 g of MS salt, and mix the solution thoroughly. Repeat **steps 3–4**.
6. Label two 2 L glass conical flasks and add 8 g of Phytagel (0.8% w/v) and a magnetic stir bar to each flask.
7. Pour the respective 1 L media from the graduated measuring cylinders into the labeled flasks.
8. Cover the mouth of the flasks with a large piece of aluminum foil and put an autoclave tape on the top of the foil (*see* **Note 8**).
9. Add enough water to an autoclave tray to cover the bottom and place the covered flasks containing the media in that tray.
10. Autoclave the flasks containing media for 30 min in the liquid cycle, and after the completion of the autoclave, place the flasks on magnetic stir plates and let the solution stir at ~250 rpm (*see* **Note 9**).
11. Turn on the UV light in the biosafety cabinet for about 15 min before pouring the media into the plates (*see* **Note 10**).
12. When the MS medium-containing flasks (with and without 150 mM NaCl) are cooled to 50–60 °C, use a pipette to add 35–40 mL of media to a 10 cm square Petri dish.
13. When the phytagel-containing plates have solidified, place them upside-down in a bag and store them in a refrigerator at 4 °C for future use.

3.3 Plating and Assessing of Seed Germination Under No Salt and 150 mM NaCl Salt Stress Conditions

1. Sterilize all the necessary equipment with 70% (v/v) ethanol (such as marker pens, plate racks, etc.) or autoclaving (tweezers, filter paper, etc.).
2. Wipe down the biosafety cabinet with 70% (v/v) ethanol. Place all the sterilized equipment inside it and treat it with UV light for 15 min (*see* **Note 11**).
3. Take one ½ strength MS plate without salt and place the filter paper with the sterilized seeds onto it (*see* **Note 12**).
4. Using a sterilized tweezer, plate 10–20 sterilized seeds onto an MS plate with 5 mm spacing between them. Prepare four to

five plates per genotype per condition and label them appropriately (*see* **Note 13**).

5. Seal each plate with two rounds of micropore surgical tape (3 M) around it (*see* **Note 14**).
6. Place the sealed plates vertically in plate racks in a controlled growth chamber under standard growth conditions (22–24 °C, 120–130 μmoL/m^2/s white light, 16 h of light/8 h of dark long-day photoperiod).
7. Using a magnifying glass, score the germination as rupture of the testa (seed coat) (Fig. 1b) and the emergence of the radicle every 12 h for 7 days. Photograph the plates every 12 h using a high-resolution camera.
8. Calculate the percentage of seed germination with and without salt stress using the following formulas (Fig. 3a, b):
   - Percentage of seed germination at a particular time point: (Number of germinated seeds at that time point/Total number of seeds planted) × 100.
9. Plot the average percentage of seed germination ± standard deviation (SD) at each time point for each seed line.

3.4 Molecular Markers to Assess Germination Under Salt Stress

The following procedure for isolating RNA from seeds has been modified from the Spectrum Total Plant RNA Kit (Sigma Aldrich) protocol, and the protocol for real-time quantitative polymerase chain reaction (qPCR) to analyze the transcript levels of some salt-responsive genes has been modified from the procedure described by Luna Universal One-Step RT-qPCR Kit (New England BioLabs).

1. Weigh 25 mg of seeds and place them on three sheets of filter paper with 5 mL of ½ strength MS solution in a clear square Petri dish (*see* **Note 15**).
2. Keep the plates with seeds in a controlled growth chamber under standard growth conditions (22–24 °C, 120–130 μmoL/m^2/s white light, 16 h of light/8 h of dark long-day photoperiod) (*see* **Note 16**).
3. Collect the seeds from the filter papers after a specific time point; freeze them in liquid nitrogen and transfer them to a −80 °C freezer for future use, or they can be processed for total RNA isolation immediately (*see* **Note 17**).
4. Label the tubes appropriately for RNA isolation.
5. Grind the harvested seed samples into powder with liquid nitrogen in a mortar and pestle (*see* **Note 18**).
6. Add 1.2 mL of extraction buffer to the grounded sample and mix it well.

7. Transfer the homogenate to a 2 mL microcentrifuge tube; vortex it well for 5 min, and spin it for 5 min at 11,000 × *g*.
8. Transfer the resulting supernatant to a fresh 2 mL microcentrifuge tube, add 500 μL of chloroform to it, and vortex it for 2 min.
9. Add 500 μL of acid phenol (water-saturated) to this mixture and vortex it for 2 min (*see* **Note 19**).
10. Centrifuge the mixture for 15 min at 11,000 × *g*.
11. Transfer carefully the upper aqueous phase (avoiding the interface) to an RNase-free 2 mL tube; add 90 μL of 3 M sodium acetate (pH 5.2) and 600 μL of isopropanol; mix it well, and incubate it for 10 min at room temperature.
12. Centrifuge this mixture for 10 min at 11,000 × *g*, discard the upper aqueous phase, and wash the pellet with 1 mL of 75% (v/v) ethanol.
13. Centrifuge this mixture for 10 min at 11,000 × *g*; discard the supernatant, and let the pellet air-dry for 4 min.
14. Add 500 μL of binding solution into the pellet; mix immediately by pipetting at least five times or vortex briefly.
15. Transfer this mixture into a binding column seated in a 2 mL collection tube.
16. Centrifuge the sample at a maximum speed for 1 min to bind RNA onto the binding column.
17. Decant the flow-through and tap the collection tube up and down on the absorbent paper to drain the residual liquid. Then, return the column to the collection tube.
18. Pipette 300 μL of wash solution 1 into the column and centrifuge at a maximum speed for 1 min.
19. Decant the flow-through and tap the collection tube up and down on an absorbent paper to drain the residual liquid. Then, return the column to the collection tube.
20. For each digestion, combine 10 μL of DNase 1 with 70 μL of DNase digestion buffer and mix gently by pipetting (*see* **Note 20**).
21. Pipette 80 μL of the mixture directly onto the center of the binding column, close the cap, and incubate the sample at room temperature for 15 min.
22. Pipette 500 μL of wash solution 1 into the column and centrifuge at maximum speed for 1 min to remove the digested DNA.
23. Decant the flow-through and tap the collection tube up and down on an absorbent paper to drain the residual liquid. Then, return the column to the collection tube.

24. Pipette 500 μL of wash solution 2 into the column and centrifuge at maximum speed for 1 min.
25. Decant the flow-through and tap the collection tube up and down on an absorbent paper to drain the residual liquid. Then, return the column to the collection tube.
26. Repeat **steps 24–25** and centrifuge for another 2 min to dry. Carefully remove the column and transfer it into a fresh 1.5 mL tube.
27. Pipette 50 μL of nuclease-free double-distilled water onto the center of the column and let the tube sit for 2 min at room temperature.
28. Centrifuge at a maximum speed for 1 min to elute the purified RNA. Discard the binding column and save the isolated DNase-treated RNA present in the flow-through eluate. It is ready for immediate use or storage in a −20 °C (short-term) or −80 °C (long-term) freezer.
29. Examine the quality and concentration of the purified RNA (2 μL per sample) on a Nanodrop spectrophotometer (*see* **Note 21**).
30. Dilute the DNase-treated RNA samples with nuclease-free double-distilled water to 40–50 ng/mL per reaction tube to determine the transcript levels of two salt-responsive markers (*RESPONSIVE TO DESICCATION* 29A: *RD29A*, At5g52310*; RESPONSIVE TO* ABA18: *RAB18*, At5g66400) using real-time qPCR analysis (Fig. 3c).
31. For the qPCR, add 10 μL of Luna Universal One-Step Reaction Mix (2×), 1 μL of Luna Warm Start Reverse-Transcriptase Enzyme Mix (20×), 0.8 μL of forward and reverse primers for each of the stress-responsive genes (the following primer pairs were used for the qRT-PCR of *RD29A* and *RAB18*: *RD29A*-Forward: 5′-GCCGACGGGATTTGACG, *RD29A*-Reverse: 5′-GCCGGAAATTTATCCTCTTCTGA, *RAB18*-Forward: 5′-TCGGTCGTTGTATTGTGCTTTTT, and *RAB18*-Reverse: 5′-CCAGATGCTCATTACACACTCATG [28]), 2 μL of DNase-treated diluted RNA (40–50 ng/mL), and the remaining nuclease-free double-distilled water (5.4 μL) to make a total volume of 20 μL. Prepare the reaction in a 96-well plate and seal the plate with an adhesive plastic seal.
32. Mix the reaction mixture briefly and spin the plate briefly to remove bubbles for approximately 2–4 min at 2500–3000 rpm.
33. Run qPCR with the following thermocycling protocol: an initial reverse transcription step for 10 min at 55 °C for 1 cycle, an initial denaturation step for 1 min at 95 °C for 1 cycle, followed by 40–45 cycles of 10 s of denaturation at

95 °C and 30 s of extension at 60 °C and a final melting curve analysis at 60–95 °C.

34. First, normalize the transcript data to *At3g12210* at each condition to obtain the relative amount of transcript levels. In addition, normalize these levels to the levels observed in the control wild-type seeds germinated under no salt condition using the method of Livak and Schmittgen (Fig. 3c) [29].
35. Test the statistical significance for all the data using a one-way ANOVA and a post hoc test such as Tukey–Kramer for multiple comparisons.

4 Notes

1. Since MS salt is hygroscopic, do not leave the reagent outside; it should always be stored in a refrigerator at 4 °C.
2. Do not leave the seeds in ethanol for too long. If overtreated, ethanol acts as a stressor for seed germination and may reduce the viability of the seeds.
3. Do not leave the seeds in the bleach for too long. It may kill the seeds or could be toxic to the seedlings.
4. Wash off the bleach entirely with water.
5. If seeds are stuck inside the tube, add more ddH_2O to wash them out.
6. The plate containing the filter paper with dried seeds can be stored for months in a closed sterile Petri dish wrapped with micropore surgical tape and kept in a refrigerator at 4 °C. If you are sterilizing multiple genotypes simultaneously, label the plates/filter paper containing seeds before storage.
7. MS salt can be purchased either in "pH adjusted and buffered" or unbuffered forms. If it is not pre-buffered, MS salt mixed with water measures an acidic pH (around 4.5–4.7); hence, it must be pH adjusted and buffered to pH 5.7.
8. The mouth of the conical flask should be properly covered to prevent spilling during the autoclave. A glass bottle with a cap can also be used instead of conical flasks.
9. To avoid cooling the media in the autoclave, take the flasks out as soon as the cycle ends and place them on the magnetic stir plate for homogeneous mixing of the solution.
10. While UV-sterilizing the biosafety cabinet, plates, pipettes, and pipette gun can also be sterilized to avoid contamination.
11. While pouring different media, use different color coding on the rim of the plates to avoid future confusion. For example, use marker pens to draw two black lines for ½ strength MS

media without 150 mM NaCl and two red lines for ½ strength MS media with 150 mM NaCl.

12. Placing the previously sterilized seeds on the filter paper on ½ MS plate will help the seeds imbibe water and make them easy to pick up with tweezers.
13. When 20 seeds of one genotype are placed on the MS plates, they should be placed in two rows (top and bottom) of 10 seeds, with a 5 mm spacing between them.
14. The plates are wrapped with a surgical tape to allow for gas exchange and prevent ethylene accumulation [30].
15. While setting up the plates, put the filter paper on the lid of the Petri dish and use its base as the lid.
16. Keep the plates horizontal and randomly distribute them in the growth chamber to minimize positional effects.
17. For this experiment, seeds were harvested for RNA isolation after 2 days.
18. You can also grind the seeds using a tissue homogenizer machine. In that case, add the extraction buffer to the seeds in a microcentrifuge tube before grinding.
19. The acid phenol needs to be kept at room temperature overnight.
20. Do not vortex the DNase 1 as it is sensitive to physical denaturation.
21. For most plant tissues, the ratio of absorbance at 260 nm to 280 nm should be between 1.8 and 2.2. Evaluate the integrity of the purified RNA by agarose gel electrophoresis or by a bioanalyzer instrument.

Acknowledgments

The author would like to thank Dr. Simon Gilroy for critical reading of the manuscript, giving valuable advice, and allowing the author to conduct the experiments in his laboratory at the University of Wisconsin, Madison, supported by NASA 80NSSC21K0577.

References

1. Abeles FB, Morgan PW, Saltveit ME, Mikal E (1992) Ethylene in plant biology 414
2. Bakshi A, Shemansky JM, Chang C, Binder BM (2015) History of research on the plant hormone ethylene. J Plant Growth Regul 34. https://doi.org/10.1007/s00344-015-9522-9
3. Chen YF, Randlett MD, Findell JL, Schallert GE (2002) Localization of the ethylene receptor ETR1 to the endoplasmic reticulum of Arabidopsis. J Biol Chem 277:19861–19866. https://doi.org/10.1074/JBC.M201286200
4. Chang C, Kwok SF, Bleecker AB, Meyerowitz EM (1993) Arabidopsis ethylene-response

gene ETR1: similarity of product to two-component regulators. Science 262:539–544. https://doi.org/10.1126/SCIENCE.8211181

5. Hua J, Sakai H, Nourizadeh S et al (1998) EIN4 and ERS2 are members of the putative ethylene receptor gene family in Arabidopsis. Plant Cell 10:1321. https://doi.org/10.1105/TPC.10.8.1321
6. Hua J, Chang C, Sun Q, Meyerowitz EM (1995) Ethylene insensitivity conferred by Arabidopsis ERS gene. Science 269:1712–1714. https://doi.org/10.1126/SCIENCE.7569898
7. Hua J, Meyerowitz EM (1998) Ethylene responses are negatively regulated by a receptor gene family in Arabidopsis thaliana. Cell 94: 261–271. https://doi.org/10.1016/S0092-8674(00)81425-7
8. Bakshi A, Binder BM (2018) Chapter 8: plant ethylene sensing and signalling
9. Shakeel SN, Wang X, Binder BM, Schaller GE (2013) Mechanisms of signal transduction by ethylene: overlapping and non-overlapping signalling roles in a receptor family. AoB Plants 5. https://doi.org/10.1093/AOBPLA/PLT010
10. Binder BM, Chang C, Schaller GE (2018) Perception of ethylene by plants – ethylene receptors. In: Annual plant reviews, pp 117–145. https://doi.org/10.1002/9781119312994.APR0477
11. Binder BM, O'Malley RC, Wang W et al (2004) Arabidopsis seedling growth response and recovery to ethylene. A kinetic analysis. Plant Physiol 136:2913. https://doi.org/10.1104/PP.104.050369
12. Binder BM, O'Malley RC, Wang W et al (2006) Ethylene stimulates nutations that are dependent on the ETR1 receptor. Plant Physiol 142:1690. https://doi.org/10.1104/PP.106.087858
13. Cho YH, Yoo SD (2007) ETHYLENE RESPONSE 1 histidine kinase activity of Arabidopsis promotes plant growth. Plant Physiol 143:612–616. https://doi.org/10.1104/PP.106.091504
14. Hall AE, Bleecker AB (2003) Analysis of combinatorial loss-of-function mutants in the Arabidopsis ethylene receptors reveals that the ers1 etr1 double mutant has severe developmental defects that are EIN2 dependent. Plant Cell 15:2032. https://doi.org/10.1105/TPC.013060
15. Qu X, Schaller GE (2004) Requirement of the histidine kinase domain for signal transduction by the ethylene receptor ETR1. Plant Physiol 136:2961–2970. https://doi.org/10.1104/PP.104.047126
16. Kim H, Helmbrecht EE, Blaine Stalans M et al (2011) ETHYLENE RECEPTOR ETHYLENE RECEPTOR1 domain requirements for ethylene responses in Arabidopsis seedlings. Plant Physiol 156:417. https://doi.org/10.1104/PP.110.170621
17. Wilson RL, Bakshi A, Binder BM (2014) Loss of the ETR1 ethylene receptor reduces the inhibitory effect of far-red light and darkness on seed germination of Arabidopsis thaliana. Front Plant Sci 5. https://doi.org/10.3389/fpls.2014.00433
18. Wilson RL, Kim H, Bakshi A, Binder BM (2014) The ethylene receptors ethylene response1 and ethylene response2 have contrasting roles in seed germination of arabidopsis during salt stress. Plant Physiol 165. https://doi.org/10.1104/pp.114.241695
19. Bakshi A, Wilson RL, Lacey RF et al (2015) Identification of regions in the receiver domain of the ETHYLENE RESPONSE1 ethylene receptor of arabidopsis important for functional divergence. Plant Physiol 169. https://doi.org/10.1104/pp.15.00626
20. Bakshi A, Piy S, Fernandez JC et al (2018) Ethylene receptors signal via a noncanonical pathway to regulate abscisic acid responses. Plant Physiol 176. https://doi.org/10.1104/pp.17.01321
21. Bentsink L, Koornneef M (2008) Seed dormancy and germination. Arabidopsis Book 6: e0119. https://doi.org/10.1199/TAB.0119
22. Weitbrecht K, Müller K, Leubner-Metzger G (2011) First off the mark: early seed germination. J Exp Bot 62:3289–3309. https://doi.org/10.1093/JXB/ERR030
23. Nawaz M, Sun J, Shabbir S et al (2023) A review of plants strategies to resist biotic and abiotic environmental stressors. Sci Total Environ 900:165832. https://doi.org/10.1016/J.SCITOTENV.2023.165832
24. Jakab G, Ton J, Flors V et al (2005) Enhancing Arabidopsis salt and drought stress tolerance by chemical priming for its Abscisic acid responses. Plant Physiol 139:267. https://doi.org/10.1104/PP.105.065698
25. Lång V, Palva ET (1992) The expression of a rab-related gene, rab18, is induced by abscisic acid during the cold acclimation process of Arabidopsis thaliana (L.) Heynh. Plant Mol Biol 20:951–962. https://doi.org/10.1007/BF00027165/METRICS
26. Murashige T, Skoog F (1962) A revised medium for rapid growth and bio assays with tobacco tissue cultures. Physiol Plant 15:473–

497. https://doi.org/10.1111/J.1399-3054.1962.TB08052.X

27. Elwell AL, Gronwall DS, Miller ND et al (2011) Separating parental environment from seed size effects on next generation growth and development in Arabidopsis. Plant Cell Environ 34:291–301. https://doi.org/10.1111/J.1365-3040.2010.02243.X

28. Fujii H, Verslues PE, Zhu JK (2007) Identification of two protein kinases required for Abscisic acid regulation of seed germination, root growth, and gene expression in Arabidopsis. Plant Cell 19:485. https://doi.org/10.1105/TPC.106.048538

29. Livak KJ, Schmittgen TD (2001) Analysis of relative gene expression data using real-time quantitative PCR and the 2−ΔΔCT method. Methods 25:402–408. https://doi.org/10.1006/METH.2001.1262

30. Buer CS, Wasteneys GO, Masle J (2003) Ethylene modulates root-wave responses in Arabidopsis. Plant Physiol 132:1085. https://doi.org/10.1104/PP.102.019182

Chapter 12

GCN2 and EIN2 Pathway Activation Assays in Arabidopsis Under Hypoxia

Hsing-Yi Cho, Mei-Yi Chou, and Ming-Che Shih

Abstract

General translational repression is a crucial process that reduces energy consumption under hypoxia. Ethylene can stimulate both GCN2 and EIN2 pathways to counteract this repression and facilitate adaptation to hypoxia upon submergence. Here, we present three rapid assays to investigate the activation of GCN2 and EIN2 pathways in translation under submergence in *Arabidopsis*, including the survival test, eIF2α phosphorylation detection, and newly synthesized protein quantification. These methodologies will contribute to understanding ethylene signaling and plant adaptation to hypoxia stress.

Key words Hypoxia, Submergence, Ethylene, GCN2, EIN2, Translation

1 Introduction

General control nonderepressible 2 (GCN2) is an evolutionarily conserved regulator that perceives nutrient starvation and stress signals [1, 2]. In yeasts and mammals, GCN2 senses amino acid starvation via its C-terminal region, histidyl tRNA synthetase, which stimulates the activity of the N-terminal kinase domain and turns on downstream signaling [3]. *Arabidopsis* has one copy of *GCN2* and two copies of *eIF2α*. The phosphorylation site in eIF2α (Ser56) recognized by GCN2 is conserved and is phosphorylated in response to submergence [4].

Ethylene is a key phytohormone involved in plant responses to hypoxic stress. In the early stage of submergence, ethylene is trapped in the submerged tissues due to restricted gas diffusion underwater [5]. The rapid increase in ethylene triggers a signaling pathway that, through the elimination of nitric oxide (NO), stabilizes the group VII ethylene response factors (ERFs), leading to the activation of HRGs [6]. In addition, ethylene and aminocyclopropane-1-carboxylic acid (ACC), the precursor of ethylene, enhance GCN2-mediated phosphorylation of eIF2α [1, 4].

Devesh Shukla (ed.), *Ethylene Sensing and Signaling*, Methods in Molecular Biology, vol. 2945,
https://doi.org/10.1007/978-1-0716-4650-2_12,

In our previous study [4], we demonstrated that under hypoxia upon submergence, the enhanced GCN2 activity is triggered through a noncanonical ethylene signaling pathway, in which the activity of GCN2 is independent of EIN2, EIN3, and EIL1. Ethylene activates both GCN2 and EIN2 to repress the general translation under submergence. We provide three rapid assays to evaluate the activation of GCN2 and EIN2 pathways under submergence in *Arabidopsis*. First, a survival test of seedlings under submergence indicates that activation of GCN2 and EIN2 is required for the submergence adaptation. Second, measuring the level of eIF2 phosphorylation provides insights into GCN2 activity under submergence. Finally, quantifying the newly synthesised protein reveals the activation levels of GCN2 and EIN2 pathways in translation under hypoxia upon submergence. These provide valuable tools for studying ethylene response in translational regulation under hypoxia upon submergence.

2 Materials

2.1 Survival Test Under Submergence

1. 2.5 L sterilized water.
2. Filter papers and Petri dishes.

2.2 eIF2α Phosphorylation Assay

1. ½ MS medium.
2. 3% oxygen (mixed with nitrogen).
3. 6× protein loading buffer: 375 mM Tris–HCl (pH 6.8), 9% SDS, 50% glycerol, 0.03% bromophenol blue, 10% β-mercaptoethanol.
4. Mini-Protein TGX precast gels (Bio-Rad).
5. Trans-Blot SD Semi-Dry Transfer Cell (Bio-Rad).
6. Transfer buffer (1 L): 5.8 g Tris-base, 2.9 g glycine, 0.25 g SDS, and 20% methanol.
7. TBST buffer: 1× TBS buffer with 0.1% Tween 20.
8. Phospho-eIF2α (Ser51) (Abcam).

2.3 Newly Synthesized Protein Assay Under Hypoxia

1. Click-IT Tetramethylrhodamine (TAMRA) Protein Analysis Detection Kit (Invitrogen): 100 μM L-azidohomoalanine (Click-IT AHA, Invitrogen).
2. Protein extraction buffer: 1% SDS, 50 mM Tris–HCl pH 8.0, 1/250 (v/v) proteinase inhibitor cocktail for plant cell lysate (Sigma-Aldrich).
3. Gas jars.
4. 3% oxygen (mixed with nitrogen).
5. 4–20% protein gel (Novex, Invitrogen).

6. Sapphire Biomolecular Imager (Azure Biosystems) or Typhoon (GE).
7. Coomassie blue stain.

3 Methods

- Seeds are sterilized with 0.5% sodium hypochlorite for 10 min, washed at least three times with sterilized water, and then incubated at 4 °C for 3 days to achieve uniform germination.
- All seeds are germinated on 5% phytagel plates with 1/2 MS and 0.5% sucrose and grown vertically for 7 days under a 16 h light (81 μmoL m^{-2} s^{-1}; at 22 °C)/8 h dark (at 18 °C) cycle (*see* **Note 1**).
- On the seventh day, seedlings are carefully transferred to new 5% phytagel plates with 1/2 MS and 0.5% sucrose and grown vertically for 2 days (*see* **Note 2**).

3.1 Survival Test Under Submergence

1. Nine-day-old seedlings on plates are submerged entirely in sterilized water in the dark for 48 h or in constant light (81 μmoL m^{-2} s^{-1}; at 22 °C) for 8 days (*see* **Note 3**).
2. After submergence treatment, plants are transferred to the filter paper with 2.5 mL sterilized water and put back into a growth chamber under a 16 h light (81 μmoL m^{-2} s^{-1}; at 22 °C)/8 h dark (at 18 °C) cycle, and the number of damaged or chlorotic leaves is counted after 2 days of recovery (Fig. 1) (*see* **Note 4**).

3.2 eIF2α Phosphorylation Assay Under Submergence

1. ½ MS medium for submergence treatments is bubbled with 3% oxygen for 1 h (the final dissolved oxygen concentration in the medium: 1.7 mg/L). ½ MS medium is kept bubbled with 3% oxygen upon seedlings that are submerged.
2. Nine-day-old seedlings are submerged in low oxygen-containing ½ MS medium in the light (81 μmoL m^{-2} s^{-1}). Seedlings kept at room temperature in the light are used as control (*see* **Note 5**).
3. After 30 min of submergence treatment, seedlings are harvested and homogenized with liquid nitrogen.
4. The homogeneous sample (eight seedlings) is mixed with 2× protein loading buffer (50 μL) and immediately cooked at 98 ° C for 10 min.
5. Centrifuge at the highest speed for 5 min.
6. Collect the supernatant carefully, and 20 μL of the supernatant is loaded into a well of the Mini-Protein TGX precast gel.
7. Run Mini-Protein TGX precast gels following the Bio-Rad instruction.

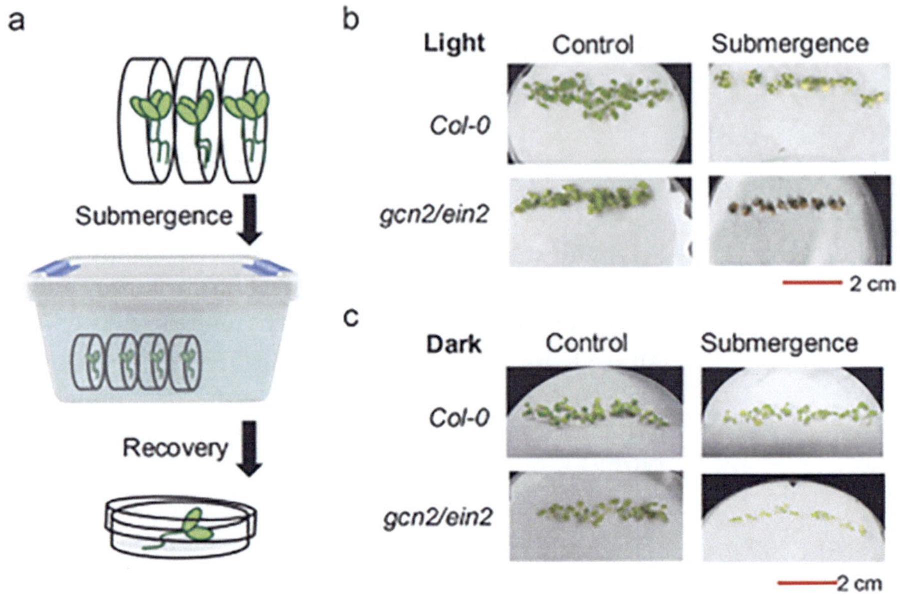

Fig. 1 GCN2 and EIN2 pathways are critical for submergence tolerance. (**a**) Vertical growth seedlings are put in the tank and submerged till the indicated time. Submerged seedlings are transferred to the filter papers, soaking in 2.5 mL of sterilized water for recovery. (**b**) and (**c**) The phenotype of *Col-0* and *gcn2/ein2* after 8 days of submergence (light) + 2 days recovery and 2 days submergence (dark) + 2 days recovery. The leaves of *gcn2/ein2* showed more serious damage than *Col-0*

8. Transfer the protein from the gel to the membrane:
   (a) Place the gel, filter paper, and methanol-activated PVDF membrane into the transfer buffer for 3 min.
   (b) Assemble the western blot transfer sandwich.
   (c) Transfer by a semi-dry transfer cell (constant 25 voltage for 30 min).
9. Incubate the membrane in 5% milk TBST buffer with agitation at room temperature for 1 h.
10. Add Phosphor-eIF2α (Ser51) antibody (titer 1:2000) to the blocking solution and then incubate the blot with agitation at room temperature for 2 h.
11. Wash the membrane three times in TBST buffer with agitation at room temperature for 10 min each.
12. Incubate the membrane with the second antibody (anti-rabbit, HRP conjugate, titer: 1/10000) in TBST buffer with agitation at room temperature for 1 h.
13. Wash the membrane three times in BST buffer with agitation at room temperature for 10 min each.

14. Examine the results by chemiluminescence assays and quantify the band signals with Image J. The Coomassie blue stain of the membrane is used as the loading control (*see* **Note 6**).

3.3 Newly Synthesized Protein Assay Under Hypoxia

1. Transfer nine-day-old seedlings to the filter paper soaked in 2 mL of ½ MS medium for 1 h.
2. Transfer seedlings to the filter paper soaked with 1.8 mL ½ MS medium containing 100 μM L-azidohomoalanine (Click-IT AHA, Invitrogen) into a jar containing 3% oxygen under light or a jar containing air under light for 6 h (Fig. 2a) (*see* **Note 7**).
3. Harvest and homogenize the seedlings with liquid nitrogen.
4. Add the ice-cold extraction buffer and incubate the mixture on ice for 10 min.
5. Spin at the highest speed for 15 min in a 4 °C precooled centrifuge to collect the supernatant. Quantify the protein concentration of the supernatant by DC protein assay (Bio-Rad).
6. 60 μg labeled proteins in a maximum volume of 50 μL solution is mixed with the 40 μL Click-iT reaction buffer.
7. Vortex for 5 s.
8. Add 4 μL 40 mM $CuSO_4$ and vortex for 5 s.

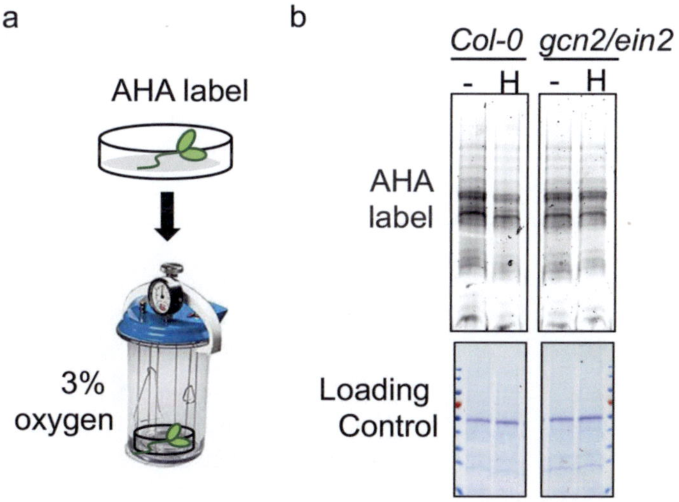

Fig. 2 Newly synthesized protein is reduced in *Col-0* under hypoxia. (**a**) Seedlings are transferred to the filter paper containing 2 mL of 1/2 MS for 1 h. Then, seedlings are transferred on the filter paper, soaking 1.8 mL of ½ MS medium containing 100 μM L-azidohomoalanine and put into a jar and equilibrated the whole jar with 3% oxygen or air for 6 h. (**b**) Newly synthesized proteins (NSP) in the whole seedlings of Col-0 and gcn2/ein2 under control (–) and 6 h hypoxia (H) in the light are visualized by the AHA label. The Coomassie-stained gel shows the total protein of two lines under control and 6 h of hypoxia (light)

9. Add 4 μL Click-iT reaction buffer with component C in the reaction mixture and vortex for 5 s. Wait for 3 min.
10. Add 8 μL Click-iT reaction buffer with component D in the reaction mixture and vortex for 5 s.
11. Rotate the reaction mixture end-over-end for 20 min in the dark.
12. Add 300 μL of methanol to the reaction mixture and vortex for 5 s.
13. Add 75 μL chloroform to the reaction mixture and vortex for 5 s.
14. Add 200 μL ultrapure water to the reaction mixture and vortex for 5 s.
15. Spin at the highest speed for 5 min and discard the upper phase.
16. Add 225 μL methanol to the mixture and vortex.
17. Spin at the highest speed for 5 min and collect the protein pellet.
18. The precipitated protein is suspended by 1× sample loading buffer and denatured at 70 °C for 10 min.
19. Load the denatured samples into the 4–20% protein gel (Novex, Invitrogen).
20. Detect the fluorescence by Sapphire Biomolecular Imager (Azure Biosystems) or Typhoon with excitation at 520 nM and emission at 565 nM.
21. Quantify the fluorescence intensity by ImageJ and normalize it with a Coomassie blue stain (Fig. 2b).

4 Notes

1. Vertical growth, which allows roots to remain on the agar surface, is essential for the following assays.
2. Because sugar content is also essential for hypoxia tolerance [7], transferring eight seedlings of similar size to each new plate on the seventh day is necessary for consistent experimental results.
3. To reduce the contamination in water under submergence, spray 70% ethanol on the plates and the tank for submergence treatment, and then let ethanol dry out before the submergence of the plates. To submerge all plants, remove the plates' caps, layer plates vertically, and prevent any air bubbles between the plates when flooding the plates (Fig. 1a). The water surface is at least 5 cm higher than the top of the plates. Ampicillin can be added to water to reduce the bacterial growth.

4. 48 h of submergence (dark) and 8 days of submergence (light) caused severe damage to 50% of *Col-0* seedlings. Thus, 48 h of submergence (dark) and 8 days of submergence (light) are used to examine the tolerance of plants.
5. Because dark–light transient activates GCN2 [8], to examine GCN2 activity under submergence (light), the treatment must always be done after 2 h of photoperiod when eIF2α phosphorylation is undetectable.
6. The tubulin or eIF2α protein can also be used to normalize the phosphorylation band density.
7. An alternative method is available for hypoxia treatment in the NSP assay.
   (a) Grow seedlings in a 6-well plate containing 1 mL of ½ MS medium + 0.5% sucrose in each well for 7 days.
   (b) On the seventh day, remove the old medium and replenish 1 mL of ½ MS medium +0.5% sucrose to grow the seedlings.
   (c) Till the ninth day, remove the medium and add 1.8 mL of ½ MS medium containing 100 μM L-azidohomoalanine (Click-IT AHA, Invitrogen) and EC-Oxyrase solution (EC-Oxyrase®, titer: 1:100). EC-Oxyrase can remove oxygen from the medium.

References

1. Sébastien L, Elodie L, Marie-Noëlle PP, Marie C, Christophe R, Jean MD, Thierry P (2008) *Arabidopsis* eIF2alpha kinase GCN2 is essential for growth in stress conditions and is activated by wounding. BMC Plant Biol 8:134
2. Immanuel TM, Greenwood DR, MacDiarmid RM (2012) A critical review of translation initiation factor eIF2α kinases in plants – regulating protein synthesis during stress. Funct Plant Biol 39(9):717
3. Hinnebusch AG (2014) The scanning mechanism of eukaryotic translation initiation. Annu Rev Biochem 83:779–812
4. Hsing-Yi C, Mei-Yi C, Hsiu-Yin H, Wan-Chieh C, Ming-Che S (2022) Ethylene modulates translation dynamics in *Arabidopsis* under submergence via GCN2 and EIN2. Sci Adv 8:eabm7863
5. Lamers J, van der Meer T, Testerink C (2020) How plants sense and respond to stressful environments. Plant Physiol 182(4):1624–1635
6. Schmidt R, Weits DA, Feulner CF, van Dongen JT (2017) Oxygen sensing and integrative stress signaling in plants. Plant Physiol 176(2): 1131–1142
7. Hsing-Yi C, Elena L, Ming-Che S, Perata P (2021) Energy and sugar signaling during hypoxia. New Phytol 229(1):57–63
8. Ansul L, Ju G, AUC R, Sung KC, Philip WM, Madison L, Masaki S, Brad D, Albrecht GVA (2020) Light activates the translational regulatory kinase GCN2 via reactive oxygen species emanating from the chloroplast. Plant Cell 32(4):1161–1178

Chapter 13

Detection of Ethylene Signaling Using Bioinformatics Tools in Plants

Yogeshwar Vikram Dhar, Priyamvada Mishra, Garima Saxena, Shahre Alam, and Mehar Hasan Asif

Abstract

Ethylene is a gaseous hormone that is vital in plant growth, development, biotic or abiotic stresses, fruit ripening, and senescence. The ethylene signaling machinery includes several genes, many of which are multigene family members and exhibit spatial and temporal expression patterns. These genes also show divergence and expansion during evolution. Recent high-throughput technologies have promoted a more comprehensive profiling of genes regulated by ethylene signaling. The availability of a large amount of sequencing data from whole genome, transcriptome, small RNA, etc., in the public databases provides an informative resource for gene mining of ethylene. However, these omics data generally result in large gene lists that make it challenging to interpret the data and extract insights into biological significance. Different platforms have different architectures and designs, and even similar studies using the same platform may obtain data with significant variances because of the highly dynamic and flexible effects of plant hormones, making comparisons across different studies and platforms difficult. Hence, the method of analysis suggests the requirement of database for their respective analysis. Some of the databases reported to find the genomics, transcriptomics, and regulatory mechanism of ethylene are *The Arabidopsis Information Resource* (TAIR), Banana Genome Hub (BGH), Citrus genome database (CGD), OneKP database, miRDeep-2, miRDP, psRNATarget, Arabidopsis Hormone Database (AHD), and gene set–level analyses of hormone responses (GSHR). The present chapter deals with the online and offline methods and basic workflow for the detection of ethylene signaling machinery.

Key words Ethylene signaling, Bioinformatics, Gene mining, Transcriptomics, miRNA analysis, Promoter analysis, Cis-regulatory elements, Plant hormones

1 Introduction

Ethylene signaling plays a pivotal role in the various steps of major biological processes in plants like fruit ripening [1, 2], stress responses [3, 4], growth, and development [5, 6]. This involvement of gaseous hormone ethylene in several processes makes them an important target to understand the gene network involved in the ethylene-driven signal transduction and other responses. The boom

Devesh Shukla (ed.), *Ethylene Sensing and Signaling*, Methods in Molecular Biology, vol. 2945,
https://doi.org/10.1007/978-1-0716-4650-2_13,

in sequencing-based methods and new-generation applications along with the infusion of high-performance supercomputing makes it an integral part of gene identification with their transcript expression, epigenetic status, evolutionary landscape, and computational protein structural analysis. Throughout the world, several researchers are working to identify and analyze these genes in accordance with their interest, to elucidate the role of ethylene-mediated gene responses in various multistep biological processes and functions. Fruit ripening is one such process, where ethylene regulated the major subprocesses like fruit growth, color, aroma, and acid-to-sugar conversion, with a number of transcription factors and other regulatory genes [1, 7–12]. Several reports showed the significant amount of ethylene signaling genes and their involvement in unripe to ripe conditions of various fruits. Apart from fruits, the ethylene machinery is also identified in biotic and abiotic stress conditions [3, 4, 13], regulating the downstream pathways. In the present time, due to the extensive use of sequencing technologies and advanced methods, huge amounts of sequencing data for plant genomes and transcriptomes are available, which require deep mining and exploration to fetch the useful information about target genes and their relative machinery. These datasets provide huge amounts of sequence and annotation data along with variations, enriched promoters, and expression during given conditions. On the other hand, exploiting these datasets for specific genes is a complex task, where the researcher needs to check several aspects of data like sequence domain (conserved protein domain), complete length, protein family, GO (gene ontology) and pathway terms, expression, protein structure, etc. For the initial identification process, there is a need of one such protocol which can provide initial genes and related details which can help in building the complete study design (Fig. 1). Here, the present chapter suggests an exploratory method for the initial identification of such genes.

Studies on the post-transcriptional controllers of fruit growth and ripening, such as small RNAs (microRNAs (miRNAs), are, nevertheless, scarce. Small RNAs (sRNA) are noncoding molecules that range in length from 18 to 34 nucleotides (nts), and it is well recognized that they are essential for the regulation of post-transcriptional gene expression. The identification of sRNAs in numerous plant species is done using the potent technology of next-generation sequencing (NGS). The regulatory mechanisms of small RNAs have been intensively studied using next-generation sequencing (NGS) over the past few decades. There are numerous bioinformatic tools available to help biologists make sense of the immense volumes of data produced by NGS platforms. Unfortunately, NGS platforms produce a huge quantity of data, and managing these data can be quite time-consuming. In order to gain useful information and address important issues, the analysis of

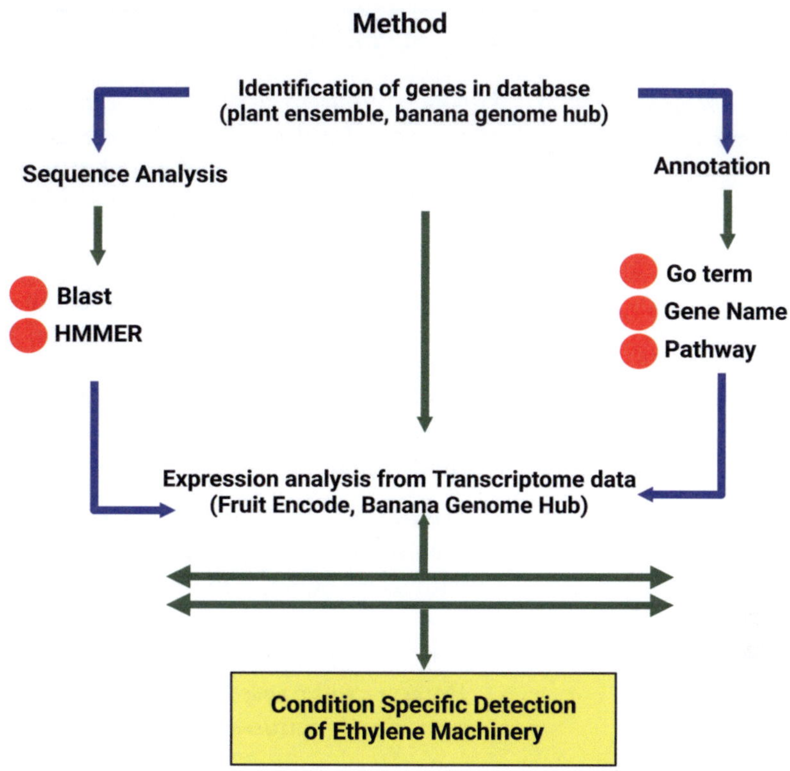

Fig. 1 A complete study design to explore condition-specific identification of ethylene machinery in plants

these data requires substantial computing and statistical expertise. To extract useful information from these data and answer challenging biological issues, examination of these data takes extensive computational and statistical expertise. In this chapter, many aspects of the sRNA-seq data analysis are introduced to the researchers. Analysis of sRNA-seq data is distinct from conventional RNA-sequencing methods. Before low-quality data are filtered and adapter sequences are eliminated, short read aligners are used to map the filtered data onto the transfer RNA (tRNA), ribosomal RNA (rRNA), small nucleolar RNA (snoRNA), and small nuclear RNA (snRNA). Because they are not byproducts of dicer-like (DCL) protein activity and are believed to have a very low possibility of being involved in short RNA pathways, the reads that map to these RNAs are ignored. A number of tools, including miRDeep-P [14], ShortStack [15], miRPlant [16], MIReNA [17], and miRkwood [18], are used to process the unmapped reads and find novel miRNAs after the filtered reads have been mapped to miRBase to identify the conserved miRNAs. These techniques concentrate on miRNA prediction by considering the fundamental characteristics of miRNAs, such as structural details, length, DCL cleavage, and their converseness among species.

2 Gene Identification and Expression Analysis

2.1 Database Selection

This step requires the selection of database from which the genes of interest can be downloaded, for example, the TAIR database of Arabidopsis (https://www.arabidopsis.org/), rice genome database (http://rice.uga.edu/), banana genome hub (https://banana-genome-hub.southgreen.fr/), citrus genome database (https://www.citrusgenomedb.org/), etc. Apart from these, there are several other whole genome databases that are available for different fruits, crop plants, medicinal plants, etc. If the whole genome is not available for the plant of interest, then onekp database (https://db.cngb.org/onekp/) can be used for transcriptomic sequences. These databases provide two types of data search method, where:

- Dataset can be downloaded offline, in the form of sequence (genomic, transcript, amino acid) and annotation file (locus id with gene name, gene function, GO terms, KEGG pathway). These data can be analyzed manually and with the help of software (i.e., offline blast) for required information.
- Dataset can be searched online with database utilities like "gene search", "function search" term search for GO and Blast.

For example:

- In the case of banana genome hub database, user can easily download the sequence and annotation data from the download tab and analyze.
- Similarly, the "search" tab can be used for gene, function, and term search where example terms can be "ethylene" "ethylene sensitive," "ethylene responsive," "hormonal," or direct by gene name which can be "AP2/ERF"[19], "ERS" [6], "ACS," "ACO" [20–22], etc.

Despite the abundance of genetic research data, a system is still required to arrange, screen, and analyze the data to be used directly in database selection. Researchers and bioinformaticians started working on the publicly available genetic and genomic data in various ways to make it usable for developing and choosing better databases; this led to the notion of creating a comprehensive database.

2.2 Offline Sequence Search and Retrieval

The sequences can be searched offline by sequence similarity search using the BLAST [23] and HMMER (http://hmmer.org/) tool. The offline blast and HMMER tools can be installed in the user's computer (windows/linuxos). This tool can be employed for sequence search (nucleotide/amino acid) with best hits and sensitivity cutoff (e-value). The input seed sequence, from any previously reported dataset, can be used in BLAST tool. In the case of HMMER tool, the pfam profile (Pfam database is now merged in Interpro database) of selected protein domains can be employed as the input, and the entire profile can be searched.

For example, the two-component system signal transduction machinery genes, which work as noncanonical pathway for ethylene signaling, from *Oryza sativa*, can be used as input, to find out the genes of the two-component system in banana, using the blast tool. The nucleotide sequence with blastn program and amino acids with blastp program can be searched. After the search finished, the id's identified can be retrieved using "fastacmd" or command line scripts.

Benefits and Drawbacks of Online Blast: Access databases stored remotely at NCBI. There is minimal to no setup necessary. However, they are unable to use a customized database. In contrast, the benefits of an offline blast include the ability to use a customized database. It works well for scripting, automation, or situations where a lot of queries need to be run (UNIX). Offline blast's other drawback is that it necessitates some computer knowledge and setup.

2.3 Sequence Identification

The identified sequences can be checked for their conserved domains using the CDD (https://www.ncbi.nlm.nih.gov/Structure/cdd/cdd.shtml) and SMART (http://smart.embl-heidelberg.de/) tools, where the retrieved amino acid sequence can be submitted to the abovementioned online software for the conserved domain. The abovementioned steps can similarly be followed for transcriptomic database. This method is essential in the identification of ethylene-responsive domains like "ERF," "GAF," "AP2," etc. The sequences of these domains can also be used to align with other sequences. Apart from that, in next steps, the comprehensive methods can be used to explore ethylene signaling.

CDD is a protein annotation resource that consists of a collection of well-annotated multiple sequence alignment models for ancient domains and full-length proteins. These are available as position-specific score matrices (PSSMs) for fast identification of conserved domains in protein sequences via RPS-BLAST. The CDD content includes NCBI-curated domains, which use 3D structure information to explicitly define domain boundaries and provide insights into sequence/structure/function relationships, as well as domain models imported from a number of external source databases (Pfam, SMART, COG, PRK, and TIGRFAMs). These resources can be used to help elucidate protein function. Note that conserved domain data continue to evolve as research progresses.

There are two modes of operation for SMART: normal and genomic. The fundamental distinction lies in the protein database that is employed. The Swiss-Prot, SP-TrEMBL, and stable Ensembl proteome databases are included in Normal SMART. Only fully sequenced genome proteomes, Ensembl for metazoans and Swiss-Prot for other organisms, are utilized in Genomic SMART. Three sections comprise the entire list of genomes in Genomic SMART: bacteria, eukaryotes, and archaea. Even after identical proteins are eliminated, Normal SMART's protein database still contains much redundant information. Consider using the genomic mode instead of SMART if you want to get precise domain counts in different genomes or if you want to study domain architectures using SMART. There will not be as many protein fragments in the architectural query results that relate to the same gene, and the numbers in the domain annotation pages will be more precise. And that is applicable for small number of genomes.

2.4 GO Term Search

The specific GO terms (gene ontology) defining "signal transduction," "hormonal machinery," etc. can be searched in specific genome databases as well as database like AgriGO (http://systemsbiology.cpolar.cn/agriGOv2), AmiGO (http://amigo.geneontology.org/amigo), and Uniprot (https://www.uniprot.org/). The annotation files (mostly in .gff and .csv format) for specific genomes can be downloaded and searched on excel for selective locus id's containing the specific term. agriGO v2.0 is a Web-based gene ontology analysis tool and database. It is easy to use and focuses mainly on agricultural species. The goal of agriGO v2.0 is to offer the agricultural community comprehensive support in ontology analysis. The following are some of the new benefits and features of agriGO v2.0: Particular attention is paid to agricultural species in agriGO v2.0. Datatypes and species are supported. agriGO was also intended to be an easy-to-use Web server. Users can conduct quick and precise studies using reference datatype priorities, single species analysis, and a new species classification method. The following analysis tools were used:

- BLAST4ID (Transfer IDs using BLAST)
- SEACOMPARE (Cross comparison of SEA)
- Parametric Analysis of Gene Set Enrichment (PAGE)
- Singular Enrichment Analysis (SEA)

With these tools' help, users can conduct systematic result exploration and data mining, improving data analysis and interpretation. Customized direct acyclic graph (DAG) trees and scatter plots were produced as part of the custom analysis tools. These technologies increase the flexibility of input. Along with the distributions of the p-values (PVD) of the critical GO terms created randomly, a Batch SEA tool with numerous inputs, including time-course samples, was supplied.

2.5 Promoter Analysis

The promoter region of genes contains important cis-regulatory motifs that govern gene activity. The presence of motifs like GCC box, ERE, DRE, and CTGG in the promoters of genes suggests the possibility of their ethylene responsiveness. Thus, the putative promoter region of selected genes can be subjected for the ethylene-specific cis-regulatory elements using Plantcare (http://bioinformatics.psb.ugent.be/webtools/plantcare/html/) and PlantPan (http://plantpan.itps.ncku.edu.tw/) databases. The identified motifs can be compared with denovo motif patterns, also using MEME tool (https://meme-suite.org/meme/). For this, the promoter regions can be extracted from databases like plant Ensemble or extracted offline.

For example:

- The biomart tab from plant Ensemble database can be selected.
- In next step, the specific plant of study will be selected along with the search term (gene name, locus id, GO term).
- In next step, the sequence feature will be selected with upstream region with 2kb length, to cover the putative promoter region.
- The extracted promoter region can be searched in Plantcare and Plantpan database with full report for the present cis-regulatory elements.

Despite the abundance of programs accessible, a lot of work still needs to be done by hand. The programs are rarely flawless; occasionally, they fail to include related motifs or their opposite complements. A thorough analysis of potential themes presented by a program can provide fresh perspectives. Running multiple programs multiple times is vital as no program can guarantee accuracy the first time. To create a final presentation, we found that copying promoters in Fasta format into Word was useful after conducting in-depth theme searches using various programs. To locate and highlight/modify the font, color, or size of cis-elements like W-boxes or recently discovered potential cis-elements, we might utilize the Word search function. The outcomes should coincide, at minimum, with the information produced from databases and promoter search programs. We reduced the amount of manual labor by using a basic algorithm that compares two lists and detects similar words. Comparing many lists would be considerably quicker to determine which genes matched which lists. These programs are frequently updated, and many of them are new which resolve many earlier issues. A list of some of the databases used to identify a gene and their respective expression analysis can be found in Table 1.

Table 1
List of some bioinformatic databases and servers, which can be used in genome-wide analysis

S. No.	Databases/ servers	URL	Description	References
1	NCBI	https://www.ncbi.nlm.nih.gov/	Provides access to biomedical and genomic information	[24]
2	Uniprot	https://www.uniprot.org/	A comprehensive resource for protein sequence and annotation data	[25]
3	EnsemblPlants	https://plants.ensembl.org/index.html	A genome-level portal for plant sciences and research	[26]
4	Phytozome	https://phytozome-next.jgi.doe.gov/	The Plant Comparative Genomics portal	[27]
5	fruitENCODE	http://www.epigenome.cuhk.edu.hk/encode.html	An encyclopedia of DNA elements for fruit ripening	[28]
6	TAIR	https://www.arabidopsis.org/	A database for genetic and molecular biology data of *Arabidopsis*	[29]
7	Banana Genome Hub	https://banana-genome-hub.southgreen.fr/	Provides centralized access for genome assemblies, annotations, and omics resources available for bananas and relatives	[30]
8	Sol Genomics Network	https://solgenomics.net/	A database and website dedicated to the genomic information of the nightshade family	[31]
9	PlantGDB	https://www.plantgdb.org/	A database of molecular sequence data for all plant species with significant sequencing efforts	[32]
10	PlantTFDB	http://planttfdb.gao-lab.org/	A portal for the functional and evolutionary study of plant transcription factors	[33]
11	GEO	https://www.ncbi.nlm.nih.gov/geo/	An NCBI database for expression values of organisms	[34]
12	CDD	https://www.ncbi.nlm.nih.gov/Structure/cdd/cdd.shtml	An NCBI database for domain prediction	[35]
13	Pfam	http://pfam.janelia.org/	To explore domains and other functional regions in a protein	[36]
14	SMART	http://smart.embl-heidelberg.de/	Database for domain analysis	[37]
15	StringDB	https://string-db.org/	A database for protein–protein interaction networks functional enrichment analysis	[38]

(continued)

Table 1 (continued)

S. No.	Databases/ servers	URL	Description	References
16	AgriGO v2	http://systemsbiology.cau.edu.cn/agriGOv2/	A tool and database for gene ontology	[39]
17	Revigo	http://revigo.irb.hr/	A tool and database for gene ontology	[40]
18	Time-Tree	http://www.timetree.org/	A database for evolutionary and divergence time of species	[41]
19	KEGG Pathway	https://www.genome.jp/kegg/pathway.html	A collection of manually drawn pathways based on molecular interactions and reactions	[42]
20	KOBAS	http://kobas.cbi.pku.edu.cn/	A collection of manually drawn pathways based on molecular interactions and reactions	[43]
21	KAAS	https://www.genome.jp/kegg/kaas/	A collection of manually drawn pathways based on molecular interactions and reactions	[44]
22	GSDS Server	http://gsds.gao-lab.org/	A server to visualize the features of genes	[45]
23	MG2C	http://mg2c.iask.in/mg2c_v2.0/	A Web tool to visualize the location of genes on chromosomes	[46]
24	chromoWIZ	http://mips.helmholtz-muenchen.de/plant/chromoWIZ/index.jsp	A Web tool to visualize the location of genes on chromosomes	[47]
25	MEME	https://meme-suite.org/meme/	A Web-based tool to predict conserved amino acid patterns	[48]
26	Phyre2 server	http://www.sbg.bio.ic.ac.uk/~phyre2/html/page.cgi?id=index	A server to recognize protein folding	[49]
27	Pathreader	http://zhanglab-bioinf.com/PAthreader/	A server to recognize protein folding pathways	[50]

3 Identification of Small RNA During Fruit Ripening

Short noncoding endogenous RNAs, or "small RNAs," are a family of molecules vital to numerous biological processes. According to recent research, plants' ethylene signaling process also involves microRNAs or miRNAs. Within the tomato, LeERF1 is one of the ethylene response factors (ERFs) located downstream of the ethylene signal transduction pathway. Small RNAs were found to target many genes, some of which were implicated in the ethylene pathway. These included AP2 TFs, auxin response factors, F-box proteins, ERF TFs, APETALA2-like proteins, and MADS-box TFs.

A regulatory model that illustrates the connections between the targets of short RNAs engaged in ethylene signaling and their regulators was established. This chapter offers fundamental data for more research on the role of short RNAs in the ethylene pathway and fruit ripening. The initial step in performing the small RNA analysis is to find data submitted to a public database [51].

Further, Small RNA Sequencing Raw Readings can be accessed as Small RNA Sequencing Fastq formatted data. The Sequence Read Archive (SRA) of the National Centre for Biotechnology Information (NCBI; sra.ncbi.nlm.nih.gov) provides publicly available data for download. The following resources are required to identify known and novel miRNAs. The installation instructions for the tool are available in the following website:

1. Trimmomatic v0.39 (http://www.usadellab.org/cms/?page=trimmomatic [52]).
2. Cutadapt v2.10 (https://cutadapt.readthedocs.io/en/stable/guide.html [53]).
3. FASTX-Toolkit v0.0.14 (http://hannonlab.cshl.edu/fastx_toolkit/) [54].
4. Bowtie v1.2.2 (http://bowtie-bio.sourceforge.net/ [55]).
5. ViennaRNA Package v2.4.15 (http://www.tbi.univie.ac.at/~ivo/RNA/ [56]).
6. randfold v2.0 (https://github.com/erbon7/randfold) [14]),
7. miRDeep-P v1.3 (https://sourceforge.net/projects/mirdp/ [57]),
8. SAMtools v1.10-2 (https://github.com/samtools/ [58]).

The ethylene production by osmotic stress triggers or maintains the signaling cascade of ethylene-sensitive (EIN) proteins. The maturity of various fruits, such as banana fruits, depends on the synthesis of ethylene, fatty acids, esters, other secondary plant products, and cell wall hydrolysis activities [59]. Our understanding of how hormone action and miRNAs govern plant development has dramatically increased with the recent revelation of the link between phytohormone responses and miRNA pathways. The development of seedlings, the senescence of leaves and flowers, the stimulation of fruit ripening, the determination of floral sex, resistance to pathogen infection, and stress adaptability are the physiological and morphological responses that ethylene regulates [60–62].

Tomato is a highly effective model system for fruit ripening and development. Furthermore, tomato fruit has been the subject of most research on the function of ethylene during fruit ripening [63, 64]. One study suggests that mature green fruits were treated with exogenous ethylene and 1-methyl cyclopropene to investigate the correlations between ethylene treatment and miRNA expression. The expression levels of several known and unknown miRNAs were examined using stem-loop RT-PCR. Exogenous ethylene therapy resulted in the downregulation of three miRNAs (miR394, miR414, and miR1917), the upregulation of five miRNAs (miR156, miR159, miR396, miR482, and miRZ7), and no effect on one miRNA (miR828), whose target is *EIN2* [65].

Fruit growth is regulated by several ripening transcription factors (TFs) and ethylene signaling elements, such as ETHYLENE RESPONSE FACTOR (ERF). Interestingly during fruit ripening of citrus fruits, wild type (WT) and mutant type (MT) showed different expression levels of four ERF TFs (ERF RAP2-3, ERF012, ERF023, and AP2/ERF-B3 domain-containing transcription repressors). ERF1 and ERF2 expression levels were decreased by ethylene, suggesting that ethylene may play a role in regulating specific features of citrus fruit peel maturity [66]. By directly controlling downstream genes (ACO, ACS, EXP, PSY, XET, PME, and PG) that impact hormone synthesis, carotenoid synthesis, and fruit softening, ERFs can also serve as critical genes in the climacteric fruit [67–71]. When differentially expressed miRNAs from WT and MT fruits were compared, several changes associated with ethylene pathways were found. These changes included miR172a, which targets the AP2 ethylene-responsive transcription factor-TOE3; miR9470-3p, which targets ERF5 and ERF021; and miR396a-5p and miR477-3p, which target ETHYLENE INSENSITIVE 3-like[72]. Tomato slymiR1917, which regulates tomato ethylene responses, breaks down SlCTR4 splice variants (SlCTR4sv) [71, 72]. On the other hand, EIN3 and EIN3-LIKE1 (EIL1), two downstream transcription factors (TFs), further activate a multitude of ethylene response factors (ERFs) that facilitate the production of stress-susceptible genes. Ethylene's effects on sRNA levels in many species are well understood, but its functions in the body's reaction to abiotic stress still need to be understood. These functions shed light on the differences in gene expression that ethylene controls. Notably, miR172a (target AP2 ethylene-responsive transcription factor-TOE3), miR9470-3p (ERF5 and ERF021), and miR396a-5p and miR477-3p (target ETHYLENE INSENSITIVE 3-like) all suggested alterations in ethylene pathways [72]. The non-conserved tomato slymiR1917 controls the degradation of the SlCTR4 splice variant (SlCTR4sv) and, in turn, the tomato ethylene responses [50, 73]. The DkERF8/16/18 genes in persimmon fruit may facilitate fruit maturity by increasing ethylene synthesis and changing the cell wall [74]. According to reports, ethylene in rice has been linked to miR159 and miR394 [75]. Although it has been discovered that MiR159 also targets a distinct non-MYB target in tomatoes, its primary target is MYB [76]. Another critical target of miR159 is 1-aminocyclopropane-1-carboxylate synthase, which is necessary for the synthesis of ethylene [77]. Consistent with our observations, climacteric plants synthesize large amounts of ethylene during the breaker stage, and miR159 expression decreases simultaneously.

Furthermore, miR394 and miR414 target F-box family proteins, essential for several plant hormones' signal transduction pathways. The ethylene signaling regulators EIN2 and EIN3 are stabilized by the F-box proteins ETP1/2 and EBF1/2, respectively [78, 79]. A target of miR828 called Ethylene Insensitive 2 (EIN2), the principal positive regulator of ethylene signal transduction and other hormones like abscisic acid, auxin, cytokinin, and jasmonate, may function as a hub for communication between many hormone signaling pathways [62, 80, 81]. CTR1, a negative regulator of the ethylene response, most likely interacts directly with receptor molecules to produce a signaling complex [82, 83]. Like ethylene receptors, CTR1 genes are expressed in every tissue studied, and the mRNAs linked to these genes differ according to the tissue. LeCTR1 induction has been confirmed as a target by MiR1917, and it is associated with tissues during developmental stages related to high ethylene, like fruit ripening [84]. miRNA expression patterns can be altered by exogenous ethylene treatment; all but miR828 of these miRNAs have a negative relationship with their targets. Some of the reported miRNAs have been identified in fruit ripening and mentioned in Table 2.

The single-end sRNA sequencing data produced by the Illumina sequencing technology can be used in the method that is illustrated below.

3.1 Methods: Preprocessing of Data

The initial quality control process for the raw read from sRNA-seq involves eliminating adaptor sequences and poor-quality reads. Trimmomatic can be used to finish this stage using the following command.

1. **java -jar /home/nbri/bin/Trimmomatic-0.39/trimmomatic-0.39.jar SE –threads 2 –SL-1.fq.gz phred33 SL-1_cleaned.fq ILLUMINACLIP:/home/nbri/bin/Trimmomatic-0.39/adapters/TruSeq3-PE-2.fa:2:30:10 SLIDINGWINDOW:4:15 MINLEN:36**

In this instance, trimmomatic includes the adapter sequences in the illumina.fa file along with trimmomatic. With the ILLUMINACLIP option, Trimmomatic is instructed to delete adapters and other Illumina-specific sequences from the reads. The SLIDINGWINDOW option offers the window size and the Phred-quality score for cutting low-quality bases. The number of threads "-threads" can be changed to speed up this operation depending on the number of cores that are available on the user's machine. Moreover, the poly-A tails on the reads are trimmed, and Cutadapt is used to reject reads that are longer than 34 nt and shorter than 18 nt by following the command.

Table 2
Some of the reported miRNAs identified in fruit ripening

Fruit biology	Classification	Species	miRNA	Target	Function	References
Fruit development	Fruit set and size	Apple	miR172p	AP2	Reduced fruit size and altered organ development	[96, 97]
		Tomato	miR156	SPL	Fruit growth, ovary, and fruit development	[98]
			miR159	*SlGAMY B2(GA biosynthesis gene)*	Large fruit	[99]
			miR172d	*AP2*	Floral organ identity and number	[100]
			miR396A/B	*GRF*	Bigger fruits	[101]
			MiR1917	*CTR4(altered ethylene response)*	Fruit size	[102]
			MiR171a	*SlGRAS24 & sl GRAS40 (altered gibberellins and auxin signaling)*	Cell number size	[103]
			miRNA164a	*NAM2/3*	Decreased fruit size	[104]
	Fruit set	Tomato	miRNA159	*SlGAMYB2* (GA biosynthesis gene)	Fruit morphology, precocious fruit initiation flattened, fruit with more locules inside	[105]
			miR160	*ARF10, ARF16* and *ARF17*	Sugar accumulation, leaf and flower development, somatic embryo development, pear-shaped fruit	[106]
			miR166	*SlREV*	Fruit formation	[107]
			miR168	*SlAGO1s*	Fruit initiation and growth	
		Pear	miR397a	*LACs*	Stone cell formation, reduced lignin content and stone cell number	[108]
		Longan	miR160	*ARF10, -16,* and *-17*	Somatic embryo development	[109]
	Seed development/ parthenocarpy	Tomato	miR159	*GAMyb-like1* and *GAMyb-like2*	Parthenocarpy	[104]
			miR166	*SlHB15A*	Parthenocarpic fruit set	[110]
			miR167	*SlARF8*	Parthenocarpy	[104]
			miR168	*SlAGO1s*	Parthenocarpy	[107]
			miR172	*AP2*	Small parthenocarpic fruit-like organ	[100]

(continued)

Table 2
(continued)

Fruit biology	Classification	Species	miRNA	Target	Function	References
Fruit ripening	Fruit color	Litchi	miR156a	*LcSPL1/2*	Anthocyanin biosynthesis	[111]
			NEW41	*CHI*	Anthocyanin accumulation	[111]
		Pear	miR156	*SPL*	Red peel coloration, anthocyanin biosynthesis	[112]
		Blueberry	miR156a	*VcSPL12*	Anthocyanin accumulation	[113]
			miR396	*FtsZs*	Coloration	[114]
			miR_n10	*BAG1*	Coloration	[115]
		Apple	miR172	*AP2-MYB10*	Flavonoidse, reduction in red coloration	[116]
			miR7125 (light-induced)	*MYB16/MYB1-CCRs*	Promoted anthocyanin synthesis, reduced lignin biosynthesis	[114]
			mdm-miR828	*TAS4-MdMYB1*	Inhibited anthocyanin synthesis	[117]
			miR858	*MYB*	Anthocyanin biosynthesis	[118]
Fruit ripening	Fruit color	Grape	miR858	*VvMYB114*	Anthocyanin and flavonol accumulation	[119]
			miR858a	*MYB2*	Anthocyanin accumulation, anthocyanin biosynthesis	[120]
			miR156	*SPL9* and *SPL15*	Anthocyanin biosynthesis	[121]
		Tomato	miR858	*SlMYB7* and *SlMYB48*	Anthocyanin accumulation	[120]
		Kiwifruit	miR858	*AaMYBC1*	Anthocyanin biosynthesis	[121]
Fruit ripening	Fruit ripening, fruit softening, and fruit quality	Persimmon	miR395	*bHLH*	Tannin biosynthesis	[122]
			miR156	*SPL*	Tannin biosynthesis	
			miR396	*Flavonoid 3-O-glucosyltransferase* (*UFGT*)	Tannin biosynthesis	
			miR858	*MYB19/20*	Reduced the content of proanthocyanidin (PA)	
			miR2991	*ADH*	Tannin biosynthesis	
		Strawberry	*FRILAIR* (lncRNA)-miR397	*LAC11a*	Delayed fruit ripening	[123]

	fan-miR73	*ABI5*	Fruit ripening	[124]
	miR399	*PHO2*	Flavor, sugar content	[125]
	miR157	*SPL-CNR*	Delayed fruit ripening	[126]
	miR156	*SPL*	Accelerates tomato fruit softening	[127]
	miR172	*AP2a*	Accelerates fruit ripening with enhanced ethylene biosynthesis	[105]
	miR166	*SlREV*	Fruit ripening	[128]
	miR828	*EIN2*	Ethylene-dependent ripening	[129]
	miR1917	*CTR4*	Enhances ethylene response and accelerates fruit ripening	[130]
Kiwifruit	miR164	*NAC6/7*	Fruit ripening	[116]
Apple	miR7125	*MYB16/MYB1-CCRs*	Reduced lignin biosynthesis	[131]

2. **cutadapt -a "A{20}" -m 18 -M 34 -o SL-1_trimmed.fq SL-1_cleaned.fq**

The -m and -M parameters in this case define the minimum and maximum lengths of the sequences to be retained after trimming. The aforementioned filtering techniques are applied to all samples, and the filtered reads from each sample are then joined together into a single fastq file using the cat command.

3. **cat .trimmed.fq>all_reads.fq**

The combined reads are divided into unique tags and compressed in fasta format. The readings are collapsed using the FastxCollapser from the FASTX-Toolkit. The unique tag file is further prepared to be compatible with miRNA prediction tools and to be used in downstream analysis using the sed programme.

4. **cat all_reads.fq | paste - - - - | sed 's/^@/>/g' | cut -f 1,2 | tr '\t' '\n' >all_reads.fa**
5. **fastx_collapser -iall_reads.fa -o all_uniquetags.fa**
6. **sed -i 's/-/_x/' all_uniquetags.fa**

The last step in the data preparation procedure is the elimination of read mapping to ribosomal RNA (rRNA), transfer RNA (tRNA), small nuclear RNA (snRNA), small nucleolar RNA (snoRNA), and repeat sequences. By using Bowtie to align reads with a database containing rRNA, tRNA, snRNA, and snoRNA sequences, rRNA, tRNA, snRNA, and snoRNA can be removed from the reads (can be downloaded from the Rfam database [85]). Similar to this, read mapping to repetition regions can be avoided by using the Repbase database (https://www.girinst.org/repbase/). Before beginning the alignment stage, users must first create an index file for the database using bowtie-build. These commands are required here:

7. **bowtie-build rfam.farfam**
8. **bowtie rfam -f all_uniquetags.fa -S unique_tags.ncRNA.sam --ununique_tags.unaligned.ncRNA.fa**
9. **bowtie-build repeats.fa repeats**
10. **bowtie repeats -f unique_tags.unaligned.ncRNA.fa -S unique_- tags.unaligned.ncRNA.repeats.sam --ununique_-tags.filtered.fa**

Here, the options -f and -S designate that the output is in sequence alignment map (SAM) format and that the input file is in fasta format, respectively. The name of the file in which the unaligned readings are stored is specified by the command option -un.

3.2 Identification of Known miRNAs

The miRBase database's (Fig. 2) [14] known plant miRNAs were matched against the filtered reads (unique_tags.filtered.fa) using Bowtie to find conserved or recognised miRNAs.

11. **bowtie-build mirbase.famirbase**
12. **bowtie mirbase -n 2 -f unique_tags.filtered.fa -S unique_-tags.filtered.mirbase.sam --unNovel.fa**

Thus, the parameter "-n" can be used to alter the number of mismatches; in this example, we have chosen two mismatches. By parsing the alignment file with the following program, you may obtain the list of recognized miRNAs and their sequences:

13. **grep -v "^@" unique_tags.filtered.mirbase.sam | awk -F"\t" '{if($3!="") print}' | cut -f 1,3 | sort -u >known_miRNAs.**

3.3 Identification of Novel miRNAs

For the prediction of novel miRNAs, the unique reads that do not align with the miRBase database are employed. In this instance, the miRDeep-P package will be used to make the prediction. In order to forecast miRNAs based on plant-specific characteristics, nine Perl scripts are successively executed in the free package known as miRDeep-P [86]. With the software, sample datasets and a thorough manual are included. For the purpose of predicting novel plant miRNAs, reads that do not correspond to known plant miRNAs are first mapped to the reference genome using Bowtie with either a 0 or 1 mismatch.

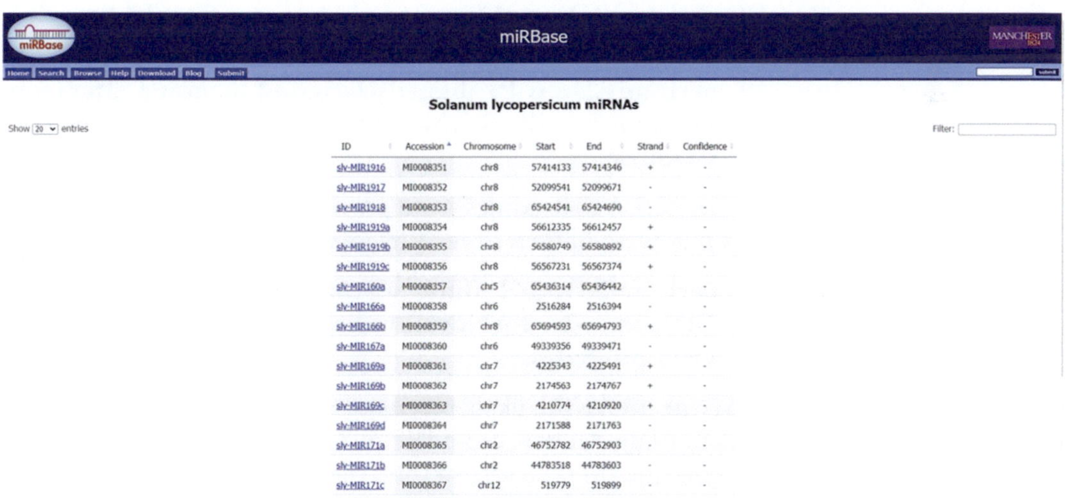

ID	Accession	Chromosome	Start	End	Strand	Confidence
sly-MIR1916	MI0008351	chr8	57414133	57414346	+	-
sly-MIR1917	MI0008352	chr8	52099541	52099671	-	-
sly-MIR1918	MI0008353	chr8	65424541	65424690	-	-
sly-MIR1919a	MI0008354	chr8	56612335	56612457	+	-
sly-MIR1919b	MI0008355	chr8	56580749	56580892	+	-
sly-MIR1919c	MI0008356	chr8	56567231	56567374	+	-
sly-MIR160a	MI0008357	chr5	65436314	65436442	-	-
sly-MIR166a	MI0008358	chr6	2516284	2516394	-	-
sly-MIR166b	MI0008359	chr8	65694593	65694793	+	-
sly-MIR167a	MI0008360	chr6	49339356	49339471	-	-
sly-MIR169a	MI0008361	chr7	4225343	4225491	+	-
sly-MIR169b	MI0008362	chr7	2174563	2174767	+	-
sly-MIR169c	MI0008363	chr7	4210774	4210920	+	-
sly-MIR169d	MI0008364	chr7	2171588	2171763	-	-
sly-MIR171a	MI0008365	chr2	46752782	46752903	-	-
sly-MIR171b	MI0008366	chr2	44783518	44783603	-	-
sly-MIR171c	MI0008367	chr12	519779	519899	-	-
sly-MIR171d	MI0008368	chr12	2882790	2882901	+	-

Fig. 2 Screenshot of the miRBase software highlighting miRNA IDs, accession numbers, start, and end positions, respectively

14. **bowtie-build Sol_genome.faSol_genome**
15. **bowtie Sol_genome -n 0 -f Novel.fa -S Novel_genome.sam**

Following that, the SAM-formatted bowtie alignments are converted to blast format using the "convert_SAM_to_blast.pl" script included with the miRDeep-P package. The reference genome (in fasta format), bowtie alignments, and unique tags must all be included in the file needed by the script:

16. **perl miRDeep-P/convert_SAM_to_blast.pl Novel_genome.sam Novel.fa Sol_genome.fa >genome.bst**

Using the "filter_alignments.pl" script, the alignments are further filtered to only keep those that have full-length alignment, 100% sequence identity, and a user-specified cutoff (c 15 in this example; this can be altered depending on the species of interest).

17. **perl miRDeep-P/filter_alignments.pl genome.bst -c 15 > genome.filter15.bst**

Reads that overlap with the exons, CDS, or other recognized annotated characteristics of the species under study are disregarded. Phytozome (https://phytozome.jgi.doe.gov/), Ensemble (https://plants.ensembl.org/index.html), and NCBI (https://www.ncbi.nlm.nih.gov/) are a few examples of public databases where the appropriate annotations can be discovered depending on the species of interest. This phase is performed by the "overlap.pl" and "alignedselected.pl" scripts:

18. **perl miRDeep-P/overlap.pl genome.filter15.bst Sol_-gene.gff -b > genome.filter15.overlap_CDS**
19. **perl miRDeep-P/alignedselected.pl genome.filter15.bst -g genome.filter15.overlap_CDS > genome.filter15.CDS.bst**

The script that comes next provides a fasta sequence for the reads that were filtered in the previous step.

20. **perl miRDeep-P/filter_alignments.pl genome.filter15.CDS.bst -b forNovelPrediction.fa> genome.filter15.CDS.filtered.fa**

We next take the likely precursor sequences from the reference genome using "excise_candidate.pl." The script receives the reference genome (in fasta format) and the filtered alignments. According to the creators of miRDeep-P, a window size of 250 bp is optimal for obtaining precursor sequences for both monocot and dicot plants.

21. **perl miRDeep-P/excise_candidate.pl Sol_genome.fa genome.filter15.CDS.bst 250 > genome.filter15.CDS.precursors.fa**

The secondary structures of the likely precursor sequences are then predicted using the RNAfold tool from the ViennaRNA package [56]. Users can turn off the graphical output by using the —noPS option.

22. **cat genome.filter15.CDS.precursors.fa | RNAfold --noPS> genome.filter15.CDS.structures**

The filtered reads are now matched to likely precursor sequences to provide miRNA signatures (Fig. 3). The alignment file is converted to blast format using the steps below, which are then used to sort the alignment file to extract signatures from the precursor sequence alignment file:

23. **bowtie-build -f genome.filter15.CDS.precursors.fa genome. filter15.CDS.precursors**
24. **bowtie genome.filter15.CDS.precursors –f genome.filter15.CDS.filtered.fa -S genome.filter15.CDS.precursors.sam**
25. **perl miRDeep-P/convert_SAM_to_blast.pl genome.filter15.CDS.precursors.sam genome.filter15.CDS.filtered.fa genome.filter15.CDS.precursors.fa >precursors.bst**
26. **sort +3 -25 precursors.bst> signatures**

After the signatures have been created, the "miRDP.pl" script uses this information along with structures identified using RNAfold to make predictions for miRNAs.

27. **perl miRDeep-P/miRDP.pl signatures genome.filter15.CDS.structures -y > predictions**

The projected miRNAs are then screened with the "rm_redundant_meet_plant.pl" script to remove redundant miRNAs and miRNAs that do not fit the criteria for plant miRNAs. This script requires the length of each chromosome from the reference genome as well as miRNA predictions. To measure chromosomal lengths, use the faidx tool in Samtools.

28. **Samtools faidx Sol.genome.fa**
29. **perl miRDeep-P/rm_redundant_meet_plant.pl Sol_genome.fa.fai genome.filter15.CDS.precursors.fa predictions nr_predictionsnovel_plant_miRNAs**

The novel miRNAs obtained in this stage, along with the known miRNAs, make up the final list of miRNAs (all miRNAs).

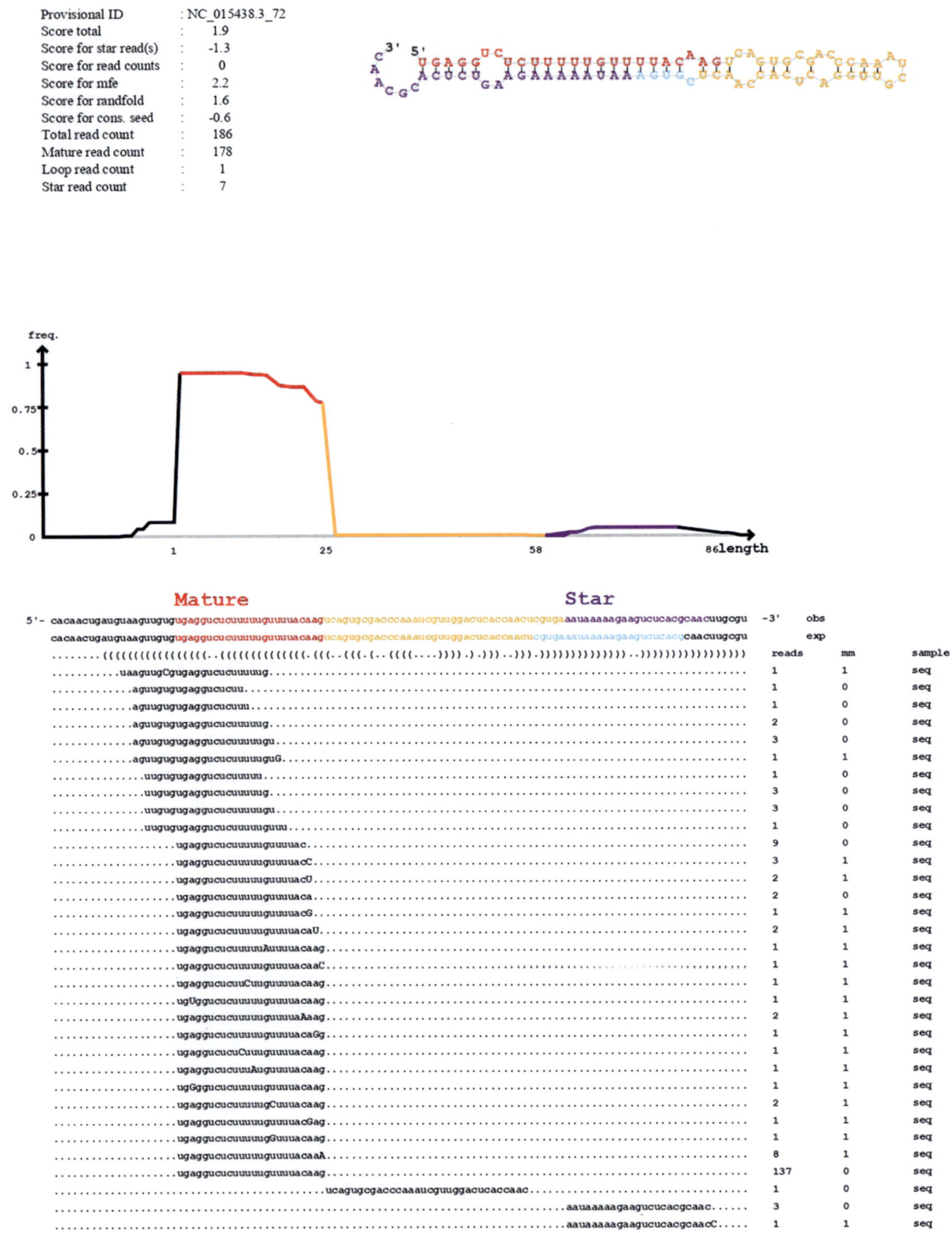

Fig. 3 Screenshot of miRDeep-P 'signature' output showing mature and star strand read piles after alignment of filtered sRNA reads to predicted precursors

3.4 miRNA Targets Prediction

To fully understand the function of miRNAs, it is essential to forecast their targets. In contrast to animal miRNAs, plant miRNAs are known to display perfect or nearly perfect complementarity with their targets. For miRNA target prediction, a variety of techniques with differing degrees of specificity and sensitivity are available using the complementarity attribute. Tools for target prediction such as TAPIR [87], psRobot [88], comTAR [89], and psRNATarget [90] are frequently utilized. In addition to computational approaches based on similarity, degradome sequencing is a highly sensitive and effective method for predicting miRNA targets [65, 91]. Several computational tools, including CleaveLand [92], SeqTar [93], sPARTA [94], and PAREsnip2 [95], can be used to evaluate the degradome sequencing data, which provides patterns of RNA degradation. psRNATarget is what we use in this demonstration. This presentation makes use of the psRNATarget server, which has an intuitive graphical user interface and can be found at (http://plantgrn.noble.org/psRNATarget/analysis) [90]. The target candidate's optimal sRNA/mRNA complementarity location is determined using an accelerated Smith–Waterman method, and it is also possible to determine if the miRNA is the cause of cleavage or translational inhibition. To produce a prediction, upload the small RNA sequences to the online site and select the relevant species database. If there are no transcript sequences available for a particular species, users can provide both the small RNA and the transcript sequences for prediction. Figure 4 shows an example of psRNATarget results.

Only 20 tomato miRNAs have been examined thus far using the aforementioned experimental techniques. The list of miRNAs discovered by some of the pipelines listed above is shown in Table 2.

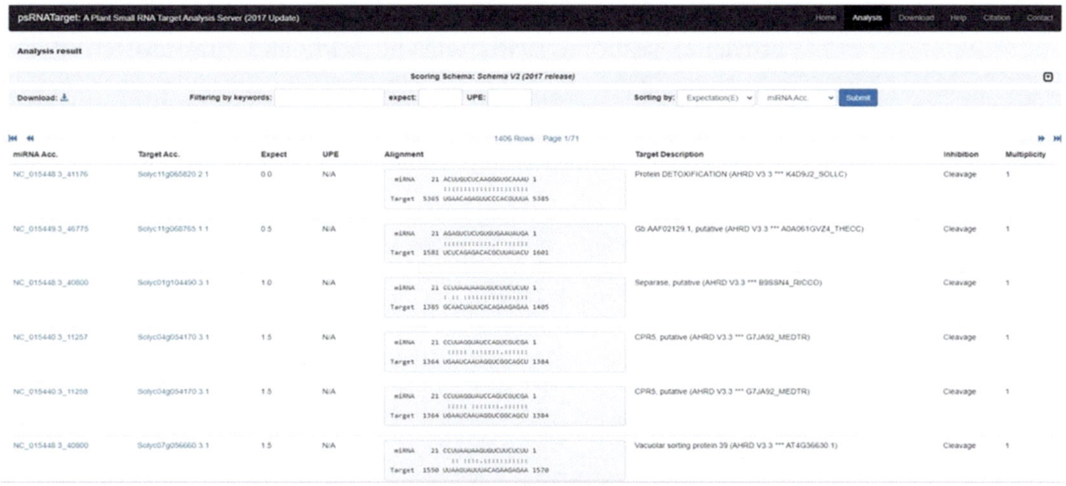

Fig. 4 Screenshot of the miRNA–mRNA pair highlighted in the psRNATarget results. miRNAs and their targets are shown in the columns "miRNA Acc." and "Target Acc.", respectively

4 Hormone Databases

In the last two to three decades, many highly advanced and high-throughput techniques for nucleotide sequencing have generated enormous data that links them to too many biological and physiological processes in plants, and these data have been available to the public all over the globe through databases and Web servers. These omic data are advantageous in studying gene regulatory relationships to understand and speculate their functionality [132]. Due to large data sets, these repositories have managed to provide comprehensive and systematic cataloging of functionally annotated hormone-related gene profile, sequences, micro-array data, transcriptome networks, phenotypic trait information, and genes related to hormone signaling [133, 134]. In this section, we will discuss two databases, namely Arabidopsis Hormone Database (AHD) and gene set–level analyses of hormone responses in Arabidopsis thaliana (GSHR), which provide a base for the studies, findings, and interpretation of gene function and regulation related to plant hormones, their phenotypic system, gene expression concerning various stimuli and deficiency, cluster analysis, co-expression network, enrichment analysis of biological pathways, GO enrichment for function and Interpro enrichment for domains, various applications, and features in plant hormonal regulation and their cascade signaling (Table 3) [134].

4.1 AHD: Arabidopsis Hormone Database

AHD is a plant hormone database consisting of extensive hormone-related genes which are based on the information from the identified Arabidopsis genes concerning plant hormone activities. The database was established on the studies of mutant, transgenic analysis, and gene ontology (GO) annotation and gathered a total of 1026 genes in the Arabidopsis genome that is related to eight plant hormones including auxin, gibberellins, cytokinin, abscisic acid, ethylene, jasmonic acid, salicylic acid, and brassinosteroid that

Table 3
List of databases

Database	Category	Link	Description
AHD and AHD2.0 Arabidopsis Hormone Database	Database	http://ahd.cbi.pku.edu.cn	A comprehensive catalog of genes playing important role in plant hormonal regulation, and morphological phenotypes managed by plant hormones
GSHR Gene set–level analyses of hormone responses in Arabidopsis thaliana	Web server	http://bioinfo.sibs.ac.cn/GSHR/	Provides a comparative, comprehensive, and systematic view of gene set–level analyses of *Arabidopsis thaliana* hormone response

show some key function in plant hormone signaling and mechanism related to plant growth, development, and responses toward environmental stress [133]. AHD has a user-friendly Web interface which can be accessed from the link provided http://ahd.cbi.pku.edu.cn and maintained by the Health Sciences Library System (HSLS) at the University of Pittsburgh, US (https://www.hsls.pitt.edu/obrc/index.php?page=URL1231258875) that allows its users to get access to particulars about the plant hormone-related genes, sequences, mutant details, phenotypic summary, functional categorization, and microarray data. Also, a phenotype ontology is provided which thoroughly and precisely explains the numerous hormone-regulated morphological activities.

4.1.1 Web Interface and Application

The graphical interface provides detailed information about hormone related genes. It features functional categories, gene descriptions, locus names, gene models, gene sequences, and gene ontology annotations sourced from the TAIR database [136]. Additionally, it includes metabolic pathway annotations from the KEGG database and data on protein-protein interactions (PPI) [136]. Users can also find a list of transgenic plants linked to each gene along with their respective phenotypes. The interface offers graphical representations of gene expression data that highlight differential regulation by various hormones and incorporates a standardized phenotypic trait classification system for mutants identified in plant hormone studies. Also, users have the facility to search for gene information by searching its name, description, or related hormones and gene sequence in NCBI BLAST search (Fig. 5).

AHD provides well-classified and organized information on hormone-regulated phenotypes and their associated genes to understand and examine the phenotypes related to the effect of various hormones via phenotype subclasses, co-regulation of different morphological features, and hormone-associated genes involved in different numerous hormone responses [133]. These classified pieces of information help in interpreting the hormone signaling pathways for a particular trait through interconnected regulatory networks of genes. The first version of AHD was released in 2008, and an updated version AHD2.0 was released in 2010 with some additional features like the update of stored Arabidopsis hormone-related genes with the predicted microRNA splicing site (954 miRNA splicing site targeting HRGs) and featured elaborated diagrams of several plant hormone biosynthesis and signaling pathways, added orthologs of sequences listed in OrthoMCL-DB to each Arabidopsis hormone-related genes for comparative and evolutionary studies, integration of genes that are genetically interconnected to each AHRG, linking of WebLab tool for online bioinformatics analysis for on-time sequence alignment, prediction of 2D and 3D protein structure and motif

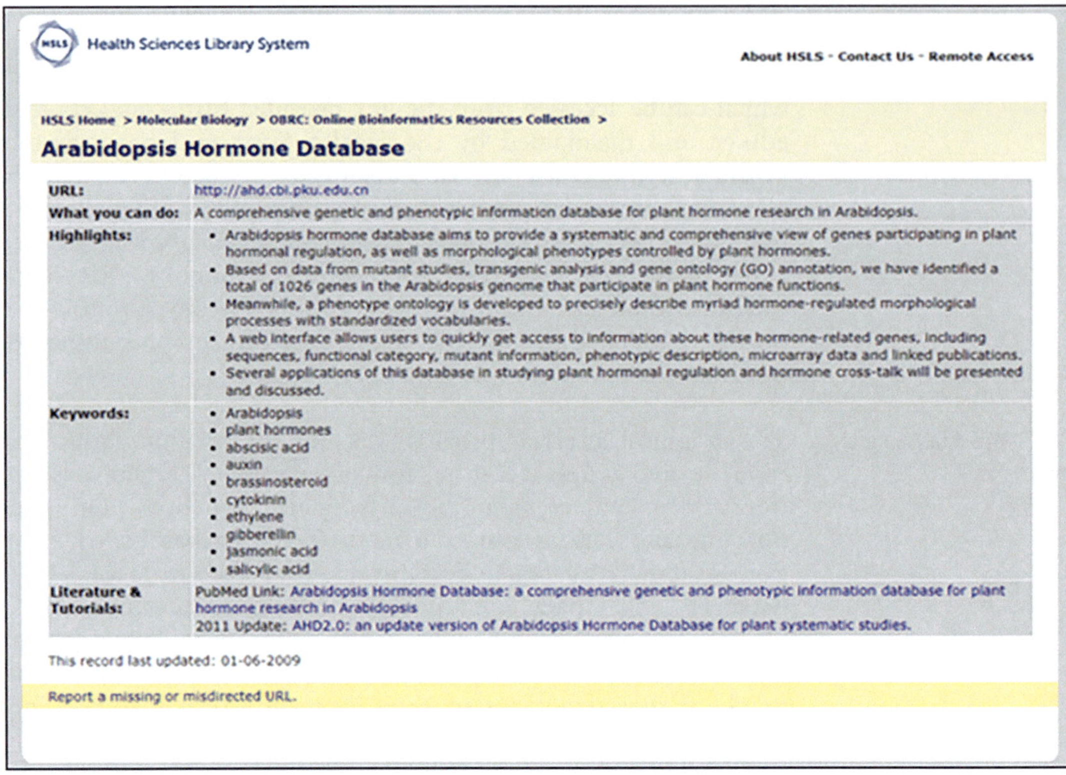

Fig. 5 Screenshot showing the Health Sciences Library System (HSLS) at the University of Pittsburgh, USA, which maintains the *Arabidopsis thaliana*hormone Database. Website source: https:/www.hsls.pitt.edu/obrc/index.php?page=URL1231258875

searches, and cross-linking of some protein databases like Pfam, PROSITE, PRINTS, SuperFamily, Gene3D, PANTHER, PIR, HAMAP, TIGRFAM, SMART, ProDom for predicting gene functionality [136]. In addition to this, there is the integration of more gene expression information which was extracted from AtGenexpress which includes data sets of several developmental stages and abiotic stresses [136].

Additionally, data obtained from AHD2.0 can be analyzed using statistical techniques. Consider the relationships that exist between genes related to hormones and microRNA. Remarkably, this database and analysis showed that 88% of the genes associated with ethylene have interactions with microRNA. This suggests that short RNA regulation likely covers many ethylene response pathways, consistent with other research findings [18–20].

We also examined primary root elongation, a phenotypic property that was reported to be regulated by several hormones, to ascertain how ethylene, in accordance with several hormones, coordinates the regulation of a common phenotypic trait. AHD database search yielded 120 mutants that, under normal or hormone-treated conditions, have longer primary roots than wild type. The phenotype is regulated by 63 genes, all of which are represented by these mutations. Twelve of them are linked to Cytokinin responses, twenty-four to ethylene, two to abscisic acid, eighteen to auxin, eleven to jasmonic acid, one to gibberellin, and one to brassinosteroid. We examined whether the encoding products of these genes exhibited protein–protein interaction (PPI) in order to gain insights into the molecular process underlying how these genes cooperate to control primary root elongation. 23 of the 63 proteins can be mapped using the Arabidopsis interactions viewer Web server (http://bar.utoronto.ca/interactions/cgi-bin/arabidopsis_interactions_viewer.cgi) to create protein–protein interaction networks. Furthermore, in a single network, 17 of the 23 hormone-related proteins are coupled either directly or through one or more intermediate proteins. It is interesting to note that 14 genes are associated with a single type of hormone, while only 3 genes respond to numerous hormones. This finding suggests that, rather than through shared genes, ethylene and other hormones govern a common process through sharing an intricate regulatory network [133].

4.2 GSHR: Gene Set–Level Analyses of Hormone Responses in Arabidopsis thaliana

GSHR is a Web server that utilized the RNA-seq and microarray datasets extracted from Gene Expression Omnibus (GEO) and gives categorical hormone-related transcriptomic variations in response to several plant hormones in *Arabidopsis thaliana*. It has 1368 gene sets which are the result of upregulated and down-regulated genes, and these sets of genes are involved mainly in hormonal signaling [134]. It combines numerous plant hormone-associated transcriptomic data sets and provides gene list-level analyses of user-input gene lists. RNA-seq reads were selected and mapped, and its differentially expressed genes were calculated through DEseq having fold change > 1 and P value < 0.05, whereas microarray data were processed by Bioconductor package affy, limma; quantile normalization. For enrichment analysis, Fisher's exact test was used to get the compared gene sets of transcriptomic data. Co-expression network was constructed by using both RNA-seq and microarray datasets in which function rcorr in R package Himsc was applied to calculate Pearson correlation coefficients above 0.70 [134].

4.2.1 Web Interface and Application

The GSHR Web server (http://bioinfo.sibs.ac.cn/GSHR/) was developed by using PHP and JavaScript for an interactive Web interface and MySQL for data management to facilitate the users to retrieve desired output related to hormones and hormone-related factors by searching gene or gene lists of *Arabidopsis thaliana* Genebank IDs. GSHR displays very detailed and statistical information about the input gene lists like lists of hormones regulated by those genes, description of hormone biosynthesis and biological processes, co-expression network, enrichment analysis of pathways, hierarchical clustering of expression patterns, gene ontology, and InterPro domains.

GSHR can be used to interpret the hormonal response, gene functionality, and signaling for a variety of stimuli in plants related to physiological, morphological, and environmental stresses. It allows and permits gene lists as input, generated from any high-throughput technique data to get clustered enriched gene sets belonging to a particular protein family which is involved in specific biological processes or hormonal signaling. This database provides a comparative study of gene sets produced from different platforms and different experiments, regulating the transcription of hormones and hormone-related factors for different biological pathways and hormonal signaling (Fig. 6) [134].

4.2.2 Application of GSHR: Hormones Associated with Plant Response to Phosphate Deficiency

We employed genes upregulated by phosphate starvation in *Arabidopsis thaliana* roots as input to demonstrate the application of GSHR [137], and according to the GSHR output, ABA, auxin, and ET significantly controlled a portion of the input genes, suggesting that these hormones may play a role in plant responses to phosphate deficit. ABA-responsive genes were increased when cells were cultivated in low-phosphate environments, consistent with

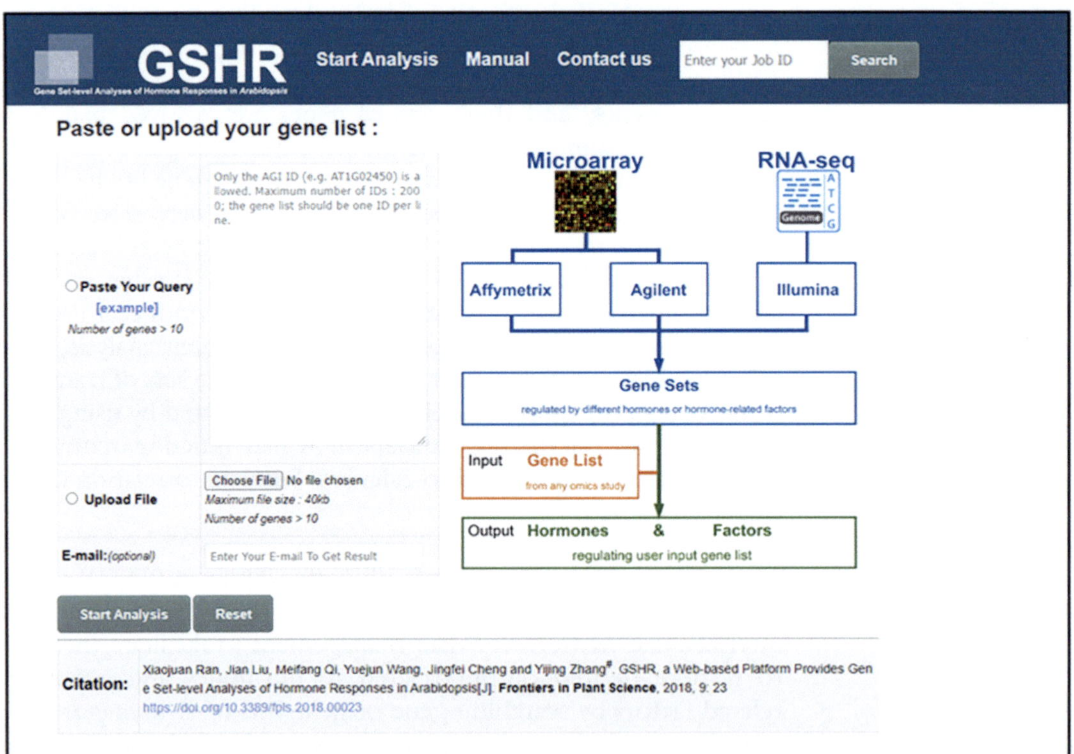

Fig. 6 Screenshot showing the homepage of GSHR database. (*Web source:* http:/bioinfo.sibs.ac.cn/GSHR/)

the findings [138, 139]. Furthermore, reduced phosphorus increased Arabidopsis roots' sensitivity to auxin and ET, changing the roots' architectural makeup [140]. After determining which hormones are relevant, the enriched gene sets may be chosen for additional functional examination. Their expression responses to particular hormones were analyzed using clustering and visualization. Further, the heatmap indicated that both up- and downregulated genes were seen upon treatment with ET and that IAA and ABA predominantly upregulated the chosen gene sets. High-degree genes in the co-expression network are also promising candidates for important mediators connecting phosphate signaling and hormone response signaling pathways. Also, the ethylene biosynthesis pathway and ABA signaling pathway were predominantly found in the enriched gene ontology terms.

Acknowledgment

The CSIR National Botanical Research Institute assigns the manuscript number as CSIR-NBRI_MS/2023/05/09.

References

1. Li X, Wang X, Zhang Y, Zhang A, You CX (2022) Regulation of fleshy fruit ripening: from transcription factors to epigenetic modifications. Hortic Res
2. Li S, Chen K, Grierson D (2021) Molecular and hormonal mechanisms regulating fleshy fruit ripening. Cells 10(5)
3. Li J, Zou X, Chen G, Meng Y, Ma Q, Chen Q, Wang Z, Li F (2022) Potential roles of 1-aminocyclopropane-1-carboxylic acid synthase genes in the response of gossypium species to abiotic stress by genome-wide identification and expression analysis. Plants (Basel) 11(11):1524
4. Parida AP, Raghuvanshi U, Pareek A, Singh V, Kumar R, Sharma AK (2018) Genome-wide analysis of genes encoding MBD domain-containing proteins from tomato suggest their role in fruit development and abiotic stress responses. Mol Biol Rep 45(6): 2653–2669
5. Iqbal N, Khan NA, Ferrante A, Trivellini A, Francini A, Khan MIR (2017) Ethylene role in plant growth, development and senescence: interaction with other phytohormones. Front Plant Sci 8:475
6. Qu X, Hall BP, Gao Z, Schaller GE (2007) A strong constitutive ethylene-response phenotype conferred on Arabidopsis plants containing null mutations in the ethylene receptors ETR1 and ERS1. BMC Plant Biol 7:3
7. Pu H, Shan S, Wang Z, Duan W, Tian J, Zhang L, Li J, Song H, Xu X (2020) Dynamic changes of DNA methylation induced by heat treatment were involved in ethylene signal transmission and delayed the postharvest ripening of tomato fruit. J Agric Food Chem 68(33):8976–8986
8. Li Z, Jiang G, Liu X, Ding X, Zhang D, Wang X, Zhou Y, Yan H, Li T, Wu K et al (2020) Histone demethylase SlJMJ6 promotes fruit ripening by removing H3K27 methylation of ripening-related genes in tomato. New Phytol 227(4):1138–1156
9. Lu P, Yu S, Zhu N, Chen YR, Zhou B, Pan Y, Tzeng D, Fabi JP, Argyris J, Garcia-Mas J et al (2018) Genome encode analyses reveal the basis of convergent evolution of fleshy fruit ripening. Nat Plants 4(10):784–791
10. Hu KD, Zhang XY, Yao GF, Rong YL, Ding C, Tang J, Yang F, Huang ZQ, Xu ZM, Chen XY et al (2020) A nuclear-localized cysteine desulfhydrase plays a role in fruit ripening in tomato. Hortic Res 7(1):211
11. Katz YS, Galili G, Amir R (2006) Regulatory role of cystathionine-gamma-synthase and de novo synthesis of methionine in ethylene production during tomato fruit ripening. Plant Mol Biol 61(1–2):255–268
12. Ardila T, Cuadrado MJ, Gallardo M, Sánchez-Calle IM, Gomez-Jimenez M (2012) The interaction between ethylene and polyamines

during ripening of olive fruit. Acta Hortic:949
13. Sharabi-Schwager M, Samach A, Porat R (2010) Overexpression of the CBF2 transcriptional activator in Arabidopsis suppresses the responsiveness of leaf tissue to the stress hormone ethylene. Plant Biol 12(4):630–638
14. Yang X, Li L (2011) miRDeep-P: a computational tool for analyzing the microRNA transcriptome in plants. Bioinformatics 27:2614–2615
15. Shahid S, Axtell MJ (2014) Identification and annotation of small RNA genes using ShortStack. Methods 67:20–27
16. An J, Lai J, Sajjanhar A, Lehman ML, Nelson CC (2014) miRPlant: an integrated tool for identification of plant miRNA from RNA sequencing data. BMC Bioinformatics 15:275
17. Mathelier A, Carbone A (2010) MIReNA: finding microRNAs with high accuracy and no learning at genome scale and from deep sequencing data. Bioinformatics 26:2226–2234
18. Guigon I, Legrand S, Berthelot J, Bini S, Lanselle D, Benmounah M et al (2019) miRkwood: a tool for the reliable identification of microRNAs in plant genomes. BMC Genomics 20:532
19. Bianchetti R, Bellora N, de Haro LA, Zuccarelli R, Rosado D, Freschi L, Rossi M, Bermudez L (2022) Phytochrome-mediated light perception affects fruit development and ripening through epigenetic mechanisms. Front Plant Sci 13:870974
20. Barry CS, Llop-Tous MI, Grierson D (2000) The regulation of 1-aminocyclopropane-1-carboxylic acid synthase gene expression during the transition from system-1 to system-2 ethylene synthesis in tomato. Plant Physiol 123(3):979–986
21. Yokotani N, Nakano R, Imanishi S, Nagata M, Inaba A, Kubo Y (2009) Ripening-associated ethylene biosynthesis in tomato fruit is autocatalytically and developmentally regulated. J Exp Bot 60(12): 3433–3442
22. Li X, Meng Z, Malik AU, Zhang S, Wang Q (2022) Maintaining the quality of postharvest broccoli by inhibiting ethylene accumulation using diacetyl. Front Nutr 9:1055651
23. Altschul SF, Gish W, Miller W, Myers EW, Lipman DJ (1990) Basic local alignment search tool. J Mol Biol 215(3):403–410
24. https://www.uniprot.org/
25. https://plants.ensembl.org/index.html
26. https://phytozome-next.jgi.doe.gov/
27. http://www.epigenome.cuhk.edu.hk/encode.html
28. https://www.arabidopsis.org/
29. https://banana-genome-hub.southgreen.fr/
30. https://solgenomics.net/
31. https://www.plantgdb.org/
32. http://planttfdb.gao-lab.org/
33. https://www.ncbi.nlm.nih.gov/geo/
34. https://www.ncbi.nlm.nih.gov/Structure/cdd/cdd.shtml
35. http://pfam.janelia.org/
36. http://smart.embl-heidelberg.de/
37. https://string-db.org/
38. http://systemsbiology.cau.edu.cn/agriGOv2/
39. http://revigo.irb.hr/
40. http://www.timetree.org/
41. https://www.genome.jp/kegg/pathway.html
42. http://kobas.cbi.pku.edu.cn/
43. https://www.genome.jp/kegg/kaas/
44. Hu B, Jin J, Guo AY, Zhang H, Luo J, Gao G (2015) GSDS 2.0: an upgraded gene feature visualization server. Bioinformatics 31(8): 1296–1297
45. http://mg2c.iask.in/mg2c_v2.0/
46. http://mips.helmholtz-muenchen.de/plant/chromoWIZ/index.jsp
47. Bailey TL, Boden M, Buske FA, Frith M, Grant CE, Clementi L, Ren J, Li WW, Noble WS (2009) MEME SUITE: tools for motif discovery and searching. Nucleic Acids Res 37(Web Server issue):W202–W208
48. http://www.sbg.bio.ic.ac.uk/~phyre2/html/page.cgi?id=index
49. Zhao K, Xia Y, Zhang F, Zhou X, Li SZ, Zhang G (2023) Protein structure and folding pathway prediction based on remote homologs recognition using PAthreader. Commun Biol 6(1):243
50. Wang Y, Wang Q, Gao L, Zhu B, Ju Z, Luo Y, Zuo J (2017) Parsing the regulatory network between small RNAs and target genes in ethylene pathway in tomato. Front Plant Sci 8. https://doi.org/10.3389/fpls.2017.00527
51. Bolger AM, Lohse M, Usadel B (2014) Trimmomatic: a flexible trimmer for Illumina sequence data. Bioinformatics 30:2114–2120
52. Martin M (2011) Cutadapt removes adapter sequences from high-throughput sequencing reads. EMBnet J 17:10–12
53. http://hannonlab.cshl.edu/fastx_ toolkit/
54. Langmead B, Trapnell C, Pop M, Salzberg SL (2009) Ultrafast and memory-efficient

alignment of short DNA sequences to the human genome. Genome Biol 10:R25

55. Lorenz R, Bernhart SH, Zu Siederdissen CH, Tafer H, Flamm C, Stadler PF et al (2011) ViennaRNA Package 2.0. Algorithms Mol Biol 6:26
56. https://github.com/erbon7/randfold
57. Li H, Handsaker B, Wysoker A, Fennell T, Ruan J, Homer N et al (2009) The sequence alignment/map (SAM) format and SAMtools. Bioinformatics 25:2078–2079
58. Dubois M, den Broeck LV, Inze D (2018) The pivotal role of ethylene in plant growth. Trends Plant Sci 23:311–323
59. Bleecker AB, Kende H (2000) Ethylene: a gaseous signal molecule in plants. Annu Rev Cell Dev Biol 16:1–18
60. Guo H, Ecker JR (2004) The ethylene signaling pathway: new insights. Curr Opin Plant Biol 7:40–49. https://doi.org/10.1016/j.pbi.2003.11.011
61. Kendrick MD, Chang C (2008) Ethylene signaling: new levels of complexity and regulation. Curr Opin Plant Biol 11:479–485
62. Zhu Z, Guo H (2008) Genetic basis of ethylene perception and signal transduction in Arabidopsis. J Integr Plant Biol 50:808–815
63. Giovannoni JJ (2004) Genetic regulation of fruit development and ripening. Plant Cell 16: 170–180. https://doi.org/10.1105/tpc.019158
64. Giovannoni JJ (2007) Fruit ripening mutants yield insights into ripening control. Curr Opin Plant Biol 10:283–289
65. Alos E, Distefano G, Rodrigo MJ, Gentile A, Zacarias L (2014) Altered sensitivity to ethylene in 'tardivo', a late-ripening mutant of clementine mandarin. Physiol Plant 151: 507–521
66. Yin XR, Allan AC, Chen KS, Ferguson IB (2010) Kiwifruit eil and erf genes involved in regulating fruit ripening. Plant Physiol 153: 1280–1292
67. Lee JM, Joung JG, McQuinn R, Chung MY, Fei Z, Tieman D, Klee H, Giovannoni J (2012) Combined transcriptome, genetic diversity and metabolite profiling in tomato fruit reveals that the ethylene response factor slerf6 plays an important role in ripening and carotenoid accumulation. Plant J 70:191–204
68. Liu M, Diretto G, Pirrello J, Roustan JP, Li Z, Giuliano G, Regad F, Bouzayen M (2014) The chimeric repressor version of an ethylene response factor (erf) family member, sl-erf.B3, shows contrasting effects on tomato fruit ripening. New Phytol 203:206–218
69. Han YC, Kuang JF, Chen JY, Liu XC, Xiao YY, Fu CC, Wang JN, Wu KQ, Lu WJ (2016) Banana transcription factor maerf11 recruits histone deacetylase mahda1 and represses the expression of maaco1 and expansins during fruit ripening. Plant Physiol 171:1070–1084
70. Fu CC, Han YC, Qi XY, Shan W, Chen JY, Lu WJ, Kuang JF (2016) Papaya cperf9 acts as a transcriptional repressor of cell-wall-modifying genes cppme1/2 and cppg5 involved in fruit ripening. Plant Cell Rep 35:2341–2352
71. Zuo J, Grierson D, Courtney LT, Wang Y, Gao L, Zhao X, Zhu B, Luo Y, Wang Q, Giovannoni JJ (2020) Relationships between genome methylation, levels of non-coding rnas, mrnas and metabolites in ripening tomato fruit. Plant J 103:980–994
72. Wang Y, Zou W, Xiao Y, Cheng L, Liu Y, Gao S, Shi Z, Jiang Y, Qi M, Xu T et al (2018) Microrna1917 targets ctr4 splice variants to regulate ethylene responses in tomato. J Exp Bot 69:1011–1025
73. He Y, Xue J, Li H, Han S, Jiao J, Rao J (2020) Ethylene response factors regulate ethylene biosynthesis and cell wall modification in persimmon (Diospyros kaki L.) fruit during ripening. Postharvest Biol Technol 168:111255
74. Liu Q, Zhang YC, Wang CY, Luo YC, Huang QJ, Chen SY, Zhou H, Qu LH, Chen YQ (2009) Expression analysis of phytohormone-regulated microRNAs in rice, implying their regulation roles in plant hormone signaling. FEBS Lett 583:723–728
75. Buxdorf K, Hendelman A, Stav R, Lapidot M, Ori N, Arazi T (2010) Identification and characterization of a novel miR159 target not related to MYB in tomato. Planta 232:1009–1022
76. Cara B, Giovannoni JJ (2008) The molecular biology of ethylene during tomato fruit development and maturation. Plant Sci 175:106–113
77. Wang X, Kong H, Ma H (2009) F-box proteins regulate ethylene signaling and more. Genes Dev 23(4):391–396
78. Qiao H, Chang KN, Yazaki J, Ecker JR (2009) Interplay between ethylene, ETP1/ETP2 F-box proteins, and degradation of EIN2 triggers ethylene responses in Arabidopsis. Genes Dev 23(4):512–521
79. Benavente LM, Alonso JM (2006) Molecular mechanisms of ethylene signaling in Arabidopsis. Mol BioSyst 2(3–4):165–173
80. Frankowski K, Kesy J, Kotarba W, Kopcewicz J (2008) Ethylene signal transduction pathway. Postepy Biochem 54(1):99–106
81. Zhong S, Lin Z, Grierson D (2008) Tomato ethylene receptor-CTR interactions: visualization of NEVER-RIPE interactions with

multiple CTRs at the endoplasmic reticulum. J Exp Bot 59(4):965–972
82. Adams-Phillips L, Barry C, Kannan P, Leclercq J, Bouzayen M, Giovannoni J (2004) Evidence that CTR1-mediated ethylene signal transduction in tomato is encoded by a multigene family whose members display distinct regulatory features. Plant Mol Biol 54(3):387–404
83. Leclercq J, Adams-Phillips LC, Zegzouti H, Jones B, Latché A, Giovannoni JJ, Pech JC, Bouzayen M (2002) LeCTR1, a tomato CTR1-like gene, demonstrates ethylene signaling ability in Arabidopsis and novel expression patterns in tomato. Plant Physiol 130: 1132–1142
84. https://rfam.org/
85. https://www.mirbase.org/
86. Bonnet E, He Y, Billiau K, Van de Peer Y (2010) TAPIR, a web server for the prediction of plant microRNA targets, including target mimics. Bioinformatics 26:1566–1568
87. Wu HJ, Ma Y, Chen T, Wang M, Wang X (2012) PsRobot: a web-based plant small RNA meta-analysis toolbox. Nucleic Acids Res 40:W22–W28
88. Chorostecki U, Palatnik JF (2014) comTAR: a web tool for the prediction and characterization of conserved microRNA targets in plants. Bioinformatics 30:2066–2067
89. Dai X, Zhuang Z, Zhao PX (2018) psRNATarget: a plant small RNA target analysis server (2017 release). Nucleic Acids Res 46: W49–W54
90. Addo-Quaye C, Eshoo TW, Bartel DP, Axtell MJ (2008) Endogenous siRNA and miRNA targets identified by sequencing of the Arabidopsis degradome. Curr Biol 18:758–762
91. Addo-Quaye C, Miller W, Axtell MJ (2009) CleaveLand: a pipeline for using degradome data to find cleaved small RNA targets. Bioinformatics 25:130–131
92. Zheng Y, Li Y, Sunkar R, Zhang W (2012) SeqTar: an effective method for identifying microRNA guided cleavage sites from degradome of polyadenylated transcripts in plants. Nucleic Acids Res 40:e28
93. Kakrana A, Hammond R, Patel P, Nakano M, Meyers BC (2014) sPARTA: a parallelized pipeline for integrated analysis of plant miRNA and cleaved mRNA data sets, including new miRNA target-identification software. Nucleic Acids Res 42:e139
94. Thody J, Folkes L, Medina-Calzada Z, Xu P, Dalmay T, Moulton V (2018) PAREsnip2: a tool for high-throughput prediction of small RNA targets from degradome sequencing data using configurable targeting rules. Nucleic Acids Res 46:8730–8739
95. Yao J-L, Tomes S, Xu J, Gleave AP (2016) How MicroRNA172 affects fruit growth in different species is dependent on fruit type. Plant Signal Behav 11:e1156833
96. Silva GFFE, Silva EM, da Silva Azevedo M, Guivin MAC, Ramiro DA, Figueiredo CR, Carrer H, Peres LEP, Nogueira FTS (2014) MicroRNA156-targeted SPL/SBP box transcription factors regulate tomato ovary and fruit development. Plant J 78:604–618
97. Zhao P, Wang F, Deng Y, Zhong F, Tian P, Lin D, Deng J, Zhang Y, Huang T (2022) Sly-miR159 regulates fruit morphology by modulating GA biosynthesis in tomato. Plant Biotechnol J 20:833–845
98. Lin W, Gupta SK, Arazi T, Spitzer-Rimon B (2021) MIR172d is required for floral organ identity and number in tomato. Int J Mol Sci 22:4659
99. Huang W, Peng S, Xian Z, Lin D, Hu G, Yang L, Ren M, Li Z (2017) Overexpression of a tomato MiR171 target gene SlGRAS24 impacts multiple agronomical traits via regulating gibberellin and auxin homeostasis. Plant Biotechnol J 15:472–488
100. Yang T, Wang Y, Liu H, Zhang W, Chai M, Tang G, Zhang Z (2020) MicroRNA1917-CTR1-LIKE PROTEIN KINASE 4 impacts fruit development via tuning ethylene synthesis and response. Plant Sci 291:110334
101. Huang W, Xian Z, Kang X, Tang N, Li Z (2015) Genome-wide identification, phylogeny and expression analysis of GRAS gene family in tomato. BMC Plant Biol 15:209
102. Gupta SK, Vishwakarma A, Kenea HD, Galsurker O, Cohen H, Aharoni A, Arazi T (2021) CRISPR/Cas9 mutants of tomato MICRORNA164 genes uncover their functional specialization in development. Plant Physiol 187:1636–1652
103. da Silva EM, Silva GFFE, Bidoia DB, da Silva Azevedo M, de Jesus FA, Pino LE, Peres LEP, Carrera E, López-Díaz I, Nogueira FTS (2017) MicroRNA159-targeted SlGAMYB transcription factors are required for fruit set in tomato. Plant J 92:95–109
104. Damodharan S, Corem S, Gupta SK, Arazi T (2018) Tuning of SlARF10A dosage by Sly-MiR160a is critical for auxin-mediated compound leaf and flower development. Plant J 96:855–868
105. Hu G, Fan J, Xian Z, Huang W, Lin D, Li Z (2014) Overexpression of SlREV alters the development of the flower pedicel abscission

zone and fruit formation in tomato. Plant Sci 229:86–95

106. Xian Z, Huang W, Yang Y, Tang N, Zhang C, Ren M, Li Z (2014) MiR168 influences phase transition, leaf epinasty, and fruit development via SlAGO1s in tomato. J Exp Bot 65: 6655–6666

107. Xue C, Yao J-L, Qin M-F, Zhang M-Y, Allan AC, Wang D-F, Wu J (2019) PbrmiR397a regulates lignification during stone cell development in pear fruit. Plant Biotechnol J 17: 103–117

108. Lin Y, Lai Z, Tian Q, Lin L, Lai R, Yang M, Zhang D, Chen Y, Zhang Z Endogenous target mimics down-regulate MiR160 mediation of ARF10, -16, and -17 cleavage during somatic embryogenesis in dimocarpus longan lour. Front Plant Sci

109. Clepet C, Devani RS, Boumlik R, Hao Y, Morin H, Marcel F, Verdenaud M, Mania B, Brisou G, Citerne S et al (2021) The MiR166–SlHB15A regulatory module controls ovule development and parthenocarpic fruit set under adverse temperatures in tomato. Mol Plant 14:1185–1198

110. Liu R, Lai B, Hu B, Qin Y, Hu G, Zhao J (2017) Identification of MicroRNAs and their target genes related to the accumulation of anthocyanins in litchi chinensis by high-throughput sequencing and degradome analysis. Front Plant Sci 7:2059

111. Qian M, Ni J, Niu Q, Bai S, Bao L, Li J, Sun Y, Zhang D, Teng Y (2017) Response of MiR156-SPL module during the red peel coloration of bagging-treated chinese sand pear (Pyrus Pyrifolia Nakai). Front Physiol 8: 550

112. Li X, Hou Y, Xie X, Li H, Li X, Zhu Y, Zhai L, Zhang C, Bian S (2020) A Blueberry MIR156a–SPL12 Module coordinates the accumulation of chlorophylls and anthocyanins during fruit ripening. J Exp Bot 71: 5976–5989

113. Hou Y, Zhai L, Li X, Xue Y, Wang J, Yang P, Cao C, Li H, Cui Y, Bian S (2017) Comparative analysis of fruit ripening-related miRNAs and their targets in blueberry using small RNA and degradome sequencing. Int J Mol Sci 18:2767

114. Ding T, Tomes S, Gleave AP, Zhang H, Dare AP, Plunkett B, Espley RV, Luo Z, Zhang R, Allan AC et al (2022) MicroRNA172 targets APETALA2 to regulate flavonoid biosynthesis in apple (Malus domestica). Hortic Res 9: uhab007

115. Yang T, Ma H, Zhang J, Wu T, Song T, Tian J, Yao Y (2019) Systematic identification of long noncoding RNAs expressed during light-induced anthocyanin accumulation in apple fruit. Plant J 100:572–590

116. Hu Y, Cheng H, Zhang Y, Zhang J, Niu S, Wang X, Li W, Zhang J, Yao Y (2021) The MdMYB16/MdMYB1-MiR7125-MdCCR module regulates the homeostasis between anthocyanin and lignin biosynthesis during light induction in apple. New Phytol 231: 1105–1122

117. Zhang B, Yang H-J, Yang Y-Z, Zhu Z-Z, Li Y-N, Qu D, Zhao Z-Y (2020) Mdm-MiR828 participates in the feedback loop to regulate anthocyanin accumulation in apple peel. Front Plant Sci 11:608109

118. Tirumalai V, Swetha C, Nair A, Pandit A, Shivaprasad PV (2019) miR828 and miR858 regulate VvMYB114 to promote anthocyanin and flavonol accumulation in grapes. J Exp Bot 70:4775–4792

119. Jia X, Shen J, Liu H, Li F, Ding N, Gao C, Pattanaik S, Patra B, Li R, Yuan L (2015) Small tandem target mimic-mediated blockage of MicroRNA858 induces anthocyanin accumulation in tomato. Planta 242:283–293

120. Li Y, Cui W, Qi X, Lin M, Qiao C, Zhong Y, Hu C, Fang J (2020) MicroRNA858 negatively regulates anthocyanin biosynthesis by repressing AaMYBC1 Expression in kiwifruit (Actinidia Arguta). Plant Sci 296:110476

121. Luo Y, Zhang X, Luo Z, Zhang Q, Liu J (2015) Identification and characterization of MicroRNAs from Chinese pollination constant non-astringent persimmon using high-throughput sequencing. BMC Plant Biol 15: 11

122. Tang Y, Qu Z, Lei J, He R, Adelson DL, Zhu Y, Yang Z, Wang D (2021) The long noncoding RNA FRILAIR regulates strawberry fruit ripening by functioning as a noncanonical target mimic. PLoS Genet 17: e1009461

123. Li D, Mou W, Luo Z, Li L, Limwachiranon J, Mao L, Ying T (2016) Developmental and stress regulation on expression of a novel miRNA, Fan-MiR73 and its target ABI5 in strawberry. Sci Rep 6:28385

124. Wang Y, Zhang J, Cui W, Guan C, Mao W, Zhang Z (2017) Improvement in fruit quality by overexpressing MiR399a in Woodland strawberry. J Agric Food Chem 65:7361–7370

125. Karlova R, van Haarst JC, Maliepaard C, van de Geest H, Bovy AG, Lammers M, Angenent GC, de Maagd RA (2013) Identification of MicroRNA targets in tomato fruit development using high-throughput sequencing and

degradome analysis. J Exp Bot 64:1863–1878

126. Chen W, Kong J, Lai T, Manning K, Wu C, Wang Y, Qin C, Li B, Yu Z, Zhang X et al (2015) Tuning LeSPL-CNR expression by SlymiR157 affects tomato fruit ripening. Sci Rep 5:7852

127. Chung M-Y, Nath UK, Vrebalov J, Gapper N, Lee JM, Lee D-J, Kim CK, Giovannoni J (2020) Ectopic expression of miRNA172 in tomato (Solanum Lycopersicum) reveals novel function in fruit development through regulation of an AP2 transcription factor. BMC Plant Biol 20:283

128. Zuo J, Zhu B, Fu D, Zhu Y, Ma Y, Chi L, Ju Z, Wang Y, Zhai B, Luo Y (2012) Sculpting the maturation, softening and ethylene pathway: the influences of MicroRNAs on tomato fruits. BMC Genomics 13:7

129. Moxon S, Jing R, Szittya G, Schwach F, Rusholme Pilcher RL, Moulton V, Dalmay T (2008) Deep sequencing of tomato short RNAs identifies MicroRNAs targeting genes involved in fruit ripening. Genome Res 18:1602–1609

130. Wang W, Wang J, Wu Y, Li D, Allan AC, Yin X (2020) Genome-wide analysis of coding and non-coding RNA reveals a conserved MiR164-NAC regulatory pathway for fruit ripening. New Phytol 225:1618–1634

131. Da L, Liu Y, Yang J, Tian T, She J, Ma X et al (2019) AppleMDO: a multi-dimensional omics database for apple co-expression networks and chromatin states. Front Plant Sci 10

132. Peng ZY, Zhou X, Li L, Yu X, Li H, Jiang Z et al (2009) Arabidopsis hormone database: a comprehensive genetic and phenotypic information database for plant hormone research in Arabidopsis. Nucleic Acids Res 37(Suppl 1)

133. Ran X, Liu J, Qi M, Wang Y, Cheng J, Zhang Y (2018) GSHR, a web-based platform provides gene set-level analyses of hormone responses in arabidopsis. Front Plant Sci 9

134. Kanehisa M, Furumichi M, Tanabe M, Sato Y, Morishima K (2017) KEGG: New perspectives on genomes, pathways, diseases and drugs. Nucleic Acids Res 45(D1):D353–D361

135. Jiang Z, Liu X, Peng Z, Wan Y, Ji Y, He W et al (2011) AHD2.0: an update version of arabidopsis hormone database for plant systematic studies. Nucleic Acids Res 39(Suppl 1)

136. Sun L, Song L, Zhang Y, Zheng Z, Liu D (2016) Arabidopsis PHL2 and PHR1 Act redundantly as the key components of the central regulatory system controlling transcriptional responses to phosphate starvation. Plant Physiol 170:499–514

137. Ciereszko I, Kleczkowski LA (2002) Effects of phosphate deficiency and sugars on expression of rab18 in Arabidopsis: hexokinase-dependent and okadaic acid-sensitive transduction of the sugar signal. Biochim Biophys Acta 1579:43–49

138. Chiou TJ, Lin SI (2011) Signaling network in sensing phosphate availability in plants. Annu Rev Plant Biol 62:185–206. https://doi.org/10.1146/annurev-arplant-042110-103849

139. Lopez-Bucio J, Hernandez-Abreu E, Sanchez-Calderon L, Nieto-Jacobo MF, Simpson J, Herrera-Estrella L (2002) Phosphate availability alters architecture and causes changes in hormone sensitivity in the Arabidopsis root system. Plant Physiol 129:244–256

140. Ma Z, Baskin TI, Brown KM, Lynch JP (2003) Regulation of root elongation under phosphorus stress involves changes in ethylene responsiveness. Plant Physiol 131:1381–1390

Chapter 14

Luciferase Complementation Imaging Assay (LCI) to Reveal the Interaction Between SPL7 and RAN1 Genes During Ethylene Signaling

Yanzhi Yang and Lei Li

Abstract

The gaseous hormone ethylene regulates multiple aspects of plant growth and development, as well as responses to biotic and abiotic factors. The copper transporter RAN1 mediates the delivery of copper ions to ETR1 and is crucial for proper ethylene receptor function and thus proper ethylene signaling. Moreover, the central regulator of copper homeostasis, SPL7 interacts with RAN1 at the endoplasmic reticulum to modulate ethylene signaling. Here, we present a detailed protocol of luciferase complementation imaging (LCI), a bioluminescence technique, to examine the interaction between membrane-bound proteins, such as SPL7 and RAN1, in ethylene signaling.

Key words Ethylene, Interaction, *Agrobacterium tumefaciens*, *Nicotiana benthamiana*, Luciferase complementation imaging

1 Introduction

The small gaseous hormone ethylene plays a pivotal role in plant growth and development [1]. Ethylene is perceived in plants by endoplasmic reticulum membrane-localized receptors, and signal transduction relays on the membrane-bound CONSTITUTIVE TRIPLE RESPONSE 1 (CTR1) and ETHYLENE INSENSITIVE 2 (EIN2) [2–4]. CTR1 interacts with the ethylene receptors and must be kinase-active to be signaling active [5]. CTR1 phosphorylates EIN2 and triggers its degradation by Ub/26S proteasome. In the presence of ethylene, CTR1 is deactivated and then leads to the dephosphorylated EIN2 [6, 7]. Thus, EIN2 is proteolytically cleaved to generate a nucleus-localized C-terminal fragment (CEND). The CEND initiates transcriptional regulation involving ETHYLENE INSENSITIVE 3 (EIN3) and EIN3-like proteins (EIL1) [8–10].

Devesh Shukla (ed.), *Ethylene Sensing and Signaling*, Methods in Molecular Biology, vol. 2945,
https://doi.org/10.1007/978-1-0716-4650-2_14,

In addition, ethylene perception is tightly connected to copper homeostasis. Binder et al. [13] have demonstrated that copper is essential for ethylene receptor biogenesis and proper function. The copper transporter RESPONSIVE-TO-ANTAGONIST1 (RAN1) mediates the delivery of copper ions to ETR1 and is crucial for proper ethylene signaling, normal plant growth, and development [11–14]. Recently, the central regulator of copper homeostasis, SQUAMOSA-PROMOTER BINDING PROTEIN-LIKE 7 (SPL7), has been reported to be localized at the endoplasmic reticulum and interact with RAN1 to modulate ethylene signaling [15, 16]. However, these proteins that are involved in ethylene perception and signaling are mostly membrane-localized. It is really difficult to detect the interaction between membrane-localized proteins using protein pull-down assay, yeast two-hybrid (Y2H), and co-immunoprecipitation (co-IP) assay.

In this chapter, we provide a comprehensive protocol for the preparation of vectors and Agrobacterium-mediated transient expression in *Nicotiana benthamiana* to test the interaction between membrane-bound proteins during ethylene signaling. It is a really simple, fast, and convenient method to detect protein interaction in vivo and in real time. This system requires limited experimental materials and instruments. These protocols are mainly derived from seminal methods from the Zhou labs [17]. While our protocols are optimized to membrane-bound proteins in ethylene signaling, they can be adapted to nuclear proteins or other proteins.

2 Materials

2.1 Vectors and Seeds

1. *Nicotiana benthamiana* seeds.
2. Vectors: pCAMBIA1300-nLUC, pCAMBIA1300-cLUC, pCAMBIA1300-nLUC-RAN1, pCAMBIA1300-nLUC-SPL7.
3. Primers for construct assembly: cLUC-SPL7-F: CGCGTCCCGGGGCGGTACCATGTCTTCTCTGTCGC; cLUC-SPL7-R: GAAAGCTCTGCAGGTCGACTCATATCCTGTCTCTTTC; RAN1-nLUC-F: GGGACGAGCTCGGTACCATGGCGCCGAGTAGAC; RAN1-nLUC-R: ACGAGATCTGGTCGACCTCCGTGGTGATTTTCAAAAC.
4. *Escherichia coli* (DH5α) competent cells.
5. *Agrobacterium* (GV3101) competent cells.

2.2 Equipment

1. 1 mL syringe.
2. 50 mL Eppendorf tubes.
3. 10 mL tubes.
4. PCR amplifier.

5. Temperature control orbital shaker.
6. Vortexer.
7. NanoDrop spectrophotometer.
8. Spectrophotometer plastic cuvettes.
9. Spray bottle.
10. Standard potting soil (potting soil: vermiculite = 2:1).
11. Fertilizer.
12. Pots.
13. Trays.
14. Plastic lids.
15. NightShade LB 985 system (Berthold).

2.3 Reagents

1. 1 M $MgCl_2$.
2. 1 M Acetosyringone (AS).
3. 50 mg/mL kanamycin.
4. 50 mg/mL rifampicin.
5. 100 mM potassium luciferin (100×).
6. LB medium: 10 g/L Tryptone, 5 g/L Yeast extract, 10 g/L NaCl.
7. Agrobacterium cell infiltration buffer: 10 mM $MgCl_2$, 200 μM AS.
8. Luciferin working solution: 1 mM potassium luciferin, 0.05% Triton X-100.

3 Methods

Here, we describe our approach for performing LCI in tobacco to test the interaction between membrane proteins, such as SPL7 and RAN1. It is also suitable for testing the interaction between nuclear proteins or other proteins.

3.1 Preparation of Nicotiana Benthamiana Plants (3 Weeks)

Day 1:

Scatter *Nicotiana benthamiana* seeds over the soil, cover the pot with a plastic lid, and place the pot in a growth chamber for 7–10 days (*see* **Note 1**) with long-day conditions (16 h: 8 h, light: dark), a light intensity of 200 μmol m^{-2} s^{-1}, relative humidity of 50%, and a temperature of 25 °C/21 °C.

Day 7:

Transfer the seedlings into individual pots. Let these seedlings grow under long-day conditions.

Fig. 1 Preparation of *Nicotiana benthamiana* plants for transient expression experiments. 3-week-old *N. benthamiana* plants with three fully expanded leaves are selected for infiltration. Replicates should be performed in at least 15 leaves from 5 plants

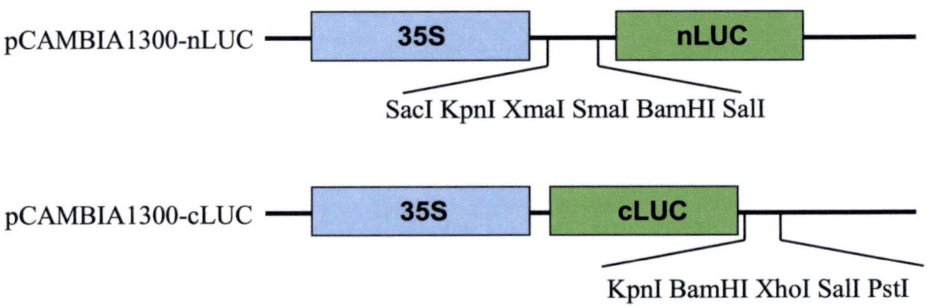

Fig. 2 Schematic diagrams of pCAMBIA1300-nLUC and pCAMBIA1300-cLUC constructs for LCI assays. The available restriction sites are listed. KpnI and SalI sites are suitable for the construct assembly in most cases. Blue rectangle: 35S promoter; green rectangle: nLUC or cLUC proteins

Day 21:

Select 3-week-old *N. benthamiana* plants with three fully expanded leaves to perform transient expression experiments (Fig. 1)

3.2 Construct Assembly (1 Week)

Day 1:

Design appropriate primers (*see* Subheading 2.1).

Day 2:

1. Amplify the coding sequences of SPL7 and RAN1 using the conventional polymerase chain reaction method.
2. Digest the pCAMBIA1300-nLUC and pCAMBIA1300-cLUC vectors using restriction enzymes KpnI and SalI (*see* **Note 2**) (Fig. 2).
3. Clone the coding sequences of SPL7 or RAN1 into pCAMBIA1300-nLUC and pCAMBIA1300-cLUC using Gibson assembly methods (*see* **Note 3**).
4. Transform the ligation solution into *Escherichia coli* (DH5α) competent cells.

Days 3–4:

Perform Colony PCR to select positive clones and confirm them by sequencing.

Day 5:

Prepare and autoclave LB liquid medium (103 kPa/121 °C/15 min). Incubate one positive single *Escherichia coli* from each plate into 20 mL of LB liquid medium containing 50 μg/mL of kanamycin in a 50 mL tube, with shaking at 220 rpm at 37 °C for 16 h.

Day 6:

Extract plasmid using commercialized plasmid extraction and purification kits.

3.3 *Agrobacterium-Mediated Transient Expression in* Nicotiana Benthamiana *(5 Days)*

Day 1:

Transform the plasmids generated in Subheading 3.2, as well as controls (cLUC-SPL7, RAN1-nLUC, nLUC, and cLUC), into the *Agrobacterium tumefaciens* (GV3101) competent cells by chemical transformation.

Day 4:

Perform Colony PCR to select positive clones. Incubate one positive single *Agrobacterium tumefaciens* from each plate into 20 mL of LB liquid medium containing 50 μg/mL of kanamycin and 50 μg/mL rifampicin in a 50 mL tube, with shaking at 220 × g at 28 °C for 16–24 h until the LB liquid medium turns orange and muddy.

Day 5:

1. Centrifuge the *Agrobacterium tumefaciens* cells at 6000 × g, room temperature, for 6 min and discard the supernatant.
2. Resuspend the *Agrobacterium tumefaciens* cells in the infiltration buffer to a final concentration of $OD_{600} = 1.0$ (*see* **Note 4**).
3. Mix equal volumes of *Agrobacterium tumefaciens* clones carrying the desired plasmids to express protein (e.g., 4 mL RAN1-nLUC and 4 mL cLUC-SPL7) to prepare the infiltration solution (Fig. 3a). Negative controls (e.g., RAN1-nLUC and cLUC; nLUC and cLUC-SPL7; nLUC and cLUC) are required (Fig. 3d).
4. Keep the infiltration solution at room temperature and in the dark for 0.5–3 h.
5. Select fully expanded leaves of 3-week-old *Nicotiana benthamiana* plants for infiltration.

Fig. 3 Illustration of the infiltration process. (**a**) Preparation of the infiltration solution. Carefully prepare the infiltration solution by mixing the *Agrobacterium tumefaciens* clone solution carrying the desired plasmids to express protein (e.g., 4 mL RAN1-nLUC and 4 mL cLUC-SPL7), and then keep the infiltration solution at room temperature and in the dark for 0.5–3 h. (**b–c**) Demonstration of infiltration. (**d**) Infiltration combination: 1. nLUC and cLUC; 2. RAN1-nLUC and cLUC; 3. nLUC and cLUC-SPL7; 4. RAN1-nLUC and cLUC-SPL7. Place the plants in the growth chamber for 24–48 h after infiltration

6. Infiltrate the *Agrobacterium tumefaciens* inoculum carrying the desired plasmids and negative controls side-by-side carefully into the abaxial side of leaves using a 1 mL needleless syringe (Fig. 3b, c).
7. Place the plants in the growth chamber for 24–48 h (*see* **Note 5**) before measuring the luciferase activity.

3.4 Measurement of Luciferase Activity with Nightshade LB 985 System (1 Day)

1. Detach the whole leaves and stick them on the blackboard (Fig. 4a) (*see* **Note 6**).
2. Spread the luciferin working solution evenly on the leaves (Fig. 4b, c) and keep them in the dark for 5 min before imaging to avoid the interference of chlorophyll luminescence.
3. Open the Nightshade LB 985 system (*see* **Note 7**) and the IndiGO software, and wait until the camera temperature is stable.
4. Capture bright-field and luminescence images with an exposure time of 5 min.

Fig. 4 Preparation for luminescence imaging. (**a**) Prepare a blackboard (about 21 cm*29 cm), and detach the whole leaves and stick them on the blackboard by using a double-sided tape. (**b**) Spread the luciferin working solution evenly on the leaves using a small spray bottle. (**c**) Keep the leaves in the dark for 5 min before imaging to avoid the interference of chlorophyll luminescence

5. Measure the relative luciferase activity using the IndiGO software (Fig. 5) (*see* **Note 8**). Briefly, open the IndiGO software and choose the images that you want to analyze. Click the "circle" icon (red arrow) to draw a circle and make sure the luminescence signals are included. Right-click on the circle, select Duplicate, duplicate the circle to the side, and drag the duplicated circle to the appropriate location. Click the "Manually" icon to obtain the luminescence data (Fig. 5).

4 Notes

1. Do not allow the seedlings to grow in one pot for more than 10 days. Otherwise, roots tend to strengthen and are easy to break when transferring the seedlings into individual pots.
2. Other restriction sites are also suitable for the construct assembly.
3. Here, we usually clone vectors using Gibson assembly methods. T4 ligation or other methods are also suitable.
4. The optimal OD_{600} should be experimentally determined. Generally, the higher the concentration, the faster the expression. It should be noted that OD_{600} must be less than 2.5. Otherwise, this would damage leaves.
5. Accumulation of the transiently expressed proteins can usually be detected at 36–60 h post-inoculation. Empirically, the smaller the protein, the faster the expression.
6. Detach the leaves when you are ready for the next step. Do not leave the detached leaves on the blackboard for too long.

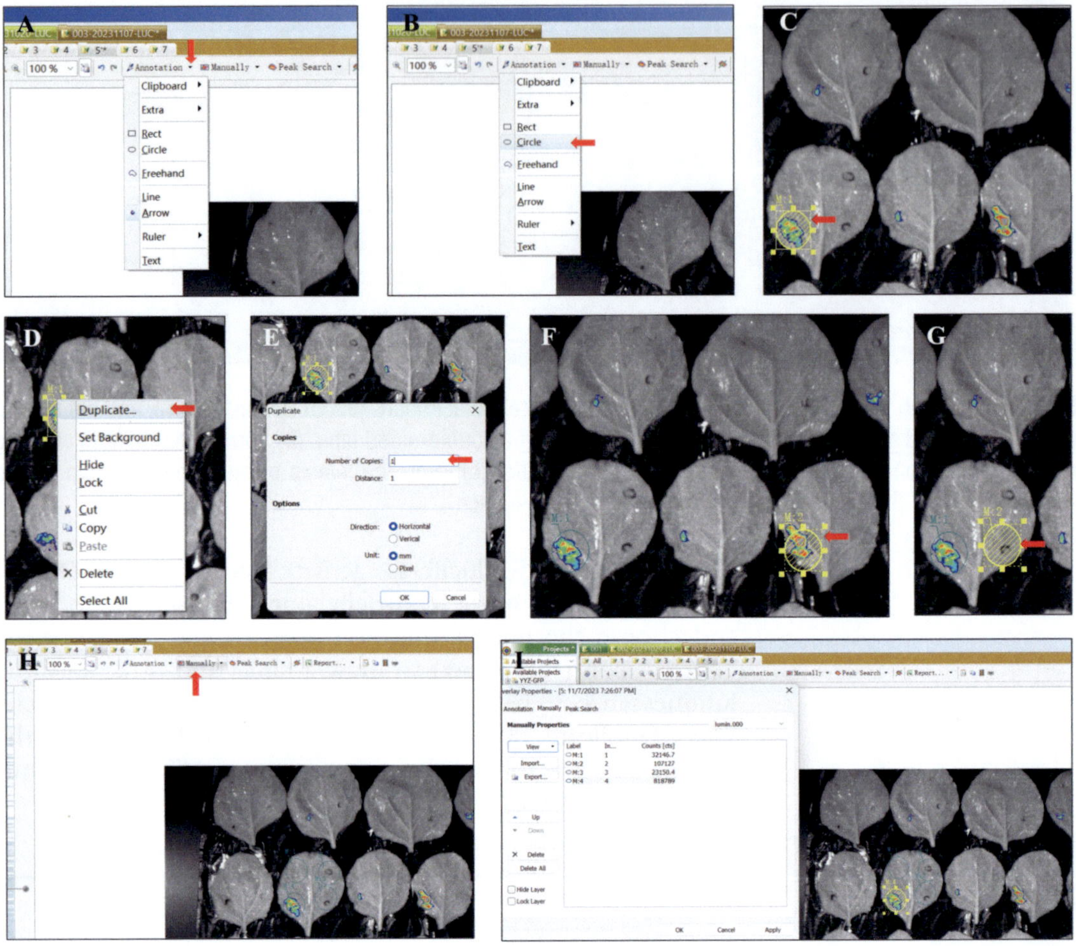

Fig. 5 Example of relative luciferase activity measurement using the IndiGO software. (**a**, **b**) Open the IndiGO software and choose the images that you want to analyze. Click the "circle" icon (red arrow). (**c**) Draw a circle to make sure the luminescence signals are included. (**d**–**g**) Right-click on the circle, select Duplicate, duplicate the circle to the side, and drag the duplicated circle to the appropriate location (**g**; red arrow). (**h**–**i**) Click the "Manually" icon (**h**) to obtain the luminescence data (**i**)

7. Other CCD imaging apparatus is also suitable for luciferase imaging.
8. We usually use the value of counts or photons to calculate the relative luminescence.

Acknowledgments

Yanzhi Yang was supported in part by the Postdoctoral Fellowship of Peking-Tsinghua Center for Life Sciences.

References

1. Abeles FB, Morgan MP, Saltveit ME (1992) Ethylene in plant biology, 2nd edn. Academic Press, New York
2. Zhu Z, Guo H (2008) Genetic basis of ethylene perception and signal transduction in Arabidopsis. J Integr Plant Biol 50(7):808–815. https://doi.org/10.1111/j.1744-7909.2008.00710.x
3. Kieber JJ, Rothenberg M, Roman G, Feldmann KA, Ecker JR (1993) CTR1, a negative regulator of the ethylene response pathway in Arabidopsis, encodes a member of the raf family of protein kinases. Cell 72(3):427–441
4. Chang C, Kwok SF, Bleecker AB, Meyerowitz EM (1993) Arabidopsis ethylene-response gene ETR1: similarity of product to two-component regulators. Science 262(5133):539–544
5. Clark KL, Larsen PB, Wang X, Chang C (1998) Association of the Arabidopsis CTR1 Raf-like kinase with the ETR1 and ERS ethylene receptors. Proc Natl Acad Sci USA 95(9):5401–5406
6. Solano R, Stepanova A, Chao Q, Ecker JR (1998) Nuclear events in ethylene signaling: a transcriptional cascade mediated by ETHYLENE-INSENSITIVE3 and ETHYLENE-RESPONSE-FACTOR1. Genes Dev 12(23):3703–3714
7. Huang Y, Li H, Hutchison CE, Laskey J, Kieber JJ (2003) Biochemical and functional analysis of CTR1, a protein kinase that negatively regulates ethylene signaling in Arabidopsis. Plant J 33(2):221–233
8. Alonso JM, Stepanova AN, Solano R, Wisman E, Ferrari S, Ausubel FM, Ecker JR (2003) Five components of the ethylene-response pathway identified in a screen for weak ethylene-insensitive mutants in Arabidopsis. Proc Natl Acad Sci USA 100(5):2992–2997. https://doi.org/10.1073/pnas.0438070100
9. Potuschak T, Lechner E, Parmentier Y, Yanagisawa S, Grava S, Koncz C, Genschik P (2003) EIN3-dependent regulation of plant ethylene hormone signaling by two arabidopsis F box proteins: EBF1 and EBF2. Cell 115(6):679–689. https://doi.org/10.1016/s0092-8674(03)00968-1
10. Binder BM, Walker JM, Gagne JM, Emborg TJ, Hemmann G, Bleecker AB, Vierstra RD (2007) The Arabidopsis EIN3 binding F-box proteins EBF1 and EBF2 have distinct but overlapping roles in ethylene signaling. Plant Cell 19(2):509–523. https://doi.org/10.1105/tpc.106.048140
11. Hirayama T, Kieber JJ, Hirayama N, Kogan M, Guzman P, Nourizadeh S, Alonso JM, Dailey WP, Dancis A, Ecker JR (1999) RESPONSIVE-TO-ANTAGONIST1, a Menkes/Wilson disease-related copper transporter, is required for ethylene signaling in Arabidopsis. Cell 97(3):383–393
12. Woeste KE, Kieber JJ (2000) A strong loss-of-function mutation in RAN1 results in constitutive activation of the ethylene response pathway as well as a rosette-lethal phenotype. Plant Cell 12(3):443–455
13. Binder BM, Rodriguez FI, Bleecker AB (2010) The copper transporter RAN1 is essential for biogenesis of ethylene receptors in Arabidopsis. J Biol Chem 285(48):37263–37270. https://doi.org/10.1074/jbc.M110.170027
14. Li W, Lacey RF, Ye Y, Lu J, Yeh KC, Xiao Y, Li L, Wen CK, Binder BM, Zhao Y (2017) Triplin, a small molecule, reveals copper ion transport in ethylene signaling from ATX1 to RAN1. PLoS Genet 13(4):e1006703. https://doi.org/10.1371/journal.pgen.1006703
15. Yang Y, Hao C, Du J, Xu L, Guo Z, Li D, Cai H, Guo H, Li L (2022) The carboxy terminal transmembrane domain of SPL7 mediates interaction with RAN1 at the endoplasmic reticulum to regulate ethylene signalling in Arabidopsis. New Phytol 236(3):878–892. https://doi.org/10.1111/nph.18376
16. Garcia-Molina A, Xing S, Huijser P (2014) Functional characterisation of Arabidopsis SPL7 conserved protein domains suggests novel regulatory mechanisms in the cu deficiency response. BMC Plant Biol 14:231. https://doi.org/10.1186/s12870-014-0231-5
17. Chen H, Zou Y, Shang Y, Lin H, Wang Y, Cai R, Tang X, Zhou JM (2008) Firefly luciferase complementation imaging assay for protein-protein interactions in plants. Plant Physiol 146(2):368–376. https://doi.org/10.1104/pp.107.111740

Part IV

Ethylene Signaling in Plant Development and Stress

Chapter 15

Ethylene Signaling in Plant Development and Stress Adaptation

Devesh Shukla

Abstract

Ethylene, often referred as "death hormone" for its role in plant abscission, senescence, and fruit ripening, is a simple olefin gas and the first gaseous hormone discovered in plants. It also regulates seed germination, root and shoot growth, flower development, fruit ripening, and responses to biotic and abiotic stresses. Ethylene biosynthesis initiates with methionine converting enzymatically to S-adenosyl-L-methionine (SAM), then to 1-aminocyclopropane-1-carboxylic acid (ACC) by ACC synthase (ACS), and finally to ethylene via ACC oxidase (ACO). Its production is regulated at transcriptional, post-transcriptional, and post-translational levels. Ethylene sensing depends on receptors such as ETR1 and ERS1, located on the endoplasmic reticulum. When ethylene binds these receptors, CTR1 becomes inactive, allowing EIN2 to relay signal to transcription factors EIN3/EIL1 and ERFs. As this pathway can interact with auxin, jasmonic acid, salicylic acid, abscisic acid, and gibberellins, it coordinates plant growth and stress management. Ethylene affects root system architecture, particularly under varying phosphate availability, modifying root hair development to enhance nutrient uptake. It also shapes plant immunity by modifying defense response against biotic and abiotic challenges. Advances such as ethylene-sensitive fluorescent reporters, genomic and proteomic tools, and computational models have clarified many aspects of ethylene signaling. However, key questions remain, including the precise activation mechanism of ethylene receptors, details of hormonal crosstalk, and post-translational modifications.

Key words Ethylene biosynthesis, Ethylene signaling, Ethylene receptors, Plant hormone crosstalk, Root system architecture, Plant immunity, Ethylene-responsive factors (ERFs), Post-translational modifications in ethylene signaling

1 Introduction

Ethylene, IUPAC name ethene, is a simple olefin gas and the first discovered gaseous hormone. It regulates many growth, development, and stress responses [1–3]. Plant produce ethylene in trace amounts, and its small size allows it to diffuse quickly within tissues and between plants, affecting multiple physiological responses. This diffusion property explain the saying "one bad apple spoils the bunch," where ethylene from ripening fruits accelerates

Devesh Shukla (ed.), *Ethylene Sensing and Signaling*, Methods in Molecular Biology, vol. 2945,
https://doi.org/10.1007/978-1-0716-4650-2_15,

ripening in nearby fruits [1–7]. Ethylene functions in seed germination, root and shoot growth, flower formation, abscission, senescence, and fruit ripening [5, 6]. It also involves in biotic and abiotic stresses including pathogen attacks, flooding, salinity, and mechanical wounding [3, 4, 8]. Its role is central in fruit ripening climacteric fruits like tomatoes, bananas, and apples [5–7]. Ethylene biosynthesis and signaling trigger transcriptional cascades that drive cell wall degradation, pigment formation, and volatile production [1, 2, 5–7].

The role of ethylene dates back to the nineteenth century when gas leaks in city lighting systems caused premature leaf fall. In 1901, Dimitry Neljubov identified ethylene as the active compound [4]. Later, researchers confirmed its endogenous synthesis and established its importance in fruit ripening and stress response pathways [1, 3]. Ethylene signaling was characterized through the "triple response" in dark-grown seedlings. This response comprises shortened hypocotyl and roots, radial swelling, and an exaggerated apical hook formation. This phenotypic became a tool for identifying the components of ethylene signaling pathways. Approaches like genetic screenings (mutant identification) and molecular characterizations have identified ethylene receptors, protein kinase CTR1, transmembrane protein EIN2, and transcription factors such as EIN3, EILs, and ethylene-responsive factors (ERFs). In canonical pathways, ethylene binds to endoplasmic reticulum-localized receptors, inactivating CTR1, which allows EIN2 and downstream transcription factors to affect ethylene responses [3, 9].

Ethylene signaling shapes plant development and responses to nutrient imbalance, including the excess phosphate response (EPiR) [8]. When phosphate levels increase, ethylene signaling affects root system architecture by reducing primary root growth and altering lateral root patterns [8]. Because nutrient excess or deficiency modulates ethylene, development and stress response genes change to maintain nutrient homeostasis [8]. In agriculture, ethylene regulates fruit ripening and senescence, critical stages for post-harvest management and crop quality [5–7]. Because of this, ethylene biosynthesis and signaling pathways are manipulated to delay ripening, improve quality, and extend shelf-life, minimizing postharvest losses [5–7]. Genetic engineering approaches, include antisense inhibition of ACC synthase and ACC oxidase. These strategies delay ripening in several climacteric fruits, helping tackle the perishability of fleshy fruits [1, 5–7]. Further, CRISPR/Cas9-based genome editing hold promise for post-harvest management of fruits and flowers.

Overall, ethylene is a multifaceted phytohormone with diverse roles in plant growth, development, stress responses, and postharvest physiology. By clarifying its mechanisms, agriculture and horticulture have achieved major advances, especially improving crop tolerance under stress and extending fruit shelf life. Ethylene signaling includes canonical (standard) and noncanonical (alternative) pathways. These pathways interacts with other plant hormone signaling pathways and highlight ethylene's impact on crop health and productivity.

2 Ethylene Biosynthesis Pathway and Its Regulation

Ethylene biosynthesis starts from methionine, a sulfur-containing amino acid, which converts to S-adenosyl-L-methionine (SAM) by the S-adenosylmethionine synthetase (Fig. 1) [6, 10]. Conversion of S-adenosyl methionine (SAM) into 1-aminocyclopropane-1-carboxylic acid (ACC) by ACC synthase (ACS) is the rate-limiting step (Fig. 1) [4, 6, 10]. Next, ACC converts to ethylene via ACC oxidase (ACO), oxygen-dependent, releasing carbon dioxide and cyanide as by-products (Fig. 1) [4, 9, 10].

Ethylene biosynthesis is regulated at multiple levels, including transcriptional, post-transcriptional, and post-translational. The ACC synthase (ACS) gene family is large and diverse. Different isoforms are expressed in response to specific developmental signals or environmental conditions [1, 4, 8–10]. In Arabidopsis, ACS genes cluster into classes based on regulatory sequences. These sequences determine responsiveness to phytohormones like auxin, gibberellins, and ethylene [9].

At the transcriptional level, homeotic genes (Developmental Regulators) and transcription factors regulate ethylene biosynthesis. They modulate ACS and ACO expression in response to environmental or developmental signals. For example, in tomato, LeHB-1 and LeMADS-RIN regulate ACS and ACO gene

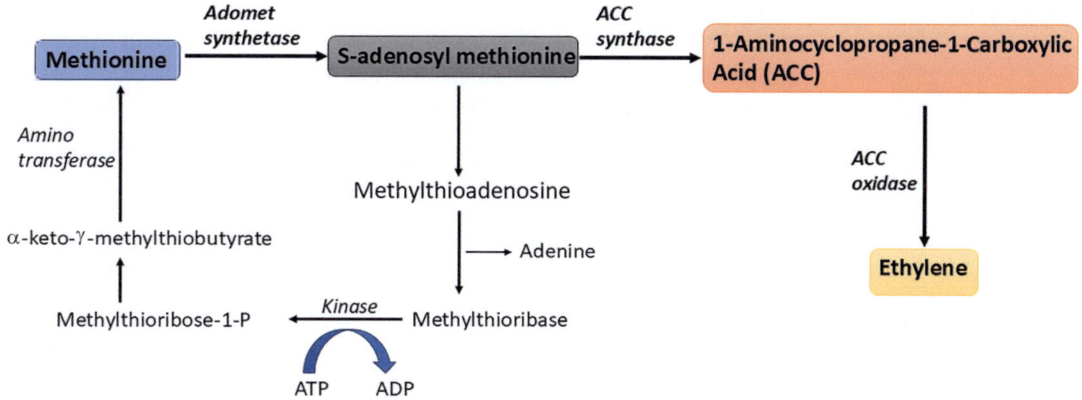

Fig. 1 Ethylene biosynthesis starts with methionine, which is converted into S-adenosyl methionine (SAM) via Adomet synthetase. SAM is then converted to 1-aminocyclopropane-1-carboxylic acid (ACC), the direct precursor of ethylene. ACC is oxidized to produce ethylene through ACC oxidase. The pathway also involves recycling methionine through intermediates like methylthioribose-1-phosphate (MTR-1-P), completing the methionine cycle

expression during fruit ripening. LeHB-1 activates LeACO1 transcription, while LeMADS-RIN binds to the promoter regions of LeACS2 to regulate its expression during ripening [4, 11]. Such regulation ensures precise control of ethylene production during critical processes such as fruit ripening [1, 4–7].

Post-transcriptionally, the stability of ACS mRNA is affected by several signaling pathways. Different ACS genes show distinct mRNA stabilities to ethylene and auxin, causing differential expression during plant stress adaptation responses [4, 12]. Post-translational regulation of ACS proteins involves phosphorylation and ubiquitination. ACS proteins are often phosphorylated by kinases such as mitogen-activated protein kinase (MAPK), which stabilizes the enzyme and increase ethylene production [9]. On the contrary, unphosphorylated ACS is tagged by the ubiquitin–proteasome system, a protein-degradation pathway, and is rapidly turned over [9]. In Arabidopsis, phosphorylation of specific serine residues stabilizes ACS6, protecting it from degradation via the 26S proteasomal pathway [9]. The C-terminal region of ACS proteins is crucial for their stability; deletions increase enzyme activity but also susceptibility to degradation [4].

ACO, also called ethylene-forming enzyme (EFE), is regulated by multiple factors [5–7]. ACO is encoded by multiple genes that are differentially expressed in response to developmental cues and environmental stress. In tomato, ACO gene family members display distinct expression during fruit ripening, senescence, and wounding, suggesting tight regulation of ethylene across development [4].

2.1 Environmental and Developmental Regulation of Ethylene Biosynthesis

Ethylene biosynthesis is responds to environmental factors, including wounding, flooding, and pathogen attack. Flooding causes hypoxia, reduced ACO activity due to limited oxygen. Whereas, wounding upregulates ACS expression and increases ethylene production, which acts as a stress signal [4]. During fruit ripening, ethylene production is autocatalytic. Ethylene stimulates further expression of ACS and ACO, producing an ethylene burst that drives the ripening [5–7]. This positive feedback loop is crucial for rapid, synchronized ripening in climacteric fruits like tomatoes and bananas [1, 5, 9]. When a young seedling grows in shade, it senses a low red:far-red light ratio through photoreceptor pigments. Consequently, this signal upregulates hormones that increase ethylene production. Ethylene and auxin then act together to elongate the hypocotyl, allowing the seedling rise above neighbouring plants and reach more light.

3 Ethylene Sensing Mechanisms

Ethylene sensing in plants starts at receptors on the endoplasmic reticulum (ER) membrane. In *Arabidopsis thaliana*, five receptors such as ETR1, ERS1, ETR2, ERS2, and EIN4 operate (Fig. 2) [4]. These receptors exhibit overlapping yet distinct functions in ethylene sensing and signaling. Structurally, these receptors resemble bacterial two-component systems and contain a histidine kinase-like domain, indicating a conserved origin for ethylene sensing across biological systems (Fig. 2) [3]. Upon ethylene binding, receptors undergo conformational changes that inactivate CTR1, a Raf-like kinase. Without ethylene, CTR1 associates with the ER membrane and negatively regulates signaling by phosphorylating EIN2 (Fig. 3). As ethylene inactivates CTR1, EIN2 can signal by releasing its C-terminal fragment (CEND) to activate ethylene-responsive genes (Fig. 3) [13, 14]. Recent findings highlight subcellular dynamics in this pathway. After ethylene perception, CTR1 can translocate from ER to nucleus (Fig. 3). In the nucleus, CTR1 stabilizes EIN3 levels, and supports responses under drought and salinity (Fig. 3) [13].

3.1 Ethylene Signal Transduction Pathways

The ethylene signaling begins when ethylene binds with receptors, such as ETHYLENE RESPONSE 1 (ETR1), ETHYLENE RESPONSE SENSOR 1 (ERS1), and ETHYLENE RESPONSE SENSOR 2 (ERS2) (Fig. 3). These receptors form subfamilies on

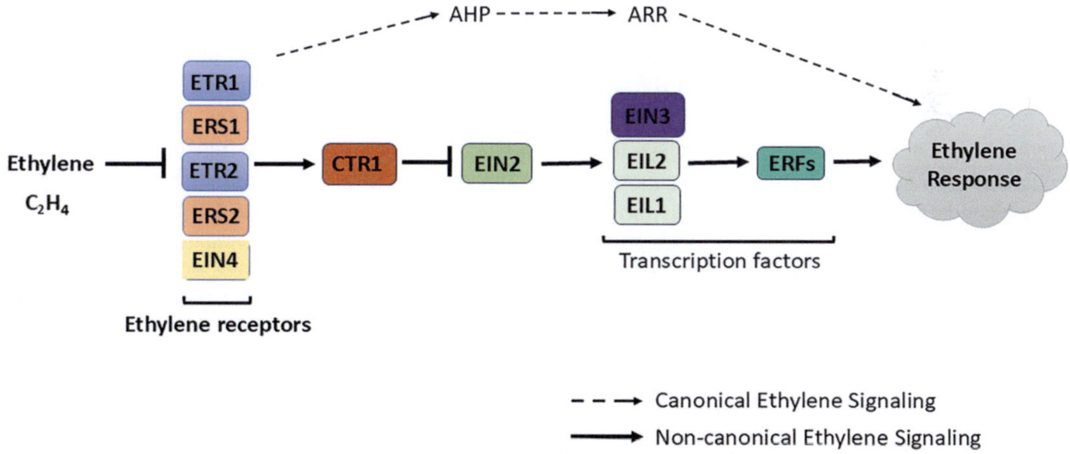

Fig. 2 In Arabidopsis, Ethylene signaling begins at receptors on the endoplasmic reticulum (ER) membrane. Ethylene binds ETR1, ERS1, ETR2, ERS2, and EIN4, inactivating the CTR1 kinase, which normally acts as a negative regulator. Consequently, EIN2 is released to activate transcription factors EIN3, EIL1, and EIL2, which activate ERFs and trigger downstream ethylene responses. The solid lines represent the canonical pathway. Dashed lines indicate a "noncanonical" pathway involving AHP (Arabidopsis Histidine Phosphotransfer Proteins) and ARR proteins (Arabidopsis Response Regulators) that modulate responses

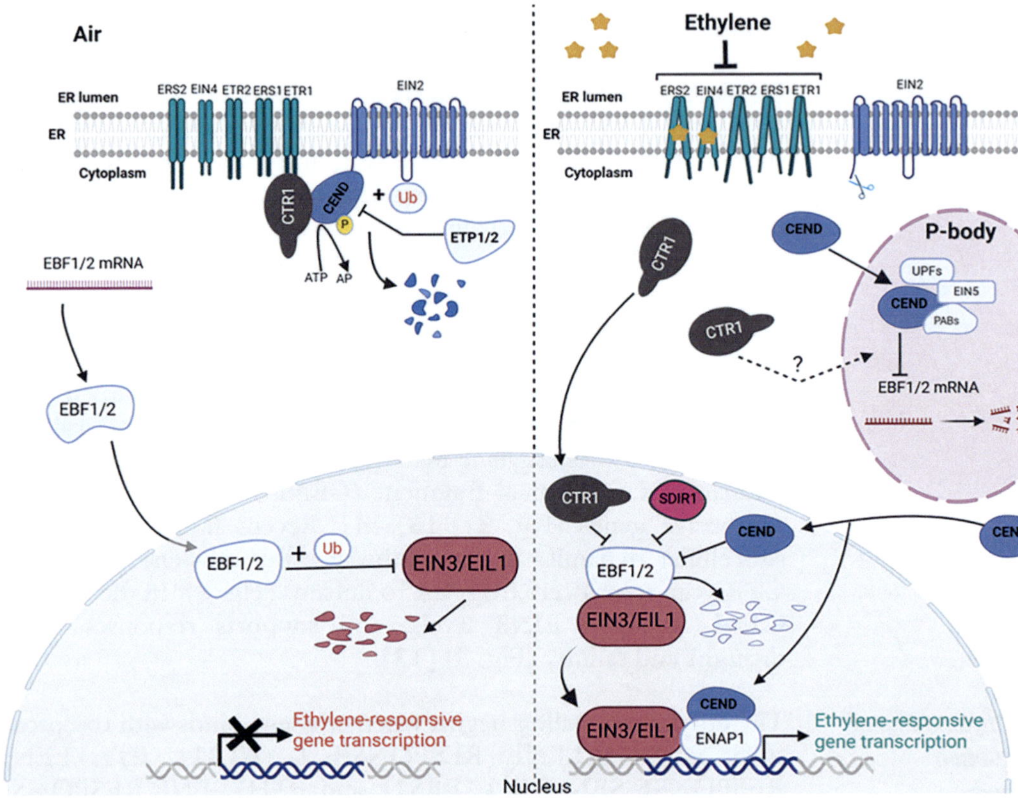

Fig. 3 Ethylene signal transduction in Arabidopsis. The figure shows the ethylene signaling pathway in *Arabidopsis thaliana*. In the absence of ethylene (left panel), active receptors at the ER membrane activate CTR1. CTR1 phosphorylates EIN2 (C-terminal end CEND), marking it for degradation through proteasomal ubiquitination by ETP1/2. This keeps EIN3/EIL1 transcription factors inactive via further ubiquitination by EBF1/2, preventing ethylene-responsive gene expression. In the presence of ethylene (right panel), ethylene binds to the receptors, and inactivates CTR1. The dephosphorylated EIN2 is cleaved, and the released C-terminal fragment (CEND) translocates to the nucleus to stabilize EIN3/EIL1, increasing gene transcription. CEND also regulates EBF1/2 mRNA via P-bodies to control EIN3/EIL1 levels. CTR1, released from the ER, may also enter the nucleus to affect EIN3 stability. SDIR1 provides additional negative regulation on EIN3/EIL1

the basis of structural and functional similarities or dissimilarities [3]. Without ethylene, receptors activate CONSTITUTIVE TRIPLE RESPONSE 1 (CTR1), a Raf-like Ser/Thr kinase and negative regulator [3]. Active CTR1 phosphorylates ETHYLENE INSENSITIVE 2 (EIN2), preventing its proteolytic cleavage and downstream signaling (Fig. 3) [14]. Upon ethylene binding, CTR1 becomes inactive, allowing EIN2 to be cleaved (Fig. 3). The released C-terminal fragment of EIN2 (CEND) moves from ER to the nucleus. There, CEND interacts with EIN3 NUCLEAR-ASSOCIATED PROTEIN 1 (ENAP1) to facilitate downstream signaling (Fig. 3). This CEND-EIN3 interaction promotes histone acetylation and enables EIN3 and EIN3-Like1 (EIL1) to bind to target DNA. As a result, ETHYLENE RESPONSE FACTOR (ERF) genes are transcribed (Fig. 3). These ERF genes regulate fruit ripening, leaf senescence, and responses to environmental stresses [3, 5–7, 9].

Recent studies highlight the importance of chromatin modifications, such as histone acetylation, in regulating the transcriptional response under ethylene treatment. These modifications alter the chromatin structure, and control DNA access for EIN3 and EIL1, modulating ethylene-responsive genes [9]. In addition to the canonical CTR1-EIN2-EIN3 pathway, ethylene signaling also involves noncanonical pathways. One such pathway is an ethylene receptor-mediated phospho-relay mechanism. It involves the ethylene receptor ETR1 interacting with the phosphotransfer protein AHP1. This pathway is thought to modulate ethylene responses independently of CTR1 [3, 9]. The ethylene signaling pathway in rice has additional features compared to Arabidopsis. In rice, a distinct phospho-relay involves ethylene receptors with regulators like MHZ1 and AHP proteins. This pathway contributes to adaptive responses, particularly to submergence stress [9].

3.2 Crosstalk of Ethylene with Other Hormones

Phytohormones interact with ethylene to coordinate plant growth, development and responses to biotic or abiotic stresses. This crosstalk helps the plant adapt to environmental and physiological conditions. Depending on context, ethylene and other hormones act together or oppose each other. Auxin and ethylene exhibit a bidirectional relationship. Ethylene promotes auxin biosynthesis, as shown by Yu et al. [15], who demonstrated how ethylene induces local auxin production to maintain cambial stem cell activity. Conversely, auxin activates certain isoforms of ACS, increasing ethylene synthesis. This mutual support steers lateral root development, root hair formation, and maintenance of the apical hook [8, 15]. When jasmonic acid (JA) interacts with ethylene, they often act synergistically to heighten defense against necrotrophic pathogens. Both hormones can elevate the biosynthesis of each one and amplify defense responses [15]. In contrast, salicylic acid (SA) often antagonizes ethylene against biotrophic pathogens but cooperates during leaf senescence [15]. Ethylene and abscisic acid (ABA) generally antagonize each other, especially during drought. ABA tends to suppress ethylene-mediated growth under adverse conditions, while ethylene can interfere with ABA effects on seed germination. This antagonism facilitates a balance between growth and stress adaptation depending on context [16]. Ethylene also crosstalks with gibberellins (GA) to regulate growth. Typically, ethylene and GA have opposing effects, such as in stem elongation and fruit ripening [16]. Altogether, these crosstalks highlight the versatile role of ethylene in integrating hormonal signals, allowing appropriate responses to developmental cues and environmental challenges.

4 Mechanisms of Ethylene-Mediated Gene Expression

Ethylene signaling regulates gene expression to drive specific physiological or environmental adaptations. A key step is stabilization and activating the EIN3 and EIL1 transcription factors after ethylene sensing. Activated EIN3 and EIL1 bind promoters of ethylene-responsive genes, including ERFs, to activate or repress expression [3, 9]. ERFs then bind GCC-box motifs in promoters of ethylene-sensitive genes. ERF family diversity supports many ethylene responses, from growth control to stress management (biotic and abiotic) [9]. Additionally, cofactors interacting with EIN3, EIL1, and ERFs sheds light on varied ethylene signaling effects across cellular and environmental context [3, 14].

5 Role in Plant Immunity and Stress Response

Ethylene plays an essential role in plant immunity and in response to biotic and abiotic stresses. It works with jasmonic acid to heighten defense against necrotrophic pathogens and elevate defense genes expression [12]. Recent findings show that ethylene adjusts the salicylic acid pathway, and vice -versa, forming interactions that improve the plant immune response [17]. When plants face drought, salinity, or mechanical injury, ethylene signaling activates and interacts with ABA and auxin to tailor stress responses [12]. For example, under drought, ethylene promotes stomata closure through crosstalk with ABA signaling, therefore, minimizing water loss. Furthermore, ethylene can improve tolerance to salt stress by managing ion balance and osmotic equilibrium.

6 Technological Advances in Ethylene Research

6.1 Novel Methodologies for Studying Ethylene Effects

The development of ethylene-sensitive fluorescent reporters and biosensors marks a significant step forward [18, 19]. These techniques enable real-time, noninvasive tracking of ethylene concentrations in living plants. Polydiacetylene-based sensors allow direct visual identification of ethylene, providing an efficient monitoring technique for agricultural practices and postharvest preservation [20]. Adoption of CRISPR/Cas9 gene editing technology has transformed ethylene research by allowing targeted manipulation of genes for production, signaling, and responses [21]. This method paves the way for creating mutant lines to probe the specific roles of particular genes in ethylene-responsive processes.

6.2 Genomic and Proteomic Approaches

Genomic and proteomic technologies have augmented our understanding of the ethylene signaling network and its links to broader cellular functions. High-throughput RNA sequencing (RNA-seq) and microarray identify ethylene-responsive genes across tissues and stages of development, revealing transcriptome complexity [8, 22]. Similarly, proteomic studies have uncovered many proteins regulated by ethylene, revealing its wide physiological roles. These methodologies also reveal post-translational modifications of ethylene signaling components, such as phosphorylation, ubiquitination, and sumoylation. These post-translational modifications affect protein function and longevity. Specifically, mass spectrometry-based proteomics detects and quantifies these modifications, clarifying regulation of the ethylene response [23, 24].

6.3 Computational Models in Ethylene Research

Computational modeling uses data from genomics, proteomics, and physiological research, giving rise to fresh perspectives on regulatory networks in ethylene signaling. These models replicate ethylene signaling dynamics across scenarios, predicting the effects of genetic alterations, and pinpointing key regulatory points within the network [25]. Agent-based models and systems biology approaches map ethylene generation, detection, and action at both the cellular and whole plant levels [26]. Computational docking and molecular dynamics simulations reveal structural details of ethylene interaction with receptors [27]. Consequently, we better understand ethylene perception and how the signaling process starts.

7 Role of Ethylene in Root Hair Development and Phosphate Availability

Ethylene also can modulate root system architecture (RSA) in response to varying phosphate (Pi) levels in the soil. Under Pi-deficient conditions, ethylene inhibits primary root elongation whereas, promoting lateral root formation and the initiation and elongation of root hairs. These changes increase root surface area, enhancing the ability of plants to absorb more Pi from the environment [28]. The interplay between ethylene and auxin is crucial here, as both hormones stimulate root hair development, under low Pi availability [8, 29, 30].

On the contrary, excess Pi triggers ethylene-mediated responses that alter RSA. High Pi reduces primary root growth. It increases length of lateral roots and root hairs, and decreases root apical meristem activity. This results in a shallower root system that resembles RSA under Pi deficiency, a phenomenon termed as the "Excess Phosphate Response" (EPiR) [8]. Ethylene-responsive factors (ERFs) are significantly induced under excess Pi conditions. These genes reprogram RSA by negatively regulating primary root growth and modifying lateral root development dynamics

[8]. Transcriptomic studies reveal a network of ethylene-induced genes driving ethylene signaling pathways for developmental and physiological functions, such as root development. These include upregulation of defensin-like genes associated with root growth [8]. The effects of ethylene on RSA are concentration-dependent and have been substantiated using ethylene mutant studies [8]. While moderate excess Pi levels may initially promote root growth, higher concentrations result in significant ethylene-mediated inhibition [8, 28]. Ethylene-insensitive mutants display a normal phenotype under excess Pi. Whereas, constitutive ethylene-sensitive mutants exhibit exaggerated responses, highlighting the role of ethylene in adapting RSA to Pi availability [8]. Overall, ethylene serves as an important mediator of root system plasticity. Therefore, it enables plants to adjust root system architecture under varying nutrient conditions [29].

8 Conclusions

Ethylene is a gaseous phytohormone that regulates development and helps plants respond to various environmental challenges. It plays a significant role in how plants develop, adapt to different conditions, and tackle stress. Ethylene biosynthesis starts from methionine. Perception then involves receptors like ETR1, leading to activation of transcription factors such as EIN3 and ERFs. Ethylene alters root system architecture when phosphate availability changes, highlighting its significance for nutrient uptake and environmental responses. It interacts with other phytohormones like auxin, jasmonic acid, and abscisic acid, revealing a complex network of interactions. Advances in methodologies, ethylene-sensitive fluorescent reporters, genomic and proteomic analyses, and computational modeling, have significantly deepened understanding of ethylene signaling networks. However, precise molecular mechanisms of ethylene receptor activation and hormonal crosstalk under varying environmental conditions remain unclear. Also, the comprehensive mapping of post-translational modifications affecting ethylene signaling components still requires investigation. To take advantage of ethylene in an agricultural context, we must address existing gaps in our knowledge. Therefore, ongoing research will help find new ways to improve plant health and increase productivity.

Acknowledgments

The Director, CSIR Central Institute of Medicinal and Aromatic Plants, Lucknow, India, is acknowledged for providing the facilities and support. The Institutional Communication No. CIMAP/ PUB/ 2025/16.

References

1. Bapat VA, Trivedi PK, Ghosh A, Sane VA, Ganapathi TR, Nath P (2010) Ripening of fleshy fruit: molecular insight and the role of ethylene. Biotechnol Adv 28:94–107. https://doi.org/10.1016/j.biotechadv.2009.10.002
2. Barry CS, Giovannoni JJ (2007) Ethylene and fruit ripening. J Plant Growth Regul 26:143. https://doi.org/10.1007/s00344-007-9002-y
3. Binder BM (2020) Ethylene signaling in plants. J Biol Chem 295:7710–7725. https://doi.org/10.1074/jbc.REV120.010854
4. Lin Z, Zhong S, Grierson D (2009) Recent advances in ethylene research. J Exp Bot 60: 3311–3336. https://doi.org/10.1093/jxb/erp204
5. Asif MH, Lakhwani D, Pathak S, Gupta P, Bag SK, Nath P, Trivedi PK (2014) Transcriptome analysis of ripe and unripe fruit tissue of banana identifies major metabolic networks involved in fruit ripening process. BMC Plant Biol 14:316. https://doi.org/10.1186/s12870-014-0316-1
6. Nath P, Trivedi PK, Sane VA, Sane AP (2006) Role of ethylene in fruit ripening. In: Khan NA (ed) Ethylene action in plants. Springer, Berlin/Heidelberg, pp 151–184
7. Lohani S, Trivedi PK, Nath P (2004) Changes in activities of cell wall hydrolases during ethylene-induced ripening in banana: effect of 1-MCP, ABA and IAA. Postharvest Biol Technol 31:119–126. https://doi.org/10.1016/j.postharvbio.2003.08.001
8. Shukla D, Rinehart CA, Sahi SV (2017) Comprehensive study of excess phosphate response reveals ethylene mediated signaling that negatively regulates plant growth and development. Sci Rep. https://doi.org/10.1038/s41598-017-03061-9
9. Zhao H, Yin C, Ma B, Chen S, Zhang J (2021) Ethylene signaling in rice and *Arabidopsis*: new regulators and mechanisms. JIPB 63:102–125. https://doi.org/10.1111/jipb.13028
10. Houben M, Van De Poel B (2019) 1-Aminocyclopropane-1-carboxylic acid oxidase (ACO): the enzyme that makes the plant hormone ethylene. Front Plant Sci 10:695. https://doi.org/10.3389/fpls.2019.00695
11. Li S, Zhu B, Pirrello J, Xu C, Zhang B, Bouzayen M, Chen K, Grierson D (2020) Roles of RIN and ethylene in tomato fruit ripening and ripening-associated traits. New Phytol 226:460–475. https://doi.org/10.1111/nph.16362
12. Fatma M, Asgher M, Iqbal N, Rasheed F, Sehar Z, Sofo A, Khan NA (2022) Ethylene Signaling under stressful environments: Analyzing collaborative knowledge. Plan Theory 11:2211. https://doi.org/10.3390/plants11172211
13. Chien Y, Yoon GM (2024) Subcellular dynamics of ethylene signaling drive plant plasticity to growth and stress: spatiotemporal control of ethylene signaling in *Arabidopsis*. BioEssays 46:2400043. https://doi.org/10.1002/bies.202400043
14. Qiao H, Shen Z, Huang SC, Schmitz RJ, Urich MA, Briggs SP, Ecker JR (2012) Processing and subcellular trafficking of ER-tethered EIN2 control response to ethylene gas. Science 338:390–393. https://doi.org/10.1126/science.1225974
15. Yu Q, Cheng C, Zhou X, Li Y, Hu Y, Yang C, Zhou Y, Soliman TMA, Zhang H, Wang Q, Wang H, Jiang C, Gan S, Gao J, Ma N (2023) Ethylene controls cambium stem cell activity via promoting local auxin biosynthesis. New Phytol 239:964–978. https://doi.org/10.1111/nph.19004
16. Yang C, Ma B, He S-J, Xiong Q, Duan K-X, Yin C-C, Chen H, Lu X, Chen S-Y, Zhang J-S (2015) *MAOHUZI6/ETHYLENE INSENSITIVE3-LIKE1* and *ETHYLENE INSENSITIVE3-LIKE2* regulate ethylene response of roots and coleoptiles and negatively affect salt tolerance in rice. Plant Physiol 169:148–165. https://doi.org/10.1104/pp.15.00353
17. Di X, Gomila J, Takken FLW (2017) Involvement of salicylic acid, ethylene and jasmonic acid signalling pathways in the susceptibility of tomato to *Fusarium oxysporum*. Mol Plant Pathol 18:1024–1035. https://doi.org/10.1111/mpp.12559
18. Fernandez-Moreno J, Stepanova AN (2020) Monitoring ethylene in plants: genetically encoded reporters and biosensors. Small Methods 4:1900260. https://doi.org/10.1002/smtd.201900260
19. Vong K, Eda S, Kadota Y, Nasibullin I, Wakatake T, Yokoshima S, Shirasu K, Tanaka K (2019) An artificial metalloenzyme biosensor can detect ethylene gas in fruits and Arabidopsis leaves. Nat Commun 10:5746. https://doi.org/10.1038/s41467-019-13758-2
20. Nguyen LH, Oveissi F, Chandrawati R, Dehghani F, Naficy S (2020) Naked-eye detection of ethylene using thiol-functionalized Polydiacetylene-based flexible sensors. ACS Sens 5:1921–1928. https://doi.org/10.1021/acssensors.0c00117

21. Nonaka S, Ito M, Ezura H (2023) Targeted modification of CmACO1 by CRISPR/Cas9 extends the shelf-life of Cucumis melo var. reticulatus melon. Front Genome Ed 5:1176125. https://doi.org/10.3389/fgeed.2023.1176125
22. Wu J, Deng Y, Hu J, Jin C, Zhu X, Li D (2020) Genome-wide analyses of direct target genes of an ERF11 transcription factor involved in plant defense against bacterial pathogens. Biochem Biophys Res Commun 532:76–81. https://doi.org/10.1016/j.bbrc.2020.07.073
23. Zolg DP, Wilhelm M, Schmidt T, Médard G, Zerweck J, Knaute T, Wenschuh H, Reimer U, Schnatbaum K, Kuster B (2018) ProteomeTools: systematic characterization of 21 post-translational protein modifications by liquid chromatography tandem mass spectrometry (LC-MS/MS) using synthetic peptides. Mol Cell Proteomics 17:1850–1863. https://doi.org/10.1074/mcp.TIR118.000783
24. Silva AMN, Vitorino R, Domingues MRM, Spickett CM, Domingues P (2013) Post-translational modifications and mass spectrometry detection. Free Radic Biol Med 65:925–941. https://doi.org/10.1016/j.freeradbiomed.2013.08.184
25. Díaz J, Álvarez-Buylla ER (2006) A model of the ethylene signaling pathway and its gene response in *Arabidopsis thaliana* : pathway cross-talk and noise-filtering properties. Chaos 16:023112. https://doi.org/10.1063/1.2189974
26. Zhang B, DeAngelis DL (2020) An overview of agent-based models in plant biology and ecology. Ann Bot 126:539–557. https://doi.org/10.1093/aob/mcaa043
27. Santos LHS, Ferreira RS, Caffarena ER (2019) Integrating molecular docking and molecular dynamics simulations. In: De Azevedo WF (ed) Docking screens for drug discovery. Springer, New York, pp 13–34
28. Roldan M, Dinh P, Leung S, McManus MT (2013) Ethylene and the responses of plants to phosphate deficiency. AoB Plants 5:plt013. https://doi.org/10.1093/aobpla/plt013
29. Masucci JD, Schiefelbein JW (1994) The rhd6 mutation of Arabidopsis thaliana alters root-hair initiation through an auxin- and ethylene-associated process. Plant Physiol 106:1335–1346. https://doi.org/10.1104/pp.106.4.1335
30. Dolan L (2001) How and where to build a root hair. Curr Opin Plant Biol 4:550–554. https://doi.org/10.1016/S1369-5266(00)00214-4

Chapter 16

Mapping of the Ethylene-Specific Root System Architecture (RSA) Traits in Arabidopsis

Devesh Shukla

Abstract

Understanding the root system architecture (RSA) is necessary for elucidation of plant growth patterns in response to environmental stimuli and hormonal signals. Ethylene, a gaseous phytohormone, modulates root developmental plasticity, including primary root elongation, lateral root formation, and root hair growth. We present a protocol for mapping ethylene-specific RSA traits in *Arabidopsis thaliana* using a hydroponic growth system. Arabidopsis seedlings grow on a polypropylene mesh supported by polycarbonate wedges in a magenta box-based setup. We treat seedlings with ethylene or its precursor, then spread root system on agar plates with an art brush. High-resolution images are recorded and analyzed with free ImageJ software. This protocol allows detailed RSA analysis under controlled ethylene treatments and can be adapted for other plant species.

Key words *Arabidopsis thaliana*, Ethylene, Root System Architecture, Hydroponic System, Root Imaging, ImageJ, Root Phenotyping, Plant Hormones, Lateral Roots, Root Hair Development

1 Introduction

The root system architecture (RSA) of plants is a plastic and adaptive tissue-organ that plays an important role in water and nutrient uptake [1–3]. It also anchors the plant and directly interacts with the soil environment [1, 2, 4]. Ethylene, a simple gaseous phytohormone, regulates plant growth and development, modulating seed germination, fruit ripening, senescence, abscission, and responses to abiotic and biotic stresses [5, 6]. In roots, ethylene affects primary root elongation, lateral root formation, root hair development, and overall root system architecture (RSA), helping plants adapt to constantly changing environment [3, 4, 7, 8]. Understanding how ethylene modulates RSA is necessary for improving crop stress tolerance and nutrient use efficiency. Traditional agar-plate-based systems have been used to study RSA; however, these methods can introduce elemental contamination from

Devesh Shukla (ed.), *Ethylene Sensing and Signaling*, Methods in Molecular Biology, vol. 2945,
https://doi.org/10.1007/978-1-0716-4650-2_16,

the gelling agents (Agar), slow root growth, and limit full root visualization due to roots being embedded in the gel [3, 7, 8]. To address these limitations, a hydroponic growth system for *Arabidopsis thaliana* is presented that allows nondestructive examination of ethylene- specific RSA traits [3, 8, 9]. Seedlings grow on a polypropylene mesh supported by polycarbonate wedges in a magenta box-based hydroponic setup [3, 7, 8]. It provides an agar-free environment, promotes rapid root growth and allows easy access to the roots. After treatment with ethylene or its precursor, roots are carefully spread on agar plates using an art brush for imaging. High-resolution images are then analyzed using free ImageJ software [8].

This protocol minimizes elemental contamination, accelerates root system growth, provides full access to the root system, and uses simple materials and freely available analysis tool [3, 8]. Although this protocol is optimized for Arabidopsis, it can be adapted for other species with adjustments [8]. The schematic workflow in Fig. 1, outlines the key steps for analyzing and visualizing the root system architecture (RSA) of plantlets.

2 Materials

2.1 Plant Material

1. *Arabidopsis thaliana* seeds (ecotype Col-0).
2. Mutant or *transgenic* lines (if applicable).

2.2 Growth Media and Solutions

1. Murashige and Skoog (MS) basal medium (½ strength).
2. Sucrose 1% w/v (10 g/L).
3. Ethylene precursor: 1-aminocyclopropane-1-carboxylic acid (ACC).
4. Seed surface sterilization solutions including 70% ethanol, 4% (v/v) diluted commercial bleach solution with a drop of Tween-20, autoclaved double-distilled water, Agar (1.2% w/v) for petri plates.

2.3 Hydroponic System Components

1. Magenta boxes (GA-7 boxes).
2. Polypropylene mesh (250 μm and 500 μm pore size).
3. Polycarbonate sheets (clear, smooth texture), cut into wedges (4 × 8 cm rectangles with midpoint notch).

2.4 Equipment

1. Laminar air flow cabinet.
2. Sterile petri dishes (150 mm × 15 mm).
3. Growth chamber with light and temperature settings (120 μmol m^{-2} s^{-1}, 16 h light/8 h dark photoperiod, 23 °C).

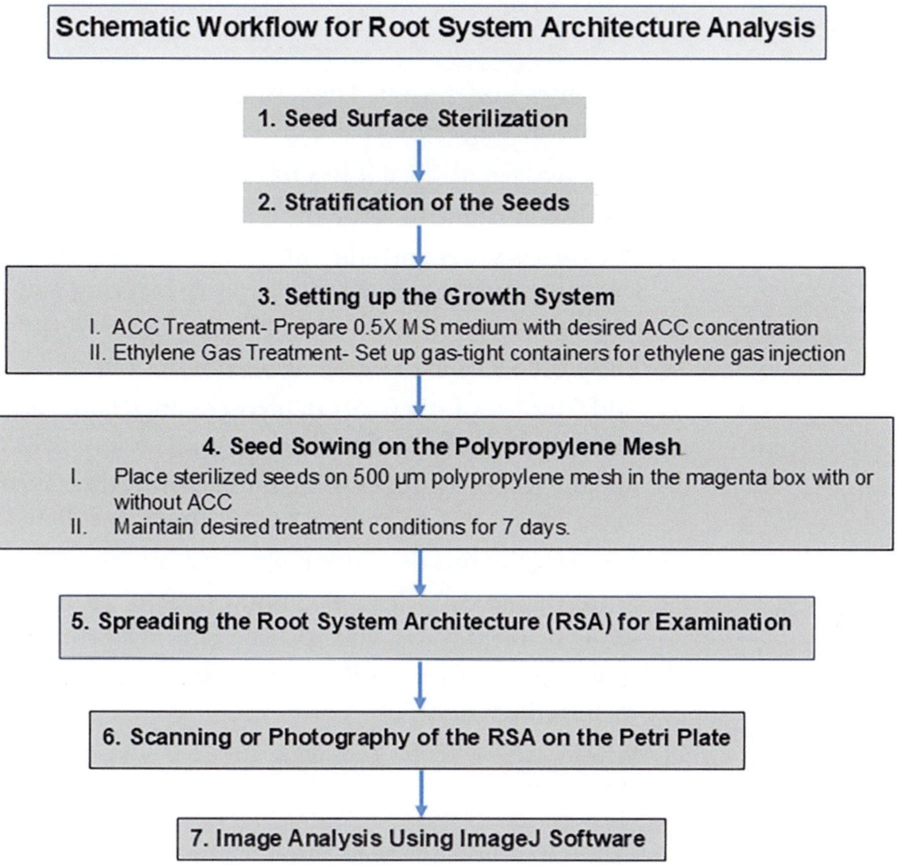

Fig. 1 Schematic workflow outlines the major steps for analyzing root system architecture. It summarizes the protocol from seed surface sterilization, and stratification, to growth setup with or without ACC or ethylene treatment. Seed are sown on a polypropylene mesh, roots are later spread for observation, imaging root systems, and the data analyzed using ImageJ software

4. Flatbed scanner (600 dpi resolution) or high-resolution digital camera.
5. Computer with ImageJ software.
6. Fine-tipped art brushes.
7. Standard laboratory supplies (pipettes, forceps, and microcentrifuge tubes).

2.5 Ethylene Treatment Setup

1. Ethylene gas cylinder and regulator (if using gaseous ethylene).
2. Air-tight containers or jars for treatment (if using gaseous ethylene).
3. Sealing materials (like Parafilm).

3 Methods

3.1 Seed Surface Sterilization and Stratification

1. Transfer approximately 100 Arabidopsis seeds (~2.5 mg) into a microcentrifuge tube.
2. Soak seeds in distilled water for 1 h at room temperature.
3. Briefly centrifuge to settle the seeds.
4. Decant the water and add 700 μL of 70% ethanol. Vortex for a few seconds and spin briefly. Repeat vortexing and spinning, but do not exceed 3 min total ethanol exposure (*see* **Note 1**).
5. Rinse the seeds with sterile distilled water once.
6. Add 700 μL of 4% (v/v) diluted commercial bleach solution having a drop of Tween-20.
7. Rapidly invert the tubes 8–12 times to mix, then briefly centrifuge. Repeat this cycle three to four times. Incubate for 7 min.
8. Take out the bleach solution using a 1 mL pipette.
9. Rinse the seeds at least five times (maybe up to seven) with sterile distilled water, and vortexing during each rinse.
10. Leave the surface-sterilized seeds in a microcentrifuge tube with sterile water.
11. Stratify at 4 °C in the dark for 2–3 days to synchronize germination.

3.2 Preparation of Hydroponic Growth System

1. Cut polycarbonate sheets into rectangles (4 × 8 cm) and notch them at the midpoint more than halfway through the rectangle (Fig. 2).
2. Assemble two notched rectangles to form an X-shaped support (Fig. 2).
3. Cut the mesh into squares (6 × 6 cm) with 250 μm pore size for initial germination (Fig. 2) (*see* **Note 2**).
4. Autoclave the polycarbonate wedges and mesh separately, wrapping the mesh in an aluminum foil to prevent wrinkles and distortion.
5. Fill magenta boxes halfway with distilled water and autoclave.
6. In a laminar flow cabinet, assemble the X-shaped polycarbonate support in the magenta box aseptically (Fig. 2).
7. Place sterile polypropylene mesh on top of the support aseptically.
8. Add sterile half-strength MS medium to each box until it reaches the bottom edge of mesh aseptically.

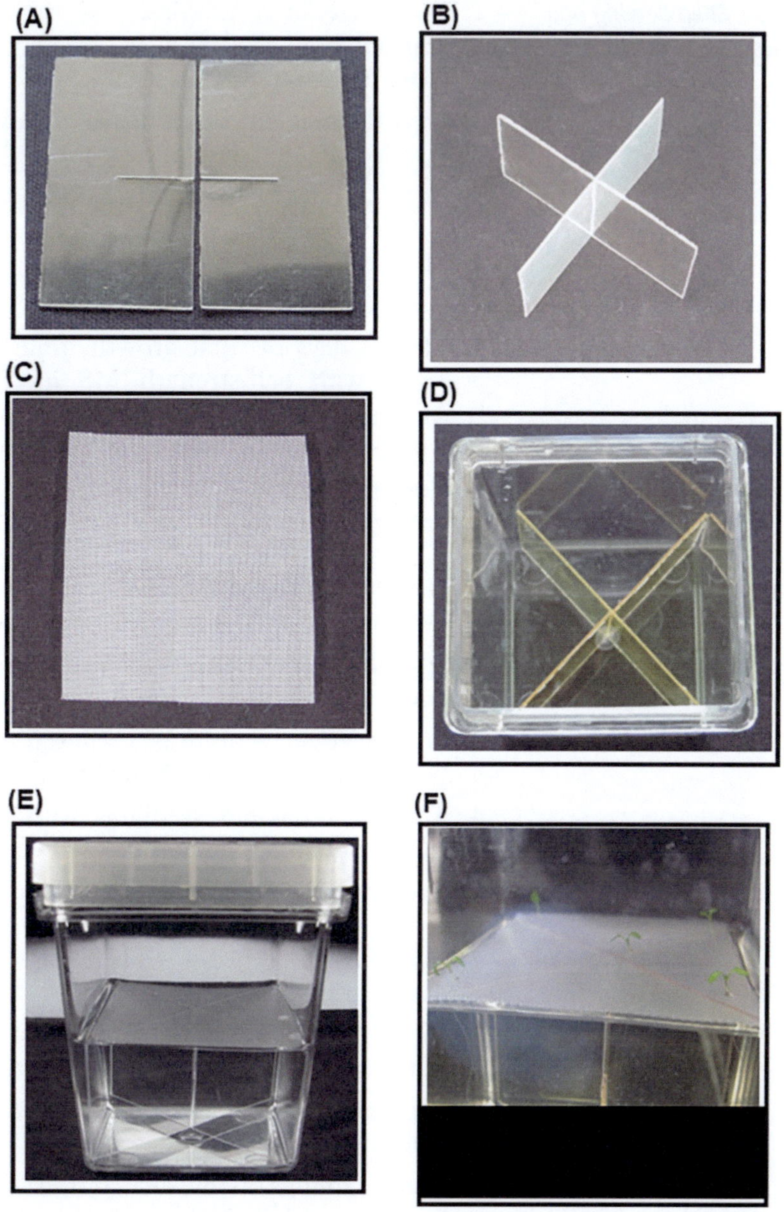

Fig. 2 Display the assembly of hydroponic growth system components (As shown in an earlier publication by Shukla et al. [8]. (**a**) Two rectangular pieces of polycarbonate wedges (4 × 8 cm); (**b**) assembly of polycarbonate wedges into an X-shaped support; (**c**) polypropylene mesh squares (6 × 6 cm) with appropriate pore sizes; (**d**) placement of the X-shaped wedge support inside the magenta box. (**e**) Assembly of the polypropylene mesh (pore size 250 μm) on the X-shaped support in the magenta box filled with half-strength MS medium up to the mesh level. (**f**) Display of seeds sown on the mesh surface

3.3 Seed Sowing and Seedling Growth

1. Under sterile conditions, sow the stratified seeds onto the 250 μm pore size mesh in the magenta boxes (Fig. 2e) (*see* **Note 3**).
2. Place the magenta boxes in a growth chamber set to 23 °C with a 16 h light/8 h dark photoperiod.
3. After 3 days, carefully transfer seedlings onto mesh with 500 μm pore size in fresh magenta boxes with half-strength MS medium.
4. Grow the seedlings for additional 2 days.
5. After 5 days of total growth, transfer seedlings to magenta boxes with half-strength MS medium supplemented with ACC (ethylene precursor) at the desired concentration (e.g., 1 μM) or set up ethylene gas treatments.
6. Continue growing seedlings under the same environmental conditions for additional 7 days.

3.4 Ethylene Treatment

3.4.1 Option A: ACC Treatment

1. Dissolve appropriate quantity of ACC (1-aminocyclopropane-1-carboxylic acid) in sterile distilled water.
2. Solution may be filter-sterilized using 0.22 μm and store aliquots at −20 °C.
3. Prepare ACC medium by adding the desired ACC concentration to half-strength MS medium.
4. Transfer 5-day-old seedlings with 500 μm mesh onto the ACC-containing medium.
5. Grow seedlings for 7 days under standard conditions.

3.4.2 Option B: Ethylene Gas Treatment (Short-Term Exposure)

1. Place magenta boxes with 5-day-old seedlings into sterile gas-tight containers or jars.
2. Inject ethylene gas into containers to reach the desired concentration (e.g., 10 μL/L or 10 ppm, v/v).
3. Seal containers tightly with Parafilm to maintain the ethylene concentration.
4. Treat seedlings briefly, such as overnight (12–16 h) or for a few hours as needed. Prolonged airtight exposure can cause physiological stress and impair seedling growth.

3.5 Harvesting and Spreading Root Systems

1. Prepare 150 mm × 15 mm petri dishes containing 1.2% (w/v) agar.
2. Add 10–20 mL of sterile distilled water to each agar plate to facilitate root spreading.
3. Gently remove the seedlings from the mesh using fine forceps or tweezers (*see* **Note 4**).
4. Submerge the seedlings in water on the agar plate.

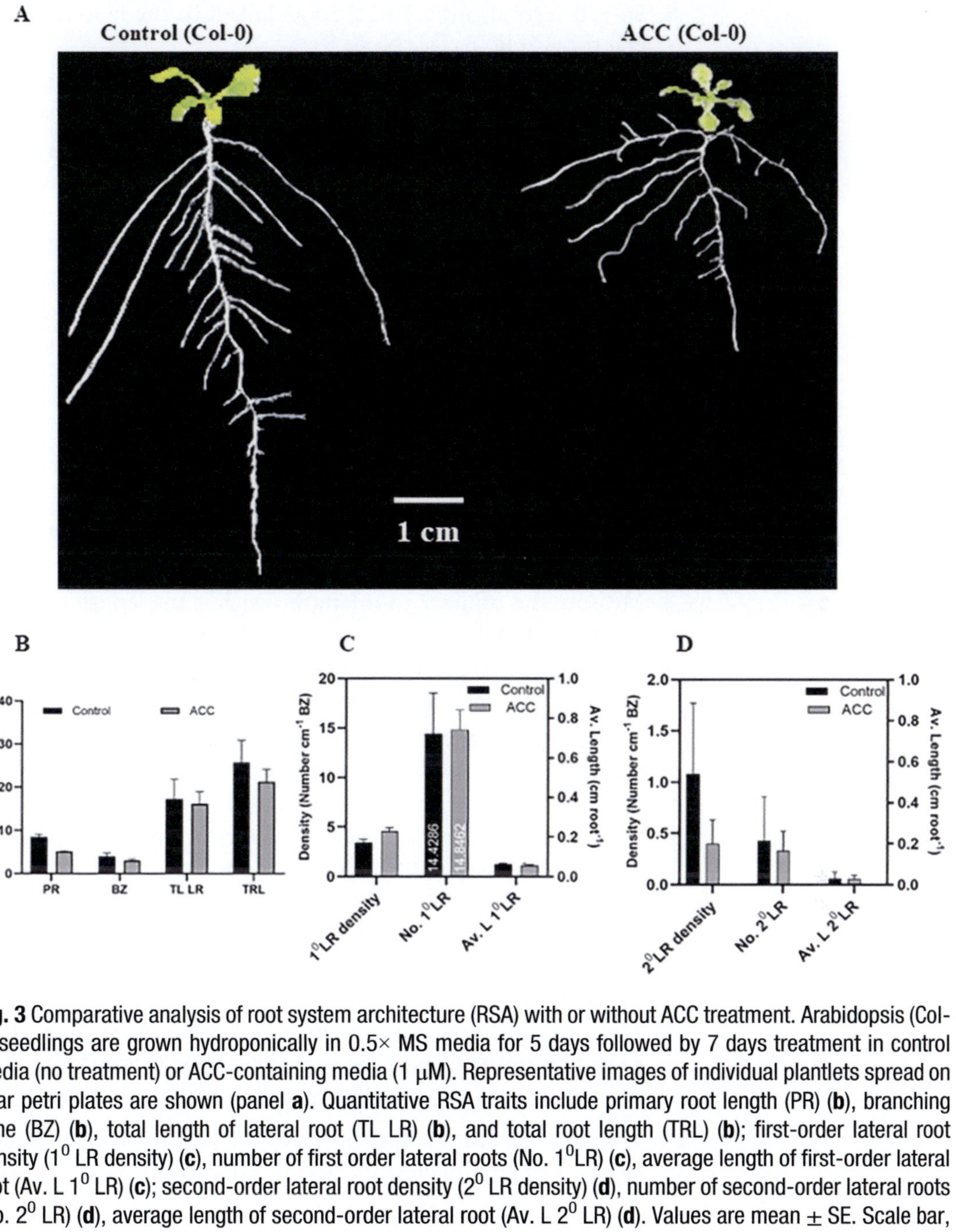

Fig. 3 Comparative analysis of root system architecture (RSA) with or without ACC treatment. Arabidopsis (Col-0) seedlings are grown hydroponically in 0.5× MS media for 5 days followed by 7 days treatment in control media (no treatment) or ACC-containing media (1 μM). Representative images of individual plantlets spread on agar petri plates are shown (panel **a**). Quantitative RSA traits include primary root length (PR) (**b**), branching zone (BZ) (**b**), total length of lateral root (TL LR) (**b**), and total root length (TRL) (**b**); first-order lateral root density (1^0 LR density) (**c**), number of first order lateral roots (No. 1^0LR) (**c**), average length of first-order lateral root (Av. L 1^0 LR) (**c**); second-order lateral root density (2^0 LR density) (**d**), number of second-order lateral roots (No. 2^0 LR) (**d**), average length of second-order lateral root (Av. L 2^0 LR) (**d**). Values are mean ± SE. Scale bar, 1 cm

5. Using a fine-tipped art brush, gently spread the root system.
6. Position the primary root in a straight line as the central axis.
7. Spread lateral roots symmetrically on either side of the primary root.

8. Spread second-order lateral roots linked to the first-order lateral roots (Fig. 3) (*see* **Note 5**).
9. Carefully tilt the plate to drain excess water without disturbing the root system.
10. If necessary, store the plates at 4 °C to pause the process before imaging (optional).

3.6 Imaging Root Systems

1. Allow the plates to return to room temperature if stored at 4 ° C.
2. Wipe any condensed water from the plate lids.
3. Capture high-resolution images of the spread root systems using a flat-bed scanner at 600 dpi or a high-resolution digital camera (Fig. 3) (*see* **Note 6**).

3.7 Quantitative Analysis of RSA Traits

1. Open the captured images in ImageJ software (refer to Shukla et al. [8] for details of ImageJ use).
2. Set the scale in ImageJ using the known distance from the scale bar (*see* **Note 7**).
3. Measure the primary root length which is the length from the hypocotyl junction to the root tip (Fig. 3).
4. Calculate the total root length which is a sum of the lengths of the primary root and all lateral roots (Fig. 3).
5. Measure the branching zone of the primary root which is the length between the first and last lateral root emergence points (Fig. 3).
6. Measure the number of first-order lateral roots by counting the lateral roots originating from the primary root within the branching zone (Fig. 3).
7. Measure the average length of first-order lateral roots, which is the total length of first-order lateral roots divided by their number (Fig. 3).
8. Measure the lateral root density, which is the number of lateral roots per unit length of the branching zone (Fig. 3).
9. Record all measurements for statistical analysis.
10. For detailed instructions on using ImageJ for root measurement, refer to Shukla et al. [8] (*see* **Note 8**).

3.8 Root Hair Phenotyping (Optional)

1. Cut a 1 cm section from the root tip of the primary root.
2. Mount the root section on a microscope slide with 10% glycerol as a mounting medium (Fig. 4).
3. Observe and capture images using a stereomicroscope (Fig. 4).
4. Analyze root hair number and length using ImageJ (Fig. 4).

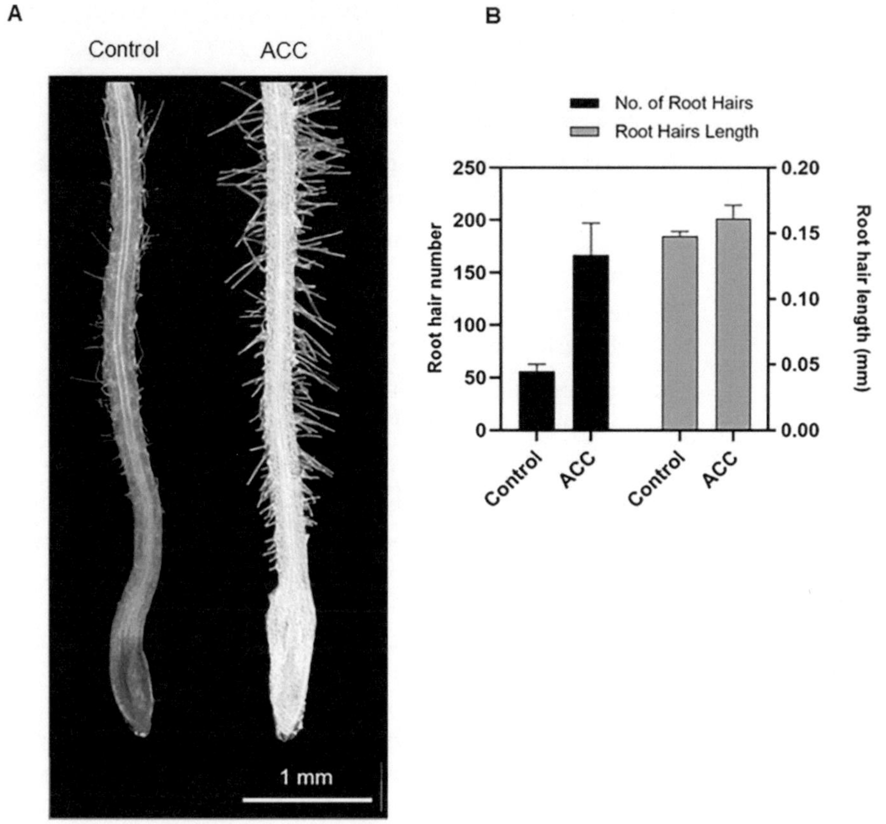

Fig. 4 Comparative analysis of root hair phenotype of *Arabidopsis thaliana* (Col-0) under control or ACC-treated condition. A 1 cm section from primary root tip was cut and mounted in 10% glycerol. A 5 mm region from the tip was documented for root hair number and length (**a**). Data are presented for root hair length (**b**) and number of root hair (**b**) in a 5 mm region. Values are mean ± SE, and scale bar is 1 mm

3.9 Data Analysis

1. Perform statistical analyses using appropriate software (GraphPad Prism, R).
2. Compare ethylene-treated samples with controls using statistical tests such as Student's *t* tests or ANOVA.

4 Notes

1. Long exposure to ethanol can reduce seed germination rates. Limit the time, not more than 3 min [8].
2. Using a larger pore size (500 μm) mesh helps easy removal of seedlings without damaging roots [8].
3. Restrict the number of seeds per mesh to avoid root entanglement. For RSA analysis, use fewer seeds (four seeds per mesh). For bulk tissue, more seeds can be sown.

4. Remove seedlings slowly and gently to avoid damaging root systems.
5. Spreading roots is an art that requires patience. Submerge roots in water to brush spreading.
6. Mark a scale on the petri plate (1 cm) or include a ruler in the image for accurate measurements.
7. Set the scale in ImageJ for each new image for accurate measurements.
8. Refer to Shukla et al. [8] for detailed ImageJ guidance on measuring lengths.
9. Hydroponic growth promotes rapid root development and minimizes elemental contamination from agar.
10. Wrap the polypropylene mesh in an aluminum foil during autoclaving to avoid bending and maintain a flat surface.
11. This method can be adapted for different plant species by adjusting the mesh pore size [8].

Acknowledgments

The Director, CSIR Central Institute of Medicinal and Aromatic Plants, Lucknow, India, is acknowledged for providing the facilities and support. The Institutional Communication No. CIMAP/ PUB/ 2025/07.

References

1. Rellán-Álvarez R, Lobet G, Dinneny JR (2016) Environmental control of root system biology. Annu Rev Plant Biol 67:619–642. https://doi.org/10.1146/annurev-arplant-043015-111848
2. Gruber BD, Giehl RFH, Friedel S, Von Wirén N (2013) Plasticity of the Arabidopsis root system under nutrient deficiencies. Plant Physiol 163: 161–179. https://doi.org/10.1104/pp.113.218453
3. Shukla D, Rinehart CA, Sahi SV (2017) Comprehensive study of excess phosphate response reveals ethylene mediated signaling that negatively regulates plant growth and development. Sci Rep. https://doi.org/10.1038/s41598-017-03061-9
4. López-Bucio J, Hernández-Abreu E, Sánchez-Calderón L, Nieto-Jacobo MF, Simpson J, Herrera-Estrella L (2002) Phosphate availability alters architecture and causes changes in hormone sensitivity in the Arabidopsis root system. Plant Physiol 129:244–256. https://doi.org/10.1104/pp.010934
5. Bapat VA, Trivedi PK, Ghosh A, Sane VA, Ganapathi TR, Nath P (2010) Ripening of fleshy fruit: molecular insight and the role of ethylene. Biotechnol Adv 28:94–107. https://doi.org/10.1016/j.biotechadv.2009.10.002
6. Nath P, Trivedi PK, Sane VA, Sane AP (2006) Role of ethylene in fruit ripening. In: Khan NA (ed) Ethylene action in plants. Springer, Berlin/Heidelberg, pp 151–184
7. Jain A, Poling MD, Smith AP, Nagarajan VK, Lahner B, Meagher RB, Raghothama KG (2009) Variations in the composition of gelling agents affect Morphophysiological and molecular responses to deficiencies of phosphate and other nutrients. Plant Physiol 150:1033–1049. https://doi.org/10.1104/pp.109.136184
8. Shukla D, Trivedi PK, Sahi S (2023) A simple protocol for mapping the plant root system architecture traits. JoVE:64876. https://doi.org/10.3791/64876

9. Jain A, Poling MD, Karthikeyan AS, Blakeslee JJ, Peer WA, Titapiwatanakun B, Murphy AS, Raghothama KG (2007) Differential effects of sucrose and auxin on localized phosphate deficiency-induced modulation of different traits of root system architecture in Arabidopsis. Plant Physiol 144:232–247. https://doi.org/10.1104/pp.106.092130

Chapter 17

Ethylene as a Central Regulator of Seed Germination, Early Seedling Development, and Stress Adaptation

Devesh Shukla and Sumya Pathak

Abstract

Ethylene, a two carbon phytohormone is required for fine-tuning the seed dormancy release, germination, and early seedling development. It interacts with abscisic acid (ABA), and alleviates dormancy by increasing endosperm softening and radicle protrusion. It works in synergy with gibberellins and helps in rapid seed germination, whereas, its cross talk with reactive oxygen species (ROS) facilitates cell wall modification. After germination of seedling, ethylene affects hypocotyl elongation and apical hook formation, protecting the shoot tip during soil emergence. It also reshapes the root system architecture by coordinating auxin transport and brassinosteroid responses, thereby optimizing the plant's search for water and nutrients. Beyond development, ethylene confers tolerance to salinity, hypoxia, and extreme temperatures, by sustaining antioxidant defenses and metabolic adjustments that keep the embryo safe. In agricultural practices, well-timed ethylene treatments can break seed dormancy, tighten germination uniformity, and improve seedling establishment under harsh conditions. Experiments in sunflower and lettuce, demonstrate that this hormone boosts seed vigor and synchronizes germination. Despite the researchers made substantial progress, we still lack answers about species-specific regulation, epigenetic control, and interplay with other hormones. Because transcriptomics, genetics, and biotechnology now provide sharper tools we can better coordinate ethylene responses for crop improvement. Consequently, adopting these strategies, we can support sustainable agriculture and strengthen global food security. This chapter discusses the critical importance of ethylene in bridging basic plant biology with practical applications.

Key words Ethylene, Seed dormancy, Germination, Early seedling development, Hormonal cross talk, ABA antagonism, Stress tolerance, ROS interactions

1 Introduction

Seeds denote a critical developmental stage, acting as small survival kits filled with genetic material that ensure species persistence. Seeds can stay dormant until conditions become favorable. Researchers first recognized ethylene in the early twentieth century for driving fruit ripening and leaf abscission, processes that facilitate leaf detachment, and resource redistribution. The significance of ethylene as a versatile regulator of plant physiology has become

Devesh Shukla (ed.), *Ethylene Sensing and Signaling*, Methods in Molecular Biology, vol. 2945,
https://doi.org/10.1007/978-1-0716-4650-2_17,

evident, particularly for seed biology [1, 2]. In 1960s, the methionine was identified as its biosynthetic precursor [1, 3]. This led to the characterization of the Yang cycle, that regenerates methionine for sustain ethylene production [1]. Consequently, many plant species show spike in ethylene during radicle emergence, signifying its essential role in seed germination [4].

Seed germination is a finely tuned process, driven by hormones where ethylene takes center stage. It counters abscisic acid (ABA), a hormone known to favour dormancy, thereby promoting germination [4]. Ethylene works with gibberellins (GAs) and reactive oxygen species (ROS) to weaken the seed coat and push radicle emergence [2, 3]. In nondormant seeds, ethylene production increases with water uptake and cell expansion during imbibition [2]. In dormant seeds, ethylene suppresses ABA and stimulates GA production that break dormancy [4, 5]. These ethylene driven functions are particularly important in species whose germination depends on environmental cues like light and temperature [2, 5]. Owing to its integrated signaling, ethylene supports radicle emergence under many conditions, confirming its role in seed germination biology (Fig. 1) [4]. After germination, ethylene affects early seedling growth (Fig. 1). It regulates apical hook formation and hypocotyl elongation, shielding the shoot apical meristem as seedling penetrates the soil [3, 5]. By modulating auxin transport, It synchronizes cell elongation and differentiation vital for seedling establishment [5]. Ethylene interacts brassinosteroids and jasmonates, to fine-tune growth patterns [5]. These hormonal interactions maintains a dynamic hormonal balance, allowing plants to adapt to changing environments [4, 5].

Ethylene also helps plants adjust to stressful environments, making it indispensable for survival [5, 6]. During flooding, heightened ethylene production triggers the formation of root aerenchyma that improve oxygen transport [6]. Under salt stress, ethylene coordinates with ABA and ROS to mitigate ionic imbalances and oxidative damage, thus enhancing plant tolerance [5]. Because of these adaptations ethylene supports plant survival in changing environments [4, 5]. Its dual ability to promote growth while mitigating stress responses showcases unique regulatory versatility throughout development [4]. By balancing dormancy and germination through interactions with ABA and GAs and by integrating environmental signals, ethylene supports adaptive growth [1, 5]. Leveraging ethylene regulatory pathways breeders can improve seed vigor, improve germination, and strengthen seedling establishment under harsh conditions [2, 4]. Consequently, researchers and agronomists are engineering crops that harness ethylene to tolerate extreme climates [4]. Under drought condition, it enhances ROS-scavenging enzymes and promotes deeper, branched roots, improving water uptake. Heat stress activates ethylene signaling which increases heat-shock protein synthesis

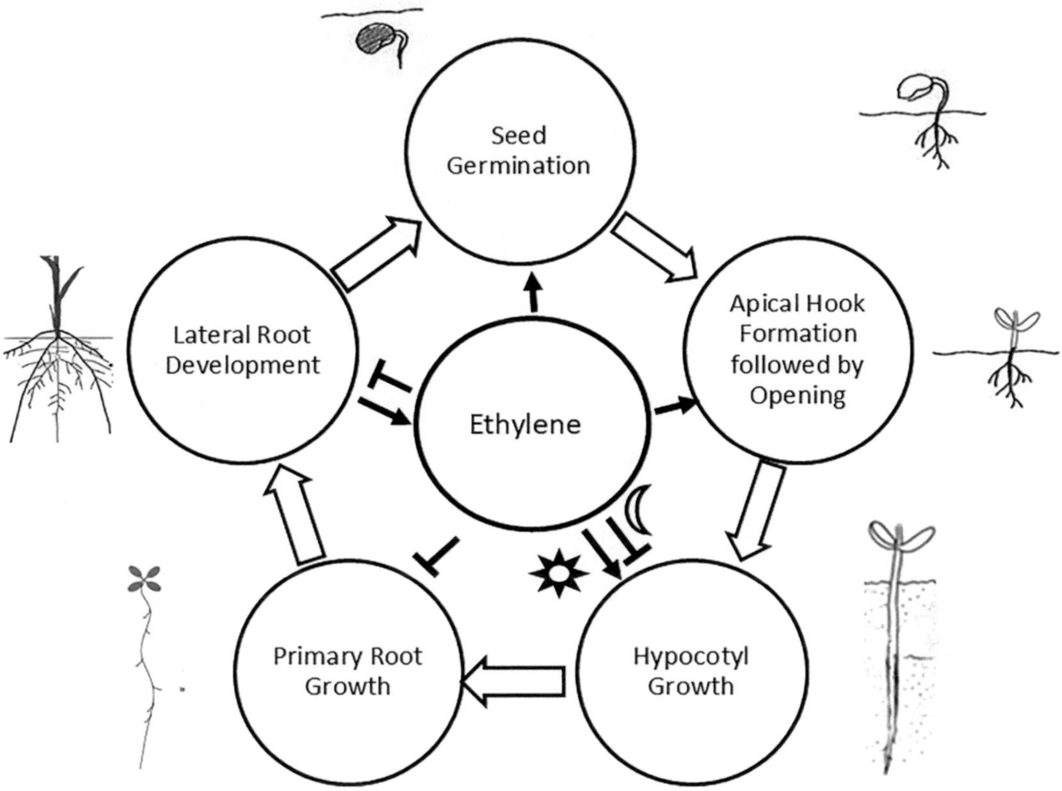

Fig. 1 Seed development transitions from dormancy to germination, seedling emergence, and finally shoot and root growth. During these stages, ethylene works with other hormones to regulate radicle protrusion, apical hook formation, root system architecture and plant maturation.

and stabilizes vital metabolic pathways. Additionally, interaction of ethylene with microbial signals offers new routes to reinforce defenses against yield impacting pathogens and pests.

Applications of ethylene reaches well beyond traditional crop production. Growers also test controlled ethylene treatments to synchronize germination in large-scale farming. As the world faces the climate change, the ethylene directed seed biology and stress adaptation gain even higher importance. Therefore, integrating ethylene strategies into farming could address food security while promoting sustainable practices. Thus, ethylene stands as a cornerstone of modern plant science, linking basic plant biology to practical tools for a resilient food future.

2 Ethylene Biosynthesis and Signaling

Ethylene biosynthesis and signaling involve tightly regulated pathways for seed germination and early seedling growth [5]. The Yang cycle converts methionine into ethylene through S-adenosyl-L-methionine (SAM) and 1-aminocyclopropane-1-carboxylic acid (ACC). SAM is synthesized by SAM synthetase and converted to ACC by ACC synthase (ACS), a rate-limiting step affected by environmental and hormonal cues [7, 8]. ACC is oxidized by ACC oxidase (ACO) to produce ethylene, releasing CO_2 and hydrogen cyanide [7, 8]. This pathway is regulated by feedback inhibition, and ACC can also be stored as malonyl-ACC (MACC) to tune ethylene levels [1, 5]. At the endoplasmic reticulum membrane, receptors ETR1 and ERS1 senses ethylene and, because of binding, deactivate the kinase CTR1. CTR1 deactivation allows EIN2, to activate and transport its cleaved C-terminal fragment to the nucleus. Inside the nucleus, the fragment stabilizes EIN3 and EIL1, therefore enabling these transcription factors to bind to ethylene-responsive promoters [4, 6].

Ethylene works with other hormones to regulate seed development. It counters abscisic acid (ABA), which normally blocks germination by altering ABA sensitivity and metabolism [7, 9]. Ethylene works with gibberellins (GA) to weaken the endosperm and promote radicle protrusion [3, 4, 8]. Understanding the ACC-to-ethylene pathway and the signaling roles of EIN2 and EIN3/EIL1 has clear agricultural value. Altering these processes may improve plant tolerance to stress. These insights lay the foundation for targeted crop management and stress mitigation strategies [4, 5].

3 Ethylene and Its Interactions in Seed Germination

Seed germination represents a crucial stage in a plant's life, governed by hormonal cross talk, environmental stimuli, and metabolic processes. Ethylene plays a key role in breaking dormancy and drives the shifts towards germination. It achieves this interacting with ABA, GA, and ROS. This chapter examines how ethylene drives dormancy release and germination while interacting with other signals to control seed physiology. During seed germination, ACC synthase (ACS) converts S-adenosyl-methionine (SAM) to 1-aminocyclopropane-1-carboxylic acid (ACC), and ACC oxidase (ACO) then oxidizes ACC to ethylene. Ethylene production increases post-imbibition, and peaks at radicle protrusion because ACO activity increases and ACS transcripts accumulate [10]. Membrane-bound receptors such as ETR1 and EIN2 perceive the ethylene and trigger downstream transcription factors EIN3/EIL1. These cascades activate gene networks that weaken the endosperm and expand cells, thus coordinating germination events. Feedback loops within the pathway adjust hormonal balance and integrate environmental responses, reflecting the signaling adaptability. Recent studies report cross-modulation by other hormones and stresses, further indicating it as a central regulator.

3.1 Interplay with ABA in Dormancy and Germination

Seed dormancy hinges on antagonism between ethylene and ABA. ABA maintains dormancy by inhibiting endosperm rupture, whereas, ethylene mitigates effect of ABA [7, 9]. Ethylene-insensitive mutants (e.g., *etr1* and *ein2*) display hypersensitivity to ABA, confirming suppressive role of ethylene. Consequently, precise interaction between both hormones is vital for timely dormancy release [8].

During dormancy, ABA curbs ACO activity and thereby lowers ethylene production. Abscisic acid (ABA) does so by suppressing *ACO* gene transcription, whereas ethylene removes this suppression, indicating reciprocal control [4, 5, 10]. External application of ethylene or its precursors can rescue germination even when ABA inhibits it, a finding with direct farm level value [4, 5, 10]. Research also indicates that ethylene reduces ABA sensitivity by switching on specific transcription factors, a clue for future genetic and biochemical studies.

3.2 Synergy with Gibberellins

Ethylene and GAs act together to start germination by weakening the endosperm and promoting embryonic growth. GAs increase ACC accumulation and ACO activity, thereby increasing ethylene biosynthesis [4]. This synergy accelerates the germination process and allows seeds respond swiftly to environmental and developmental cues. In Arabidopsis, the ethylene-responsive protein EIN2 links GAs and ethylene signals, coordinating germination gene expression [11]. Ethylene also facilitates the degradation of DELLA proteins, the main repressors of GA signaling [12]. Therefore, ethylene and GAs cooperate to degrade DELLA protein that releases growth-promoting signals [12]. Strength of this interaction may differs among species, hinting at the evolutionary adaptability of seed germination mechanisms.

3.3 Role of ROS in Ethylene-Mediated Germination

Reactive oxygen species (ROS) act as secondary messengers during ethylene-driven germination [5]. Ethylene elevates ROS production, which then softens the seed coat and endosperm, facilitating radicle to emerge. Because ROS modify hormonal pathways, including ethylene and ABA, they create a feedback loop that accelerates germination [5, 13, 14]. Hydrogen peroxide (H_2O_2), a key ROS, interacts with ethylene to amplify signaling cascades. Accumulating ROS within the embryonic axis matches rising ethylene levels, highlighting their interdependence during dormancy release and germination. As manipulating ROS levels alters effectiveness of ethylene, researchers see clear opportunities for seed technology and crop management.

3.4 Environmental Interactions

Ethylene modulates germination when seeds face salinity, hypoxia, or extremes temperatures. As ethylene alleviates ABA sensitivity and promoting ROS scavenging, it eases salt-induced dormancy [5, 13]. Interaction with light through phytochrome pathways further shows adaptability of ethylene in decoding environmental cues. Under hypoxic con, ethylene production increases and drives metabolic shifts for successful germination. Adjusting ethylene pathways under stress holds clear promise for agriculture. Enhancing ethylene production or sensitivity during adverse conditions raises germination rates and strengthens seedling establishment. Continued study of ethylene's role in stress adaptation could spur innovative agricultural practices. As ethylene interacts with ABA, GA, and ROS, it takes centrality in seed dormancy release and germination [5, 13, 14]. Its ability to modulate hormonal networks and decode environmental signals confirms it as a key regulator of germination.

4 Early Seedling Development

Early seedling development depends on hormonal signals and environmental factors, including temperature, light availability, water supply, and soil composition [15]. Ethylene drives hypocotyl growth, apical hook formation, and root system architecture via cross talk with auxins, gibberellins (GAs), brassinosteroids (BRs), and abscisic acid (ABA) [5, 11, 16]. Therefore, this early stage builds a foundation for plant establishment through integrated genetic, hormonal, and environmental signals.

4.1 Hypocotyl Growth

Hypocotyl elongation shifts with light intensity and with interaction of ethylene with other hormones (Fig. 2). In darkness, ethylene inhibits hypocotyl elongation by triggering cortical microtubule rearrangement via ETHYLENE RESPONSE FACTOR1 (ERF1) and WAVE-DAMPENED 5 (WDL5) [16]. This restraint helps seedlings push through soil. Once light reaches the shoot, ethylene instead increases elongation by stabilizing CONSTITUTIVE PHOTOMORPHOGENIC1 (COP1). COP1 then degrades LONG HYPOCOTYL5 (HY5), a repressor of hypocotyl elongation [17]. Consequently, ethylene and auxin act together as ethylene promotes auxin biosynthesis and transport by upregulating YUCCA1, YUCCA5, and PIN transporters [18]. A shifting balance among these parallel pathways lets seedlings match growth to their environment [11, 16]. Light also feeds back into ethylene signaling pathway through the photoreceptor PHYTOCHROME B (phyB). In the presence of light, phyB inhibits ethylene signaling by the degradation of EIN3. Finally, ethylene activates cell wall–loosening enzymes, allowing turgor pressure to expand cells and complete elongation [11].

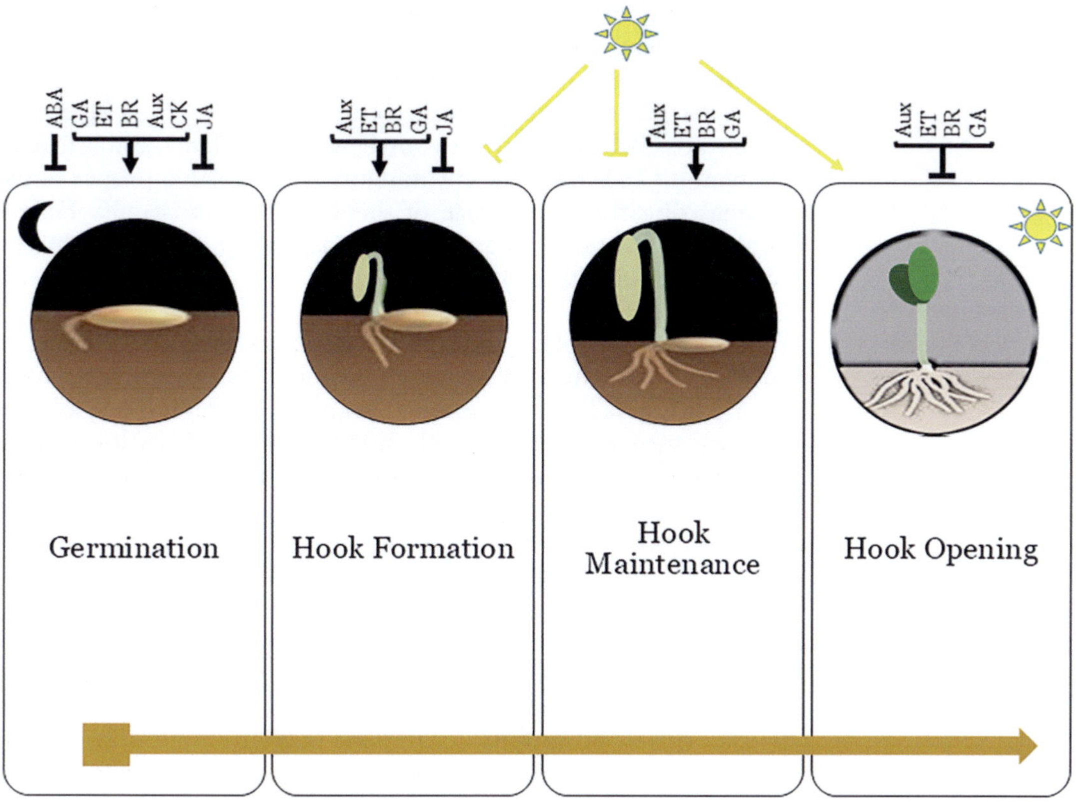

Fig. 2 Seed germination and early seedling growth unfold in five ordered steps, converting a dormant seed to a young plantlet ready for photosynthesis. 1. Water absorption breaks dormancy and starts the germination. 2. Radicle pierces the seed coat, begins root development. 3. Hypocotyl elongation and apical hook formation, the stem (hypocotyl) elongates and forms an apical hook that shields the shoot apex as it penetrates soil. 4. Light exposure and cotyledon expansion, light causes the hook to straighten, and the cotyledons (seed leaves) to expand. 5. Root system and true leaves, a strong root system develops, and true leaves emerge, enabling photosynthesis and independent growth. Environmental cues, especially light, and hormonal signals like ethylene coordinate these stages so the seedling successfully emerges and establishes

4.2 Apical Hook Formation

The apical hook protects the delicate shoot apical meristem during soil emergence (Fig. 2). Most dicot plants share this mechanism, yet species differ in timing and hormone sensitivity during hook opening. For example, Arabidopsis depends mainly on ethylene signaling whereas pea (*Pisum sativum*) shows gibberellin signaling. Such differences highlight the species-specific adaptations that ensure optimal emergence in different conditions. Ethylene leads hook development in three stages: formation, maintenance, and opening. During hook formation, ethylene elevates auxin on the concave hypocotyl side by upregulating TRYPTOPHAN AMINOTRANSFERASE1 (TAA1) and PIN3. This auxin peak bends the tissue, creating the hook formation. Maintenance depends on ethylene interplay with GAs, which stabilize DELLA growth repressors that prevent excessive hook opening [12]. The opening begins when ethylene and auxin decline, allowing symmetric growth to resume. Consequently, cotyledons and apical meristem avoid mechanical damage during emergence (Fig. 2) [19]. Brassinosteroids (BRs) interact with ethylene to modulate the apical hook formation [20]. BR-deficient mutants fail to form a proper hook, emphasizing the importance of this hormonal cross talk. Finally, EIN3/EIL1 activate targets like HLS1, adding a further ethylene regulated layer [19, 21].

4.3 Root System Architecture

Roots secure the plant and uptake water, and ethylene shapes this root system architecture in two opposing ways. At low concentrations it triggers lateral root initiation, whereas, high levels inhibits primary root elongation [22]. Ethylene upregulates local auxin biosynthesis in root tip by activating ASA1 and YUC genes, creating an auxin gradient essential for lateral root formation. This gradient propels pericycle cells to divide and develop into new lateral roots. Simultaneously, it inhibits primary root elongation by disrupting the auxin gradient via PIN transporters and by increasing ROS. This process balances root system architecture for efficient water and nutrient uptake [4, 14]. ABA channels ethylene signaling during drought, ensuring optimal growth. BRs counterbalance this by promoting cell division in the root meristem, easing the inhibitory effects of ethylene [4, 19]. Ethylene also reshapes root system architecture (RSA) when soil phosphate (Pi) levels varies. Under Pi-deficiency, ethylene inhibits primary root elongation but promotes the formation of lateral roots and the initiation and elongation of root hairs. These changes increase the root surface area, making plants to absorb more Pi from the environment [22, 23]. The interplay between ethylene and auxin is crucial in this response, their synergy drives root hair development, a key adaptation under low Pi [23–26]. On the other hand, excess Pi also triggers ethylene-mediated changes in RSA. High Pi reduces the primary root, extends lateral roots and root hairs, and decreases activity of the root apical meristem. This results in a shallower root system that resembles the RSA observed under Pi deficiency, a phenotype called as the "Excess Phosphate Response" (EPiR) [22]. Ethylene-responsive factors (ERFs) are significantly induced under excess Pi conditions, triggers the reprogramming of RSA by inhibiting primary root growth and modifying lateral root dynamics [22]. Transcriptomic studies reveal many ethylene-induced genes, including defensin-like genes that coordinate developmental and physiological traits in roots [22, 27]. Ethylene shapes RSA in a concentration-dependent manner, a fact confirmed by ethylene mutant studies [22]. At moderate excess Pi, roots may still grow, but higher Pi concentrations result in significant ethylene-mediated inhibition [22, 23]. Ethylene-insensitive mutants display a normal phenotype under excess Pi, whereas constitutive ethylene-sensitive mutants exhibit exaggerated responses, highlighting the central role of ethylene in adapting RSA to Pi availability [22]. Overall, ethylene acts as a flexible signal that allow roots remodel themselves to match varying nutrient conditions.

4.4 Hormonal Cross talk

1. **Ethylene and Auxin**: Ethylene boosts auxin biosynthesis and transport, critical for hypocotyl elongation and root development [11]. Through EIN2 signaling, ethylene activates auxin-responsive genes, reinforcing the regulatory role of auxin over cell elongation and division [11]. In Arabidopsis, ethylene increases the expression of auxin transport genes such as AUX1 and PIN3, steering auxin flow needed for both root and shoot growth [11]. Additionally, ethylene-insensitive mutants, like ein2, lose proper auxin gradients, proving that ethylene–auxin cross talk establishes accurate growth patterns [28]. This interaction underlines the auxin–ethylene synergy in seedling growth [16, 28].
2. **Ethylene and GAs**: Ethylene modulates GA synthesis and signaling during apical hook development [16]. DELLA growth-repressor proteins integrate these ethylene and GA signals, and ethylene-activated EIN3/EIL1 stabilizes DELLA activity [12]. Consequently, this cross talk keeps elongation repression in balance, ensuring coordinated seedling development [19].
3. **Ethylene and BRs**: Brassinosteroids modulate how ethylene shapes apical hook formation and root growth. BR-deficient mutants respond poorly to ethylene, showing that many ethylene-regulated growth changes rely on BR signaling [11, 20, 29]. Under stress conditions, BRs further increase the ability of ethylene to adapt growth patterns ensuring flexible adaptation [4, 19].

The coordination of hypocotyl growth, apical hook formation, and root growth shows the indispensable role of ethylene. It interacts with auxins, GAs, BRs, and ABA, creates a dynamic hormonal network that adapts growth to external signals. Knowing these pathways will help researchers develop crops that stand up to stress tolerance and thrive in different environments.

5 Different Role of Ethylene in Response to Environmental Stresses

Ethylene works flexibly with light, temperature, and salinity cues. This hormone coordinates multiple plant processes, helping seedling grow and withstand stress. Through complex cross talk with hormones and rapid responses to external signals, ethylene triggers different growth and defense pathways.

5.1 Light Stress

Ethylene regulates lightrelated development, altering plant morphology and adaptation. It drives apical hook formation, protecting the seedling during soil emergence. The hook arises from unequal hypocotyl cell growth led by hormone gradients and external cues (Fig. 3). In light, ethylene increases hypocotyl elongation through the COP1-HY5 system, expanding cells and tissues to maximize light capture (Fig. 3) [16, 17, 30]. In darkness, it inhibits elongation by stabilizing cortical microtubules and suppressing HY5, conserving energy (Fig. 3) [5]. Light intensity directly shape ethylene synthesis and signaling. As Far-red light reduces

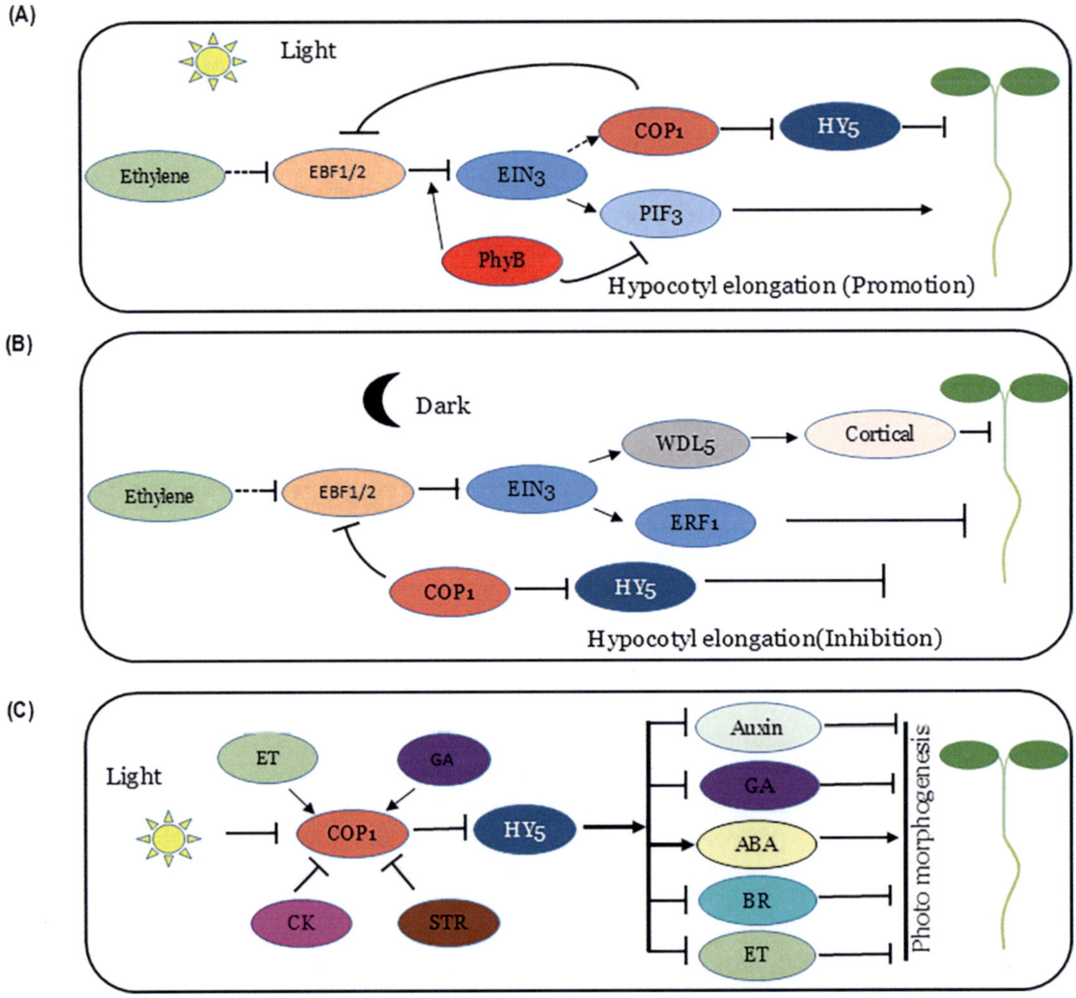

Fig. 3 Ethylene coordinates with light–dark, hormones, and developmental regulators during seed germination and hypocotyl growth. (**a**) Under light, ethylene signaling (via EIN3, EBF1/2) integrates with photoreceptors (PhyB, PIF3) and regulators (COP1, HY5) to balance growth. (**b**) Dark conditions alter ethylene effects on cortical microtubule (WDL5) and transcriptional factors (ERF1, HY5), fine-tuning hypocotyl extension. (**c**) In light condition, ethylene (ET) interacts with auxin, GA, ABA, BR, and CK and regulators STR, COP1, HY to drive seed germination and hypocotyl elongation

the control of ethylene in ABA metabolism, red and blue light instead augment its growth-modulating roles (Fig. 3) [19]. This adaptability enables plants to tune hormonal responses to shifting light conditions and save energy. Studies indicate that light and ethylene together elevate photosynthetic efficiency and lead photomorphogenesis, highlighting their joint role in developmental plasticity (Fig. 3) [30].

5.2 Temperature Stress

Ethylene assists plants cope with heat stress, maintaining germination and subsequent growth. High temperatures inhibit ethylene biosynthesis, disrupting the hormone balance required for seed germination and prolonging dormancy [4]. Ethylene acts with gibberellins to counteract thermo-inhibition by weakening the endosperm and allowing the radicle protrusion germination [4]. The influence of ethylene varies among species. In chickpea and lettuce high temperatures slow germination due to reduced ethylene production, which exogenous ethylene can reverse [13]. Adaptability of ethylene in coping with temperature changes suggests it could enhance crop tolerance under climate change [13]. Additionally, ethylene works with stress hormone ABA to balance growth and dormancy when thermal conditions change [7].

5.3 Salinity Stress

Salinity exposes plants to osmotic and ionic stress, and ethylene helps them keep growing and survive [4]. Ethylene alleviates salinity-induced dormancy by countering ABA imposed dormancy, allowing seed germination even under saline conditions [4]. Because of this mechanism, plants can grow in saline soils that would otherwise stunt them. In root development, ethylene boosts auxin synthesis and transport, facilitating lateral root initiation that improves nutrient and water uptake under salinity [13]. As ethylene and auxin act together, plants build root system architecture suited to harsh soils, supporting overall growth [13]. However, excessive ethylene disrupts auxin gradients, inhibiting root elongation [11, 13]. Ethylene works with other ROS-regulating pathways, such as ABA and auxins, to strengthen cellular defenses [14, 29]. Enhanced ROS scavenging under ethylene ensures cellular homeostasis and stress tolerance [14, 29].

5.4 Integration of Stress Responses

Ethylene acts as a master regulator, working with hormones such as ABA, GAs, and auxins to shape adaptive responses [13]. Ethylene interacts with environmental signals, such as light, temperature, and salinity, demonstrating its versatility as a central player in plant adaptation [19]. Through its signaling networks, ethylene drives cell expansion, resource allocation, and stress management [5, 29]. Approaches that leverage the role of ethylene in hormone cross talk and environmental adaptation could boost yield and stress tolerance [13]. By understanding and optimizing different functions of ethylene, we can unlock new pathways to safeguard food supply in an era of climate unpredictability.

6 Transformative Role of Ethylene in Agriculture

6.1 Conifers: Scots Pine (*Pinus sylvestris*)

Ethylene plays a pivotal role in somatic embryogenesis of Scots pine [5, 29]. During the early somatic embryo maturation, expression of ACC synthase (PsACS2) rises with ethylene production, peaking at stage 3 of embryo development. Therefore, PsACS2 expression can serve as a genetic marker to optimize embryogenic culture and improve somatic embryo production for forest regeneration. Understanding the role of ethylene allows researchers tailor hormone treatments, improving quantity and quality of somatic embryos, and their nursery establishment [6].

6.2 Crop Plants: Sunflower (*Helianthus annuus*)

Ethylene triggers germination of dormant sunflower seeds, particularly when water is scare or temperatures runs high [4]. As it lowers ABA sensitivity and balances reactive oxygen species (ROS), ethylene or its precursors boost seedling vigor [4, 5]. Growers harness this effect in seed priming treatments that help crops establish in challenging environments. Furthermore, applying ethylene synchronizes germination within a batch, a clear advantage for large-scale farming that needs uniform growth [13]. Consequently, ethylene-mediated approaches allow sunflower cultivation to expand into marginal soils [2, 4].

6.3 Horticulture: Lettuce (*Lactuca sativa*)

In lettuce, thermodormancy (failure to germinate at high temperatures) is alleviated by ethylene treatments [4]. Consequently, consistent germination is maintained in commercial seed production and on farms facing continuous variable temperature regimes [5]. Pre-sowing treatments with ethylene-releasing compounds also boost marketable yield by synchronizing germination across the field. This ensures a steady supply of high-quality lettuce, benefiting both growers and consumers [4].

6.4 Stress Mitigation: Abiotic Stresses

Ethylene strengthens seed tolerance to salt and drought during germination. It interacts with hormones ABA and GAs to regulate seedling growth and ROS scavenging. Recent studies show that ethylene sparks antioxidant defenses, protecting tissues from oxidative damage under stress [29]. By leveraging the cross talk of ethylene with auxins and cytokinins can build deeper root system architecture and raise water-use efficiency [13]. Such advances matter in arid and semi-arid regions, where drought and salty soils sharply reduce yields [4, 5].

6.5 Industrial and Research Use: *Brassica rapa* and *Arabidopsis*

Study of ethylene biosynthesis and signaling in *Brassica rapa* and Arabidopsis reveal how the hormone controls zygotic embryogenesis and seed maturation [5]. These findings have yielded practical methods that improve seed quality and storage, ensuring robust germination [5]. Arabidopsis remain the popular model for dissecting ethylene-related molecular pathways, driving new strategies in crop improvement [5]. In Brassica crops, knowing the effect of ethylene on seed viability and germination supports improved oilseed production and higher tolerance to external changes [1, 4].

7 Future Prospects in Ethylene Research on Seed Biology

Future research must close gaps in ethylene interactions with different hormones, varies among species, and senses environmental cues during dormancy and germination. Within networks of abscisic acid (ABA), gibberellins (GAs), reactive oxygen species (ROS), and brassinosteroids (BRs), ethylene acts as a key regulator needing deeper study [1, 4]. Researchers also need to examine molecular mechanisms of ethylene sensitivity under abiotic stress to understand impacts on seed metabolism and tolerance [5]. Transcriptomics, epigenetics, and synthetic biology now offer tools to fine-tune ethylene biosynthesis and signaling pathways. Transcriptomic analysis can map gene networks that control dormancy and germination, whereas, epigenetic work may uncover heritable marks guiding ethylene regulation [2, 5]. Findings from Arabidopsis should translate to crops like rice and wheat, and studies in gymnosperms could broaden knowledge of ethylene in somatic embryogenesis [6]. Translational research already uses ethylene-releasing compounds to improve germination in challenging conditions. Addressing the remaining gaps and applying interdisciplinary tools will unlock potential of ethylene as a driver of sustainable seed development and agriculture [4, 5].

8 Conclusion

Ethylene regulates seed germination and early seedling growth by weaving with different hormones and environmental cues. As it antagonizes ABA, a dormancy-promoting hormone, ethylene helps break dormancy. At the same time, it works with GA to weaken the endosperm cap and allowing radicle emergence. This hormone interplay becomes vital under salinity, hypoxia, and extreme temperature stress [1, 2, 4]. Ethylene also interacts with auxins, cytokinins, jasmonates, to drive hypocotyl elongation and apical hook formation. These processes safeguard young seedlings as they push through soil towards light, demonstrating role of ethylene in early development [4, 5]. Finally, ethylene increases germination under challenging environmental conditions by mediating reactive oxygen species (ROS) signaling, that activate stress defenses.

Acknowledgments

The Director, CSIR Central Institute of Medicinal and Aromatic Plants, Lucknow, India, is acknowledged for providing the facilities and support. The Institutional Communication No. CIMAP/PUB/ 2025/09.

References

1. Rodriguez-Gacio MDC, Matilla AJ (2001) The last step of the ethylene biosynthesis pathway in turnip tops (*Brassica rapa*) seeds: alterations related to development and germination and its inhibition during desiccation. Physiol Plant 112:273–279. https://doi.org/10.1034/j.1399-3054.2001.1120216.x
2. Oracz K, El-Maarouf-Bouteau H, Kranner I, Bogatek R, Corbineau F, Bailly C (2009) The mechanisms involved in seed dormancy alleviation by hydrogen cyanide unravel the role of reactive oxygen species as key factors of cellular signaling during germination. Plant Physiol 150:494–505. https://doi.org/10.1104/pp.109.138107
3. Matilla AJ (2000) Ethylene in seed formation and germination. Seed Sci Res 10:111–126. https://doi.org/10.1017/S096025850000012X
4. Corbineau F, Xia Q, Bailly C, El-Maarouf-Bouteau H (2014) Ethylene, a key factor in the regulation of seed dormancy. Front Plant Sci 5. https://doi.org/10.3389/fpls.2014.00539
5. Ahammed GJ, Gantait S, Mitra M, Yang Y, Li X (2020) Role of ethylene crosstalk in seed germination and early seedling development: a review. Plant Physiol Biochem 151:124–131. https://doi.org/10.1016/j.plaphy.2020.03.016
6. Lu J, Vahala J, Pappinen A (2011) Involvement of ethylene in somatic embryogenesis in scots pine (Pinus sylvestris L.). Plant Cell Tissue Organ Cult 107:25–33. https://doi.org/10.1007/s11240-011-9952-4
7. Linkies A, Müller K, Morris K, Turečková V, Wenk M, Cadman CSC, Corbineau F, Strnad M, Lynn JR, Finch-Savage WE, Leubner-Metzger G (2010) Ethylene interacts with Abscisic acid to regulate endosperm rupture during germination: a comparative approach using *Lepidium sativum* and *Arabidopsis thaliana*. Plant Cell 21:3803–3822. https://doi.org/10.1105/tpc.109.070201
8. Linkies A, Leubner-Metzger G (2012) Beyond gibberellins and abscisic acid: how ethylene and jasmonates control seed germination. Plant Cell Rep 31:253–270. https://doi.org/10.1007/s00299-011-1180-1
9. Beaudoin N, Serizet C, Gosti F, Giraudat J (2000) Interactions between Abscisic acid and ethylene signaling cascades. Plant Cell 12: 1103–1115. https://doi.org/10.1105/tpc.12.7.1103
10. Miransari M, Smith DL (2014) Plant hormones and seed germination. Environ Exp Bot 99:110–121. https://doi.org/10.1016/j.envexpbot.2013.11.005
11. Vandenbussche F, Petrášek J, Žádníková P, Hoyerová K, Pešek B, Raz V, Swarup R, Bennett M, Zažímalová E, Benková E, Van Der Straeten D (2010) The auxin influx carriers AUX1 and LAX3 are involved in auxin-ethylene interactions during apical hook development in *Arabidopsis thaliana* seedlings. Development 137:597–606. https://doi.org/10.1242/dev.040790
12. An F, Zhang X, Zhu Z, Ji Y, He W, Jiang Z, Li M, Guo H (2012) Coordinated regulation of apical hook development by gibberellins and ethylene in etiolated Arabidopsis seedlings. Cell Res 22:915–927. https://doi.org/10.1038/cr.2012.29
13. Wilson RL, Kim H, Bakshi A, Binder BM (2014) The ETHYLENE receptors ETHYLENE RESPONSE1 and ETHYLENE RESPONSE2 have contrasting roles in seed germination of Arabidopsis during salt stress. Plant Physiol 165:1353–1366. https://doi.org/10.1104/pp.114.241695

14. Subbiah V, Reddy KJ (2010) Interactions between ethylene, abscisic acid and cytokinin during germination and seedling establishment in Arabidopsis. J Biosci 35:451–458. https://doi.org/10.1007/s12038-010-0050-2
15. Žádníková P, Smet D, Zhu Q, Straeten DVD, Benková E (2015) Strategies of seedlings to overcome their sessile nature: auxin in mobility control. Front Plant Sci 6. https://doi.org/10.3389/fpls.2015.00218
16. Gallego-Bartolomé J, Arana MV, Vandenbussche F, Žádníková P, Minguet EG, Guardiola V, Van Der Straeten D, Benkova E, Alabadí D, Blázquez MA (2011) Hierarchy of hormone action controlling apical hook development in Arabidopsis. Plant J 67:622–634. https://doi.org/10.1111/j.1365-313X.2011.04621.x
17. Liang X, Wang H, Mao L, Hu Y, Dong T, Zhang Y, Wang X, Bi Y (2012) Involvement of COP1 in ethylene- and light-regulated hypocotyl elongation. Planta 236:1791–1802. https://doi.org/10.1007/s00425-012-1730-y
18. Yu Y, Wang J, Zhang Z, Quan R, Zhang H, Deng XW, Ma L, Huang R (2013) Ethylene promotes Hypocotyl growth and HY5 degradation by enhancing the movement of COP1 to the nucleus in the light. PLoS Genet 9: e1004025. https://doi.org/10.1371/journal.pgen.1004025
19. Smet D, Žádníková P, Vandenbussche F, Benková E, Van Der Straeten D (2014) Dynamic infrared imaging analysis of apical hook development in *Arabidopsis*: the case of brassinosteroids. New Phytol 202:1398–1411. https://doi.org/10.1111/nph.12751
20. Zhang Y, Liu Z, Wang J, Chen Y, Bi Y, He J (2015) Brassinosteroid is required for sugar promotion of hypocotyl elongation in Arabidopsis in darkness. Planta 242:881–893. https://doi.org/10.1007/s00425-015-2328-y
21. Li H, Johnson P, Stepanova A, Alonso JM, Ecker JR (2004) Convergence of signaling pathways in the control of differential cell growth in Arabidopsis. Dev Cell 7:193–204. https://doi.org/10.1016/j.devcel.2004.07.002
22. Shukla D, Rinehart CA, Sahi SV (2017) Comprehensive study of excess phosphate response reveals ethylene mediated signaling that negatively regulates plant growth and development. Sci Rep. https://doi.org/10.1038/s41598-017-03061-9
23. Shukla D, Trivedi PK, Sahi S (2023) A simple protocol for mapping the plant root system architecture traits. JoVE e64876. https://doi.org/10.3791/64876
24. Linkohr BI, Williamson LC, Fitter AH, Leyser HMO (2002) Nitrate and phosphate availability and distribution have different effects on root system architecture of *Arabidopsis*. Plant J 29:751–760. https://doi.org/10.1046/j.1365-313X.2002.01251.x
25. López-Bucio J, Hernández-Abreu E, Sánchez-Calderón L, Nieto-Jacobo MF, Simpson J, Herrera-Estrella L (2002) Phosphate availability alters architecture and causes changes in hormone sensitivity in the Arabidopsis root system. Plant Physiol 129:244–256. https://doi.org/10.1104/pp.010934
26. Williamson LC, Ribrioux SPCP, Fitter AH, Leyser HMO (2001) Phosphate availability regulates root system architecture in Arabidopsis. Plant Physiol 126:875–882. https://doi.org/10.1104/pp.126.2.875
27. Shukla D, Waigel S, Rouchka EC, Sandhu G, Trivedi PK, Sahi SV (2021) Genome-wide expression analysis reveals contrasting regulation of phosphate starvation response (PSR) in root and shoot of Arabidopsis and its association with biotic stress. Environ Exp Bot 188: 104483. https://doi.org/10.1016/j.envexpbot.2021.104483
28. Liu J, Mehdi S, Topping J, Friml J, Lindsey K (2013) Interaction of PLS and PIN and hormonal crosstalk in Arabidopsis root development. Front Plant Sci 4. https://doi.org/10.3389/fpls.2013.00075
29. Dubois M, Van Den Broeck L, Inzé D (2018) The pivotal role of ethylene in plant growth. Trends Plant Sci 23:311–323. https://doi.org/10.1016/j.tplants.2018.01.003
30. Yu Y, Huang R (2017) Integration of ethylene and light Signaling affects hypocotyl growth in Arabidopsis. Front Plant Sci 08. https://doi.org/10.3389/fpls.2017.00057

Chapter 18

Role of Ethylene in Plant Tissue Culture

Devesh Shukla

Abstract

Ethylene affects in-vitro plant morphogenesis in many ways, depending upon species, genotypes, and developmental stages. It can stimulate or inhibit the development, and a minute change in concentration could lead either the successful regeneration or failure. Studies in different plant taxa such as Datura, barley, coffee, Arabidopsis, tomato, and pepper show that the fine-tuning of ethylene levels could enhance embryogenesis, shoot organogenesis, whereas, changes in these optimum concentrations could lead to stunted morphogenesis. Adding ACC or Ethephon increases ethylene level, whereas silver salts or cobalt chloride diminish its synthesis or reception, showing its regulatory feature. Genetic studies depend on ethylene-insensitive mutants, ethylene-overproducing mutants, and transgenic lines altering ACC synthase or ACC oxidase. These studies reveal that ethylene works with auxins, cytokinins, and polyamine to lead developmental switches during morphogenesis. The tissue and developmental stage further shape ethylene action, callus induction, embryo maturation, and shoot regeneration each demand distinct ethylene levels. Combining ethylene management with molecular, physiological, and epigenetic approaches, could help regeneration challenges in difficult Capsicum species. Taken together, literature positions ethylene as a pivotal, context-dependent regulator of plant tissue culture morphogenesis.

Key words Ethylene, Plant tissue culture, Androgenesis, Callus induction, Shoot organogenesis, Somatic embryogenesis, Ethylene inhibitors, Phytohormone interactions, Genotype specificity, Reactive oxygen species (ROS)

1 Introduction

Ethylene is a phytohormone that strongly but selectively shapes plant growth, development, and in -vitro differentiation [1, 2]. Early studies linked ethylene to anther cultures and microspore embryogenesis, though supporting data were often inconsistent. For example, adding ethylene via Ethrel, an ethylene-releasing source, triggered extra nuclear divisions in wheat microspores (immature pollen) [3]. This response indicated at androgenesis induction in otherwise recalcitrant (hard to respond) species [2]. In contrast, similar treatments failed to enhance embryo formation in other plant systems [4]. In rice, pre-exposure to ethylene favored androgenesis [5–7], although differing protocols

Devesh Shukla (ed.), *Ethylene Sensing and Signaling*, Methods in Molecular Biology, vol. 2945,
https://doi.org/10.1007/978-1-0716-4650-2_18,

sometimes showed opposite results within the species [8]. Model organisms still emphasize complex effect of ethylene. In *Nicotiana tabacum*, anther culture releases ethylene continuously [9], yet removing or flushing it barely changes plantlet yield. Therefore, ethylene concentration inside anther locules may outweigh headspace levels in determining response [2, 9, 10]. Since findings stayed unclear, researchers turned to chemical tools. They applied inhibitors (cobalt), antagonists (silver), precursors (L-methionine), and the releaser Ethrel to reveal role of ethylene in androgenesis and morphogenesis [2, 11–14].

Ethylene regulates many in-vitro processes beyond androgenesis. the regulatory influence of ethylene extends widely across diverse in- vitro responses. Accumulation of the gas in sealed culture vessels controls shoot or root regeneration, somatic embryogenesis, and organogenesis [15–17]. Researchers adjust internal ethylene using inhibitors such as silver nitrate, silver thiosulfate, AVG, or by adding ACC or Ethrel that supplements the hormone directly. Both approaches significantly affect regeneration efficiency in cereals, legumes, and ornamentals [18–29]. Genotype strongly affects the outcome. For example, barley cultivars with high basal ethylene often suppress embryogenesis, whereas low-producing lines promote embryogenic development [27, 30, 31]. Ethylene either promotes or blocks development in Coffea, Datura, Zea, Brassica, and Capsicum, depending on developmental stage [2, 17, 26, 32–36]. Ethylene does not function in isolation. Its crosstalk with auxins, cytokinins, ABA, gibberellins, and polyamines leads in-vitro responses (Figs. 1 and 2) [37–41]. Because auxins raise endogenous ethylene while inducing callus, these combined signals can aid or impede organogenesis and embryogenesis. Competition for SAM between ethylene and polyamine pathways further complicate interpretation of the role of ethylene [26, 34].

Studies using transgenic lines and mutants show that altering ethylene sensitivity or biosynthesis can either stabilize or disrupt regeneration pathways [42–48]. These findings indicate that ethylene's impact depends not only on species or genotype but also on the precise developmental stage.

Ethylene plays many roles in plant tissue culture shaped by complex links to other hormones, environmental factors, and genetics. Although researchers have made much progress, many details remain unclear. Continued advances in analytical tools, targeted genetic methods, and whole-system studies of hormones and metabolism should clarify the function of ethylene. Such knowledge may improve the consistency and effectiveness of in- vitro regeneration across different plant species.

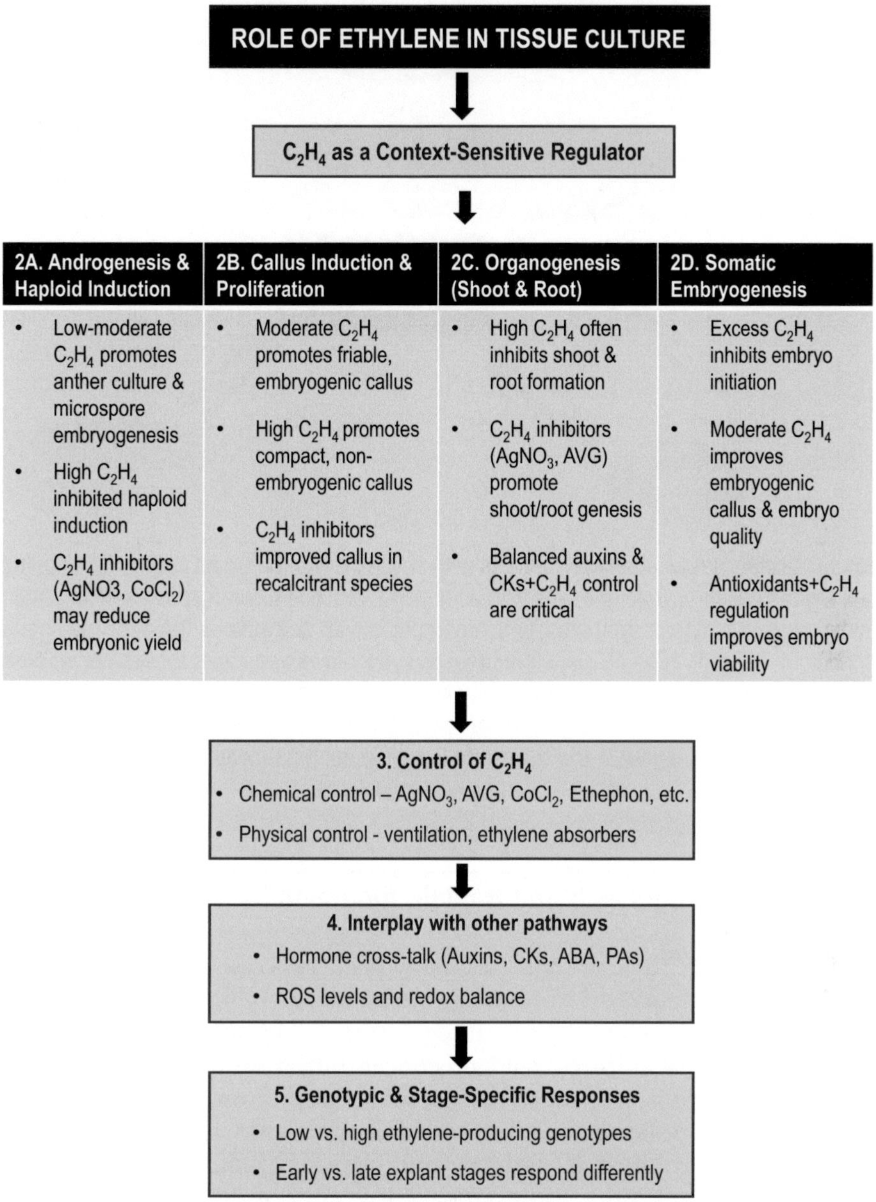

Fig. 1 Flow diagram summarizes ethylene's context-sensitive role in plant tissue culture. It links varying ethylene levels to shifts in callus formation, shoot/root organogenesis, and somatic embryogenesis. Androgenesis/haploid induction likewise depend on genotype and developmental stage. C_2H_4 ethylene, *CKs* cytokinins, *ABA* abscisic acid, *AVG* aminoethoxyvinylglycine, *ROS* reactive oxygen species, *GC* gas chromatography, *SAM* S-adenosyl-methionine, *PAs* polyamines

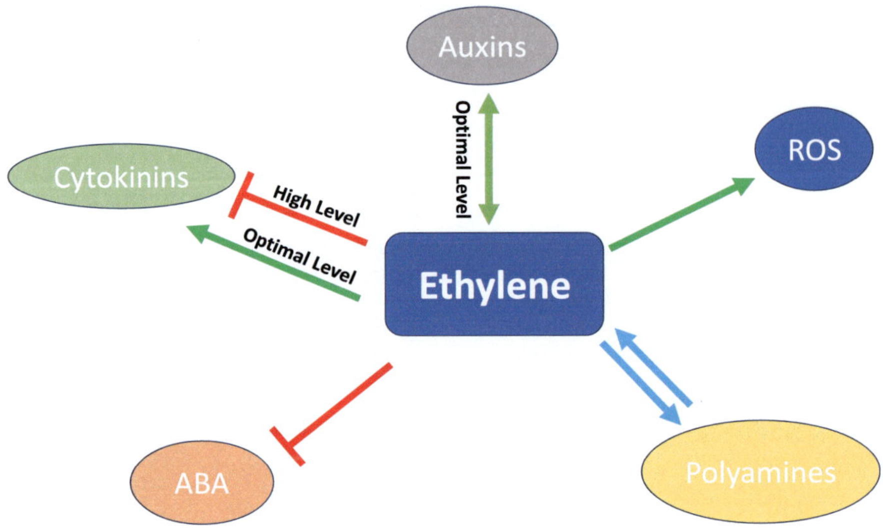

Fig. 2 Schematic display of how ethylene interacts with several phytohormones during tissue culture. It shows the cross talk among auxins, cytokinins, abscisic acid (ABA), reactive oxygen species (ROS), and polyamines (PAs). Auxins stimulates ethylene synthesis, which competes with polyamines (PAs) for the shared precursor S-adenosylmethionine (SAM). High ethylene suppresses cytokinin-driven shoot formation, whereas a moderate amount supports overall morphogenesis. Ethylene also tunes reactive oxygen species (ROS) levels, shaping stress responses and subsequent developmental outcomes. Positive (→), inhibitory (⊥), and competitive (⇄) relationships highlight the necessity to adjust each pathway precisely

2 Modulation of Androgenesis and Haploid Induction by Ethylene

Ethylene regulates many plant processes, such as androgenesis and haploid induction in tissue culture (Fig. 1). It acts by interacting with other growth regulators and by adjusting cellular metabolism. In culture, ethylene concentration significantly impacts androgenesis, which forms haploid plants from microspores or immature pollen grains. Low ethylene promotes androgenesis; high ethylene inhibits it. Babbar and Gupta [2] showed Ethrel, an ethylene-releasing compound, at low concentrations (10^{-6} M or in micromolar range) increased androgenic response in *Datura metel* (Table 1). At the same concentration, anther productivity and efficiency rose too. Androgenesis decreased at higher concentration and inhibited completely at 10^{-3} M (in millimolar range) (Table 1). Thus, culture success depends on maintaining optimal ethylene levels. Inhibitors of ethylene such as silver nitrate ($AgNO_3$) and cobalt chloride ($CoCl_2$), reduced androgenic responses. They lowered the number of embryos or plantlets per anther, suggesting endogenous ethylene supports early microspore development. Conversely, methionine, a ethylene, stimulated androgenesis, confirming the hormone role.

Table 1
Concentration-specific effects of ethylene modulators

Stage	Ethylene production	Impact on culture
Callus induction	Low to moderate	Supports callus growth
Shoot regeneration	Variable	Regulates shoot formation
Somatic embryogenesis	Elevated in early stages	Critical for embryo differentiation
Maturation	Peaks in cotyledon stage	Essential for maturation
Germination	Declines as germination progresses	Minimal effect

Ethylene may steer androgenesis by altering redox states, changing membranes, and modulating auxin and cytokinin pathways (Figs. 1 and 2). As ethylene regulates auxin transport and distribution, it shapes cellular differentiation and morphogenesis. Ethylene also interacts with reactive oxygen species (ROS), which can impact microspore viability and embryogenic potential. Inhibitors such as silver ions and cobalt ions disrupt ethylene signaling, so anther productivity and efficiency decrease. Therefore, cultures must keep ethylene within an optimal range to support androgenesis and successful haploid induction [2]. Understanding role of ethylene offers improvements in haploid induction techniques for plant tissue culture. Controlled use of ethylene precursors or inhibitors allows researchers optimize the development in anther and microspore cultures. Low concentrations of Ethrel, an ethylene-releasing compound, can increase haploid plant production in species that respond positively to ethylene [2]. In cereals such as barley, ethylene effects depend genotype [31]. Cho and Kasha [30] showed that low-ethylene genotypes benefitted from precursors. High-ethylene genotypes required inhibitors for efficient embryogenesis.

3 Callus Induction and Proliferation

Ethylene strongly regulates callus induction and proliferation (Fig. 1). Its effect on callus quality, growth rate, and embryogenic potential is substantial. The hormone acts in dose-dependent manner. Moderate ethylene levels promote friable, embryogenic callus needed for successful regeneration (Fig. 1) (Table 2). In *Datura metel*, methionine and Ethrel, both ethylene precursors enhanced callus proliferation while preserving regenerative capacity [2]. In contrast, excess ethylene produced compact, nonembryogenic callus with low morphogenic potential [49]. Callus induction in recalcitrant species like *Capsicum annuum* is particularly sensitive to ethylene. High endogenous ethylene there inhibits cell division and differentiation, hindering callus induction. Ethylene inhibitors,

Table 2
Ethylene production across culture stages

Compound	Low concentration effect	Optimal concentration effect	High concentration effect	Remarks
Ethrel	Promotes anther response and productivity	Maximal enhancement at 10^{-5} M	Complete inhibition at 10^{-3} M	Requires precise dosing to avoid inhibitory effects
ACC	Enhances bud development	Improves callus and shoot production	Inhibits organogenesis and reduces productivity	Highly dose-dependent effects
$CoCl_2$	No significant impact	Slight reduction in embryogenesis	Strong inhibition of ethylene biosynthesis	Variable outcomes; optimize per tissue type
Silver nitrate ($AgNO_3$)	Promotes shoot regeneration or minimal inhibition of organogenesis (Context-dependent)	Highly effective for shoot regeneration in most genotypes; partial inhibition in sensitive cases	Drastic reduction in regeneration rates or toxic effects	Effect is genotype and species specific. Effective for overcoming ethylene-related inhibitions

Table 3
Effects of ethylene and its modulators on plant tissue culture

Compound	Effect on callus formation	Effect on shoot regeneration	Notes
$AgNO_3$	i. Enhanced in some genotypes Ii. Complete inhibition in sensitive genotypes	Promoted shoot regeneration	Useful in recalcitrant cultures
ACC	Variable by concentration	Improved bud development	Dose-dependent effects
Ethephon	Inhibited at high levels	Inhibited at higher concentrations	Promotes buds at low levels
$CoCl_2$,	Mild reduction	No significant effect	Needs optimization for genotype

such as silver nitrate ($AgNO_3$) and cobalt chloride ($CoCl_2$), improve callus quality and proliferation (Table 3). *Capsicum* cultures treated with $AgNO_3$ showed enhanced callus growth and avoided browning and tissue necrosis, common barriers in tissue culture [35].

Physical methods help control ethylene during callus induction. Controlled ventilation and ethylene-absorbing potassium permanganate, reduce internal hormone levels (Fig. 1). Enhanced aeration in culture vessels prevents inhibitory buildup, promoting healthy callus formation [50]. Ethylene also interacts with plant growth regulators like auxins and cytokinins, adding chemical complexity. Auxin-induced ethylene increases cell division and expansion, whereas an auxin-cytokinin imbalance limits callus development (Fig. 2). Therefore, carefully combining growth regulators with ethylene modulators yields high-quality, regenerative callus [51]. Molecular work has clarified role of ethylene in callus proliferation. Biosynthetic genes ACS (ACC synthase) and ACO (ACC oxidase), and signaling components EIN2, are key players. Transcriptomic studies reveal differential expression of ethylene-responsive genes during callus induction, illuminating regulatory pathways [2]. These findings point to genetic or biochemical interventions to optimize callus induction and proliferation.

4 Organogenesis

4.1 Shoot

Ethylene affects shoot organogenesis in several ways; it can facilitate or inhibit, depending on concentration tissue state. High endogenous ethylene often reduces shoot regeneration across many species. This inhibition is pronounced in recalcitrant plants such as *Capsicum annuum* and *Brassica napus* [52] (Table 4) (Fig. 1). Elevated ethylene there upsets the hormone balance needed for shoot induction. Therefore, researchers use ethylene inhibitors

Table 4
Genotype-dependent responses to ethylene modulators

Genotype	Ethylene production	Callus formation	Shoot regeneration	Sensitivity to modulators
Golden Promise (Barley)	High	Robust	Highest	Tolerant to $AgNO_3$
Morex (Barley)	Low	Poor	Lowest	Improved by ACC
S. pennellii	Moderate	Good	Improved with ACC	Responsive to Ethephon
Fl hybrid (*Solanum lycopersicum* cv. Anl27 and *S. pennellii*)	Lower than *S. pennellii*	Moderate	Sensitive to SN (sliver nitrate)	Highly sensitive to SN
Brassica napus	Highly variable	Highly sensitive	Improved by $AgNO_3$	Highly genotype specific

such as silver nitrate ($AgNO_3$) and aminoethoxyvinylglycine (AVG). Addition of $AgNO_3$ to culture media increases shoot induction frequency and improves shoot elongation and quality (Table 1) [35] (Fig. 1). Ethylene interacts with auxins and cytokinins, and this crosstalk shapes shoot organogenesis [39–41] (Fig. 2). Auxins stimulate ethylene production, or vice versa, and together with cytokinins support shoot formation under controlled ethylene [39–41]. However, excess ethylene disrupts this synergy and favors callus over shoots. Optimizing the auxin-to-cytokinin ratio while regulating ethylene is key for successful shoot regeneration [39–41, 51]. Ethylene also acts at the cellular level, affecting division and elongation. Moderate ethylene levels enhance cellular expansion and promote meristematic activity, initiating shoot bud formation. In contrast, high ethylene concentrations induce stress responses, including reactive oxygen species (ROS), and impair shoot regeneration. Using antioxidants together with ethylene inhibitors limits these stresses and further improves shoot organogenesis [53]. Species respond differently, so culture protocols must be tailored. For example, in *Capsicum annuum*, silver nitrate increases shoot regeneration by modulating ethylene levels and balancing tissue redox. Similarly, *Brassica species* culture use ethylene-absorbing materials (potassium permanganate, activated carbon, and biochar) to reduce ethylene accumulation and raise shoot induction [52].

4.2 Root

Ethylene controls many steps of root and shoot organogenesis. During root and shoot formation, EIN2 and ERF1 adjust cellular responses to ethylene [29, 39, 40]. Transcriptomic studies indicate stage-specific expression of ethylene-responsive genes throughout root and shoot organogenesis [54–56]. Ventilating culture vessels stops ethylene buildup and therefore, encourages root or shoot elongation. Combining ethylene control with micropropagation or somatic embryogenesis improves regeneration across different plant species.

High endogenous ethylene levels often block root regeneration, especially in hard to regenerate *Capsicum annuum* and *Brassica napus* [52]. Researchers temper ethylene with silver nitrate ($AgNO_3$) or aminoethoxyvinylglycine (AVG) in the medium (Table 3). $AgNO_3$ frequently increases root induction frequency and length, producing sturdier roots [35]. Ethylene works alongside auxins and cytokinins, to regulate root organogenesis [39–41]. Auxins can synergize with cytokinins to promote root formation under controlled ethylene levels [39–41] (Fig. 2). However, too much ethylene disrupts this balance, reducing root development. Thus, optimal auxin-to-cytokinin ratio remain essential [39–41, 51]. Ethylene also steers cell division and elongation during root organogenesis (Table 2). Moderate ethylene level promotes meristematic activity, supporting early root formation (Table 2). Conversely, high ethylene concentrations can evoke reactive oxygen species (ROS), and stunt roots. Adding antioxidants together with ethylene inhibitors reduces stress and improves root organogenesis [53].

Species respond differently, that necessitate tailored tissue culture protocols (Table 2). In *Capsicum annuum*, silver nitrate moderates ethylene levels and maintains redox balance, which favors rooting. Similarly, in *Brassica species*, culture vessels lined with ethylene-absorbing materials likewise show higher root induction [52]. At the genetic level, EIN2 and ERF1 modulate ethylene responses during root development. Stage-specific expression of these genes offers useful biomarkers for optimizing protocols [54–56].

5 Somatic Embryogenesis

Somatic embryogenesis, embryos develop from somatic cells, depends heavily on ethylene regulation (Fig. 1). Ethylene tunes callus induction, embryo initiation, and maturation (Table 2). Its effect varies with concentration and species, requiring precise modulation [53]. High concentrations of ethylene inhibit somatic embryo initiation in *Daucus carota* (carrot) suspension cultures. This excessive shifts cellular redox balance and induces stress, hindering embryogenic development [53]. In contrast, moderate ethylene level supports somatic embryo initiation and later stages. For example, moderate levels boost embryogenic callus in *Coffea canephora*, yielding more somatic embryos (Table 4) [28].

Ethylene also interacts with auxins and cytokinins during somatic embryogenesis [39–41] (Fig. 2). Auxin induces ethylene early, whereas cytokinins temper the harm caused by excess ethylene [39–41]. Balancing these regulators helps cells switch into embryogenic pathways [2, 39–41]. Ethylene interacts with reactive oxygen species (ROS) as well. Low levels of ROS aid cellular differentiation, yet ethylene-induced ROS bursts cause oxidative stress, and lower embryo quality. Adding antioxidants, such as ascorbic acid or glutathione alongside ethylene regulation mitigates oxidative damage, and improves somatic embryo formation [57].

Recalcitrant species like *Capsicum annuum* face extra barriers. Excess ethylene can malform shoot apical meristems (SAMs) and cotyledons (Fig. 1). These defects arise from disrupted hormonal signaling and cell organization. Using ethylene inhibitors like silver nitrate, and better aeration and light exposure, creates conditions favorable for normal embryo formation [35].

Recent molecular studies clarify how ethylene genes shape somatic embryogenesis. Biosynthetic genes ACS and ACO, and signaling components EIN2 and ERF1, drive these pathways

[58, 59]. Transcriptomic analyses shows their stage-specific expression, offering targets for chemical or genetic fine-tuning [2]. Researchers now combine chemical inhibitors, antioxidants, and environmental adjustments to fine-tune ethylene level. These steps have improved somatic embryogenesis in a variety of plant species.

6 Ethylene Measurement and Control in Culture

Accurate measurement and regulation of ethylene is vital for successful plant tissue culture. Sealed culture vessels allow ethylene build up, so its level must be monitored and controlled (Fig. 1) (Table 2). Gas chromatography, still the standard, detects trace ethylene with high sensitivity and accuracy (Fig. 1) [17]. Portable ethylene sensors now permit real-time monitoring, useful in dynamic culture setups. The impact of ethylene on culture outcomes depends on maintaining it within suitable limits at each developmental stage. Excessive ethylene blocks shoot and root organogenesis; too little may not stimulate cell division and differentiation.

Controlled ventilation uses vented caps or perforated lids to let ethylene escape, preventing buildup (Fig. 1). Potassium permanganate or activated charcoal can absorb ethylene, keeping gas level balanced in culture system [50]. Such physical treatments minimize ethylene accumulation without adding chemicals that could harm developing tissues. Chemical inhibitors, such as silver nitrate ($AgNO_3$), aminoethoxyvinylglycine (AVG), and cobalt chloride ($CoCl_2$), are used to suppress ethylene synthesis or action (Table 3). Silver nitrate functions as an ethylene receptor antagonist and counters ethylene inhibition of shoot and root regeneration. However, too much inhibitors becomes phytotoxicity and reduces tissue viability [52]. Light, temperature, and humidity also shape ethylene production and its effects on cultures. High temperatures enhance ethylene synthesis; cooler conditions suppress it. Adjusting these parameters alongside ethylene regulation offers an integrated approach to improve culture outcomes [17].

Recent molecular work enables transgenic plants with modified ethylene biosynthesis or signaling. Lowering ACC synthase (ACS) or ACC oxidase (ACO) expression reduces ethylene production, providing genetic control in -vitro [58, 59]. Modifying ethylene receptors like ETR1, alters sensitivity, enabling cultures respond precisely to changing conditions [54, 55]. Such genetic tools promise solutions for ethylene regulation across many species [58, 59]. Combining accurate measurement with physical, chemical, environmental, and molecular controls, ensures tight ethylene regulation in culture.

7 Interactions with Other Hormonal and Metabolic Pathways

Ethylene rarely acts alone; it links closely to other hormones and metabolic pathways in plant tissue cultures (Fig. 2). Auxins increase activity of ACC synthase (ACS), raising ethylene synthesis [39–41, 58]. Such auxin-induced ethylene may aid root initiation yet limit elongation if levels rise too high [39–41, 58, 60].

The interplay between ethylene and cytokinins regulates shoot and root development (Fig. 1) [39–41]. As cytokinins drive shoot organogenesis, but elevated ethylene that skews the cytokinin-to-auxin ratio reduces shoot and enlarge callus [39–41]. Applying ethylene inhibitors alongside cytokinins restores balance and improves shoot regeneration in recalcitrant Brassica and Capsicum [52]. Ethylene also shows an antagonistic relationship with abscisic acid (ABA), which shapes stress responses and development. In somatic embryogenesis and seed germination, adjusting both hormones improves culture performance [28]. Polyamines, including putrescine, spermidine, spermine, and ethylene share a common precursor, S-adenosylmethionine (SAM). This metabolic link creates a competitive dynamics where increased ethylene biosynthesis can suppress polyamine production and vice versa. Elevated ethylene can lower polyamines pools, whereas added polyamines offset ethylene driven inhibition of somatic embryogenesis and shoot organogenesis [54, 55]. ROS act as signal for cell division and differentiation, yet excessive ethylene amplifies ROS damages cells. Combining antioxidants such as ascorbic acid or glutathione, with ethylene regulation balances ROS levels and improves culture outcomes [57].

8 Genotypic Variations and Developmental Stage Factors

Genotype and the developmental stage of explant strongly shape how plant tissues respond to ethylene in culture (Fig. 1). Different plant species and genotypes vary in ethylene production and sensitivity, altering culture outcomes (Table 4) [2, 30]. Endogenous ethylene effects are therefore highly genotype-dependent. Barley anther cultures with low endogenous ethylene production formed more embryos after ethylene precursors such as ACC or Ethephon supplementation (Table 1) [30]. Conversely, high ethylene genotypes showed poor androgenesis, which improved after $AgNO_3$ or silver thiosulfate treatment [31]. Similar patterns appeared in Brussels sprouts, high ethylene production suppressed embryogenesis, and blocking biosynthesis restored responsiveness [61]. The explant developmental stage also dictates ethylene action. In *Nicotiana tabacum* anther cultures, ethylene blocked shoot primordia early but promoted it in later stages revealing stage-specific effects [62]. Younger anthers generally respond better to ethylene manipulation than older ones [30]. Optimal ethylene levels differ across phases and genotypes. Removing ethylene with Acepack reduced somatic embryogenesis in *Coffea canephora* by 40%, proving its promotive role at moderate concentrations [28]. Sunflower callus lines that failed to regenerate produced more ethylene than embryogenic lines, indicating a threshold for benefit [29].

9 Practical Implications

Ethylene inhibitors, such as $AgNO_3$, help overcome genotype limits and increase regeneration across many species (Fig. 1) [31]. However, the optimal ethylene concentration varies with genotype and developmental stage, requiring strict control of synthesis and signaling (Fig. 1).

Because of this variability, ethylene serves both as promoter and inhibitor, depending on genetic and developmental context (Figs. 1 and 2). Thus ethylene stands as a critical regulator in plant tissue culture (Fig. 1). Adding ethylene control steps to tissue culture protocols improves regeneration, morphogenesis, and embryogenesis. For example, ethylene inhibitors like $AgNO_3$ or silver thiosulfate (STS) enhance shoot regeneration in recalcitrant *Brassica oleracea* and maize [31, 32]. Carefully lowering ethylene also stimulates somatic embryogenesis in *Coffea canephora* and carrot [28, 53]. Moderate doses of Ethephon, an ethylene releaser, have positive results in specific culture systems. Researchers also manage ethylene accumulation by adjusting culture vessel seals and gas exchange. Gas-permeable membranes or partially sealed vessels allow ethylene escape, reducing growth problems [50]. These physical fixes suit species highly sensitive to trapped ethylene, including *Ficus lyrata* and sunflower.

Breeders exploit ethylene tuning in anther or microspore cultures to recover homozygous lines. In barley, adding ACC or inhibitors increased embryogenesis in low-responding genotypes [30].

10 Conclusion

Evidence shows ethylene shapes tissue culture morphogenesis in a complex, context-dependent way. Species, genotypes, culture systems, and developmental phases decide whether it promotes embryos or inhibits organogenesis. Callus induction, shoot initiation, and shoot elongation each seem to need unique ethylene conditions. Even one genotype may require partial inhibition or precise supplementation at specific developmental stages. Ethylene also interacts with other hormones, metabolites, and environmental variables such as media composition, light, temperature, and gaseous atmosphere. These complex connections make its net effect hard to predict. Brief experiments show cobalt or silver ions modulate ethylene activity. However, prolonged treatments or subtle shifts in concentration still alter growth.

Acknowledgments

The Director, CSIR Central Institute of Medicinal and Aromatic Plants, Lucknow, India, is acknowledged for providing the facilities and support. The Institutional Communication No. is CIMAP/ PUB/ 2025/06.

References

1. Stoddart JL, Venis MA (1980) Molecular and subcellular aspects of hormone action. In: MacMillan J (ed) Hormonal regulation of development I. Springer, Berlin/Heidelberg, pp 445–510
2. Babbar SB, Gupta SC (1986) Putative role of ethylene in *Datura metel* microspore embryogenesis. Physiol Plant 68:141–144. https://doi.org/10.1111/j.1399-3054.1986.tb06609.x
3. Bennett MD, Hughes WG (1972) Additional mitosis in wheat pollen induced by Ethrel. Nature 240:566–568. https://doi.org/10.1038/240566a0
4. MacDonald IM, Grant WF (1974) Anther culture of pollen containing Ethrel-induced micronuclei. Zeitschrift für Pflanzenzüchtung 73:292–297
5. Fouletier B (1974) Conditions favorisant la néoformation de cals haploïdes à partir d'anthères de riz cultivées in vitro. C R Acad Sci 278:2917–2920
6. Wang C-C, Sun C-S, Chu Z-C (1974) On the conditions for the induction of rice pollen plantlets and certain factors affecting the frequency of induction. Acta Bot Sin 16:43–54
7. Song HG, Li SN, Li GR, Yun SG, Li JW (1981) Studies on increasing the induction rate of callus tissue and pollen plants from anthers of Oryza sativa cultured in vitro. In: Plant tissue culture: proceedings of the Beijing (Peking) symposium. Pitman, London/Peking, pp 97–106
8. Chaleff RS, Stolarz A (1981) Factors influencing the frequency of callus formation among cultured rice (Oryza sativa) anthers. Physiol Plant 51:201–206. https://doi.org/10.1111/j.1399-3054.1981.tb02699.x
9. Horner M, McCOMB JA, MoCOMB AJ, Street HE (1977) Ethylene production and plantlet formation by *Nicotiana* anthers cultured in the presence and absence of charcoal. J Exp Bot 28:1365–1372. https://doi.org/10.1093/jxb/28.6.1365
10. Dunwell JM (1979) Anther culture in *Nicotiana tabacum*: the role of the culture vessel atmosphere in pollen embryo induction and growth. J Exp Bot 30:419–428. https://doi.org/10.1093/jxb/30.3.419
11. Lau O-L, Yang SF (1976) Inhibition of ethylene production by cobaltous ion. Plant Physiol 58:114–117. https://doi.org/10.1104/pp.58.1.114
12. Beyer EM (1976) A potent inhibitor of ethylene action in plants. Plant Physiol 58:268–271. https://doi.org/10.1104/pp.58.3.268
13. Sembdner G, Gross D, Liebisch H-W, Schneider G (1980) Biosynthesis and metabolism of plant hormones. In: MacMillan J (ed) Hormonal regulation of development I. Springer, Berlin/Heidelberg, pp 281–444
14. Abeles FB (1973) Ethylene in plant biology. Academic Press, New York
15. Thomas DDS, Murashige T (1979) Volatile emissions of plant tissue cultures: I. Identification of the major components. In Vitro 15:654–658. https://doi.org/10.1007/BF02618242
16. Hughes KW (1981) In vitro ecology: exogenous factors affecting growth and morphogenesis in plant culture systems. Environ Exp Bot 21:281–288. https://doi.org/10.1016/0098-8472(81)90038-1
17. Biddington NL (1992) The influence of ethylene in plant tissue culture. Plant Growth Regul

11:173–187. https://doi.org/10.1007/BF00024072

18. Purnhauser L, Medgyesy P, Czak M, Dix PJ, Mrton L (1987) Stimulation of shoot regeneration in Triticum aestivum and Nicotiana plumbaginifolia Viv. Tissue cultures using the ethylene inhibitor AgNO3. Plant Cell Rep 6:1–4. https://doi.org/10.1007/BF00269725
19. Perl A, Aviv D, Galun E (1988) Ethylene and in vitro culture of potato: suppression of ethylene generation vastly improves protoplast yield, plating efficiency and transient expression of an alien gene. Plant Cell Rep 7:403–406. https://doi.org/10.1007/BF00269523
20. De Block M (1988) Genotype-independent leaf disc transformation of potato (Solanum tuberosum) using Agrobacterium tumefaciens. Theoret Appl Genetics 76:767–774. https://doi.org/10.1007/BF00303524
21. Vain P, Flament P, Soudain P (1989) Role of ethylene in embryogenic callus initiation and regeneration in Zea mays. J Plant Physiol 135: 537–540
22. Songstad DD, Armstrong CL, Petersen WL (1991) AgNO3 increases type II callus production from immature embryos of maize inbred B73 and its derivatives. Plant Cell Rep 9:699–702. https://doi.org/10.1007/BF00235361
23. Palmer CE (1992) Enhanced shoot regeneration from Brassica campestris by silver nitrate. Plant Cell Rep 11. https://doi.org/10.1007/BF00233088
24. Radke Sharon E, Turner Joann C, Facciotti D (1992) Transformation and regeneration of Brassica rapa using Agrobacterium tumefaciens. Plant Cell Rep 11. https://doi.org/10.1007/BF00236265
25. Plus J, George L, Eapen S, Rao PS (1993) Enhanced plant regeneration in pearl millet (Pennisetum americanum) by ethylene inhibitors and cefotaxime. Plant Cell Tissue Organ Cult 32:91–96. https://doi.org/10.1007/BF00040121
26. Hu W, Gong H, Pua E (2006) Modulation of SAMDC expression in *Arabidopsis thaliana* alters in vitro shoot organogenesis. Physiol Plant 128:740–750. https://doi.org/10.1111/j.1399-3054.2006.00799.x
27. Jha AK, Dahleen LS, Suttle JC (2007) Ethylene influences green plant regeneration from barley callus. Plant Cell Rep 26:285–290. https://doi.org/10.1007/s00299-006-0252-0
28. Hatanaka T, Sawabe E, Azuma T, Uchida N, Yasuda T (1995) The role of ethylene in somatic embryogenesis from leaf discs of Coffea canephora. Plant Sci 107:199–204. https://doi.org/10.1016/0168-9452(95)04103-2
29. Kumar PP, Lakshmanan P, Thorpe TA (1998) Regulation of morphogenesis in plant tissue culture by ethylene. In Vitro Cell Dev Biol Plant 34:94–103. https://doi.org/10.1007/BF02822771
30. Cho U-H, Kasha KJ (1989) Ethylene production and embyogenesis from anther cultures of barley (Hordeum vulgare). Plant Cell Rep 8: 415–417. https://doi.org/10.1007/BF00270082
31. Evans Jean M, Batty Nigel P (1994) Ethylene precursors and antagonists increase embryogenesis of Hordeum vulgare L. anther culture. Plant Cell Rep 13:10.1007/BF00231622
32. Pua EC, Chi GL (1993) De novo shoot regeneration from cotyledon explants of Brassica napus and B. juncea: Influence of silver nitrate and genotype. Plant Cell Rep 12:195–198
33. Pua E-C, JustinEE L (1995) Enhanced de novo shoot morphogenesis in vitro by expression of antisense 1-aminocyclopropane-1-carboxylate oxidase gene in transgenic mustard plants. Planta 196. https://doi.org/10.1007/BF00193219
34. Batista DS, Dias LLC, Macedo AF, Do Rêgo MM, Do Rêgo ER, Floh EIS, Finger FL, Otoni WC (2013) Suppression of ethylene levels promotes morphogenesis in pepper (Capsicum annuum L.). In Vitro Cell Dev Biol Plant 49: 759–764. https://doi.org/10.1007/s11627-013-9559-z
35. Sharma A, Ravishankar GA, Giridhar P (2017) Silver nitrate and 2-(N-morpholine) ethane sulphonic acid in culture medium promotes rapid shoot regeneration from the proximal zone of the leaf of Capsicum frutescens Mill. Plant Cell Tissue Organ Cult 129:175–180. https://doi.org/10.1007/s11240-016-1157-4
36. Pijeira-Fernández G, Santana-Buzzy N (2024) Capsicum recalcitrance: physiological and molecular challenges of pepper tissue culture. In Vitro Cell Dev Biol Plant. https://doi.org/10.1007/s11627-024-10445-w
37. Yang SF, Hoffman NE (1984) Ethylene biosynthesis and its regulation in higher plants. Annu Rev Plant Physiol 35:155–189. https://doi.org/10.1146/annurev.pp.35.060184.001103
38. Salisbury FB, Ross CW (1992) Plant physiology, 4th edn. Wadsworth, Belmont
39. Harpaz-Saad S, Yoon GM, Mattoo AK, Kieber JJ (2012) The formation of ACC and competition between polyamines and ethylene for

SAM. In: McManus MT (ed) Annual plant reviews, vol 44, 1st edn. Wiley, pp 53–81

40. Dias LLC, Santa-Catarina C, Ribeiro DM, Barros RS, Floh EIS, Otoni WC (2009) Ethylene and polyamine production patterns during in vitro shoot organogenesis of two passion fruit species as affected by polyamines and their inhibitor. Plant Cell Tissue Organ Cult 99:199–208. https://doi.org/10.1007/s11240-009-9594-y
41. Kumar V, Parvatam G, Ravishankar GA (2009) AgNO3 – a potential regulator of ethylene activity and plant growth modulator. Electron J Biotechnol 12. https://doi.org/10.2225/vol12-issue2-fulltext-1
42. Wilkinson JQ, Lanahan MB, Clark DG, Bleecker AB, Chang C, Meyerowitz EM, Klee HJ (1997) A dominant mutant receptor from Arabidopsis confers ethylene insensitivity in heterologous plants. Nat Biotechnol 15:444–447. https://doi.org/10.1038/nbt0597-444
43. Clark DG, Gubrium EK, Barrett JE, Nell TA, Klee HJ (1999) Root formation in ethylene-insensitive plants. Plant Physiol 121:53–60. https://doi.org/10.1104/pp.121.1.53
44. Bovy AG, Angenent GC, Dons HJM, Van Altvorst A-C (1999) Heterologous expression of the Arabidopsis etr1-1 allele inhibits the senescence of carnation flowers. Mol Breed 5:301–308. https://doi.org/10.1023/A:1009617804359
45. Hee Kim J, Botella JR (2004) Etr1-1 gene expression alters regeneration patterns in transgenic lettuce stimulating root formation. Plant Cell Tissue Organ Cult 78:69–73. https://doi.org/10.1023/B:TICU.0000020396.64257.c1
46. Cui M-L, Takada K, Ma B, Ezura H (2004) Overexpression of a mutated melon ethylene receptor gene Cm-ETR1/H69A confers reduced ethylene sensitivity in a heterologous plant, Nemesia strumosa. Plant Sci 167:253–258. https://doi.org/10.1016/j.plantsci.2004.03.024
47. Sriskandarajah S, Mibus H, Serek M (2007) Transgenic Campanula carpatica plants with reduced ethylene sensitivity. Plant Cell Rep 26:805–813. https://doi.org/10.1007/s00299-006-0291-6
48. Chatfield SP, Raizada MN (2008) Ethylene and shoot regeneration: hookless1 modulates de novo shoot organogenesis in Arabidopsis thaliana. Plant Cell Rep 27:655–666. https://doi.org/10.1007/s00299-007-0496-3
49. Khalid M, Chraibi B, Latche A, Roustan J-P, Fallot J (1991) Stimulation of shoot regeneration from cotyledons of Helianthus annuus by the ethylene inhibitors, silver and cobalt. Plant Cell Rep 10. https://doi.org/10.1007/BF00234296
50. Jackson MB, Mantell SH, British Plant Growth Regulator Group, Association of Applied Biologists, Unit for Advanced Propagation Systems, Wye College, University of London (1987) Advances in the chemical manipulation of plant tissue cultures: proc. of a meeting org. by The British Plant Growth Regulator Group ... on 6. - 7. July 1987 at Wye College (Univ. of London). British Plant Growth Regulator Group, Bristol
51. Prez-Bermdez P, Cornejo MJ, Segura J (1985) A morphogenetic role for ethylene in hypocotyl cultures of Digitalis obscura L. Plant Cell Rep 4:188–190. https://doi.org/10.1007/BF00269285
52. Chi GL, Pua EC (1989) Ethylene inhibitors enhanced de novo shoot regeneration from cotyledon explants of Brassica campestris ssp. Chinensis (Chinese cabbage) in vitro. Plant Sci 64:243–250
53. Roustan J-P, Latche A, Fallot J (1990) Control of carrot somatic embryogenesis by AgNO3, an inhibitor of ethylene action: effect on arginine decarboxylase activity. Plant Sci 67:89–95. https://doi.org/10.1016/0168-9452(90)90054-R
54. Mehta RA, Cassol T, Li N, Ali N, Handa AK, Mattoo AK (2002) Engineered polyamine accumulation in tomato enhances phytonutrient content, juice quality, and vine life. Nat Biotechnol 20:613–618. https://doi.org/10.1038/nbt0602-613
55. Santos AM, Oliver MJ, Sánchez AM, Payton PR, Gomes JP, Miguel C, Oliveira MM (2009) An integrated strategy to identify key genes in almond adventitious shoot regeneration. J Exp Bot 60:4159–4173. https://doi.org/10.1093/jxb/erp250
56. Fujii H, Shimada T, Sugiyama A, Nishikawa F, Endo T, Nakano M, Ikoma Y, Shimizu T, Omura M (2007) Profiling ethylene-responsive genes in mature mandarin fruit using a citrus 22K oligoarray. Plant Sci 173:340–348. https://doi.org/10.1016/j.plantsci.2007.06.006
57. Tiainen T (1992) The role of ethylene and reducing agents on anther culture response of tetraploid potato (Solanum tuberosum L.). Plant Cell Rep 10:10.1007/BF00232379
58. Binder BM (2020) Ethylene signaling in plants. J Biol Chem 295:7710–7725. https://doi.org/10.1074/jbc.REV120.010854
59. Bapat VA, Trivedi PK, Ghosh A, Sane VA, Ganapathi TR, Nath P (2010) Ripening of

fleshy fruit: molecular insight and the role of ethylene. Biotechnol Adv 28:94–107. https://doi.org/10.1016/j.biotechadv.2009.10.002

60. Roberts LW, Baba S (1978) Exogenous methionine as a nutrient supplement for the induction of Xylogenesis in lettuce pith explants. Ann Bot 42:375–379. https://doi.org/10.1093/oxfordjournals.aob.a085470

61. Biddington NL, Robinson HT (1991) Ethylene production during anther culture of Brussels sprouts (Brassica oleracea var gemmifera) and its relationship with factors that affect embryo production. Plant Cell Tissue Organ Cult 25:169–177. https://doi.org/10.1007/BF00036207

62. Huxter TJ, Thorpe TA, Reid DM (1981) Shoot initiation in light- and dark-grown tobacco callus: the role of ethylene. Physiol Plant 53:319–326. https://doi.org/10.1111/j.1399-3054.1981.tb04507.x

INDEX

Devesh Shukla (ed.), *Ethylene Sensing and Signaling*, Methods in Molecular Biology, vol. 2945, https://doi.org/10.1007/978-1-0716-4650-2,

F

G

H

I

J

L

M

N

O

P

Q

R

S

T

MIX
Papier aus verantwortungsvollen Quellen
Paper from responsible sources
FSC® C105338

If you have any concerns about our products,
you can contact us on
ProductSafety@springernature.com

In case Publisher is established outside the EU,
the EU authorized representative is:
Springer Nature Customer Service Center GmbH
Europaplatz 3, 69115 Heidelberg, Germany

Printed by Libri Plureos GmbH
in Hamburg, Germany